Fuzzy Logic and Soft Computing – Dedicated to the Centenary of the Birth of Lotfi A. Zadeh

Fuzzy Logic and Soft Computing – Dedicated to the Centenary of the Birth of Lotfi A. Zadeh

Editors

Ioan Dzitac
Sorin Nadaban

MDPI • Basel • Beijing • Wuhan • Barcelona • Belgrade • Manchester • Tokyo • Cluj • Tianjin

Editors
Ioan Dzitac
Aurel Vlaicu University of Arad and Agora University of Oradea
Romania

Sorin Nadaban
Aurel Vlaicu University of Arad
Romania

Editorial Office
MDPI
St. Alban-Anlage 66
4052 Basel, Switzerland

This is a reprint of articles from the Special Issue published online in the open access journal *Mathematics* (ISSN 2227-7390) (available at: https://www.mdpi.com/si/mathematics/fuzzy_logic_soft_computing_dedicated_centenary_birth_Lotfi_A_Zadeh).

For citation purposes, cite each article independently as indicated on the article page online and as indicated below:

LastName, A.A.; LastName, B.B.; LastName, C.C. Article Title. *Journal Name* **Year**, *Volume Number*, Page Range.

ISBN 978-3-0365-5587-4 (Hbk)
ISBN 978-3-0365-5588-1 (PDF)

Contents

About the Editors . . . . **vii**

Preface to "Fuzzy Logic and Soft Computing – Dedicated to the Centenary of the Birth of Lotfi A. Zadeh" . . . . **ix**

Sorin Nădăban
Fuzzy Logic and Soft Computing—Dedicated to the Centenary of the Birth of Lotfi A. Zadeh (1921–2017)
Reprinted from: *Mathematics* **2022**, *10*, 3216, doi:10.3390/math10173216 . . . . **1**

Lorena Popa and Lavinia Sida
Fuzzy Inner Product Space: Literature Review and a New Approach
Reprinted from: *Mathematics* **2021**, *9*, 765, doi:10.3390/math9070765 . . . . **5**

Faisal Mehmood and Fu-Gui Shi
M-Hazy Vector Spaces over M-Hazy Field
Reprinted from: *Mathematics* **2021**, *9*, 1118, doi:10.3390/math9101118 . . . . **15**

Bogdana Stanojević, Milan Stanojević, Sorin Nădăban
Reinstatement of the Extension Principle in Approaching Mathematical Programming with Fuzzy Numbers
Reprinted from: *Mathematics* **2021**, *9*, 1272, doi:10.3390/math9111272 . . . . **29**

Qingsong Mao and Huan Huang
Interval Ranges of Fuzzy Sets Induced by Arithmetic Operations Using Gradual Numbers
Reprinted from: *Mathematics* **2021**, *9*, 1351, doi:10.3390/math9121351 . . . . **45**

Mabruka Ali, Adem Kılıçman and Azadeh Zahedi Khameneh
Application of Induced Preorderings in Score Function-Based Method for Solving Decision-Making with Interval-Valued Fuzzy Soft Information
Reprinted from: *Mathematics* **2021**, *9*, 1575, doi:10.3390/math9131575 . . . . **61**

Yan-Yan Dong and Fu-Gui Shi
L-Fuzzy Sub-Effect Algebras
Reprinted from: *Mathematics* **2021**, *9*, 1596, doi:10.3390/math9141596 . . . . **81**

Simona Dzitac and Sorin Nădăban
Soft Computing for Decision-Making in Fuzzy Environments: A Tribute to Professor Ioan Dzitac
Reprinted from: *Mathematics* **2021**, *9*, 1701, doi:10.3390/math9141701 . . . . **95**

Alina Alb Lupaş and Georgia Irina Oros
New Applications of Sălăgean and Ruscheweyh Operators forObtaining Fuzzy Differential Subordinations
Reprinted from: *Mathematics* **2021**, *9*, 2000, doi:10.3390/math9162000 . . . . **107**

Shahida Bashir, Sundas Shahzadi, Ahmad N. Al-Kenani and Muhammad Shabir
Regular and Intra-Regular Semigroups in Terms of m-Polar FuzzyEnvironment
Reprinted from: *Mathematics* **2021**, *9*, 2031, doi:10.3390/math9172031 . . . . **119**

Miguel A. Sainz, Remei Calm, Lambert Jorba, Ivan Contreras and Josep Vehi
Marks: A New Interval Tool for Uncertainty, Vagueness and Indiscernibility
Reprinted from: *Mathematics* **2021**, *9*, 2116, doi:10.3390/math9172116 . . . . **137**

Carolina Nicolas, Javiera Müller and Francisco-Javier Arroyo-Cañada
A Fuzzy Inference System for Management Control Tools
Reprinted from: *Mathematics* **2021**, *9*, 2145, doi:10.3390/math9172145 . . . . **153**

Georgia Irina Oros
Fuzzy Differential Subordinations Obtained Using a HypergeometricIntegral Operator
Reprinted from: *Mathematics* **2021**, *9*, 2539, doi:10.3390/math9202539 . . . . **173**

Muhammad Zishan Anwar, Shahida Bashir, Muhammad Shabir and Majed G. Alharbi
Multigranulation Roughness of Intuitionistic Fuzzy Sets by SoftRelations and Their Applications in Decision Making
Reprinted from: *Mathematics* **2021**, *9*, 2587, doi:10.3390/math9202587 . . . . **187**

Alina Alb Lupaş
Applications of the Fractional Calculus in Fuzzy Differential Subordinations and Superordinations
Reprinted from: *Mathematics* **2021**, *92*, 2601, doi:10.3390/math9202601 . . . . **209**

About the Editors

Sorin Nădăban

Sorin Nădăban is a professor in the Department of Mathematics and Computer Science of Aurel Vlaicu University of Arad. He received his bachelor's in Mathematics (1991) and PhD in Mathematics (2000) from the Western University of Timişoara, Romania. His research interests are in the areas of fuzzy mathematics and operator theory. He is the Editor-in-Chief of the journal Theory and Applications of Mathematics & Computer Science and a reviewer of the American Mathematical Society.

Preface to "Fuzzy Logic and Soft Computing – Dedicated to the Centenary of the Birth of Lotfi A. Zadeh"

In 1965, Lotfi A. Zadeh published "Fuzzy Sets", his pioneering and controversial paper, which has now had over 115,000 citations. Altogether, Zadeh's papers have been cited over 248,000 times. Starting from the ideas presented in that paper, Zadeh later founded the Fuzzy Logic Theory, which has useful applications from consumer to industrial intelligent products. In accordance with Zadeh's definition, soft computing (SC) consists of computational techniques in computer science, machine learning, and some engineering disciplines to study, model, and analyze very complex reality, for which more traditional methods have been either unusable or inefficient. SC uses soft techniques, contrasting it with classical artificial intelligence hard computing (HC) techniques, and includes fuzzy logic, neural computing, evolutionary computation, machine learning, and probabilistic reasoning. HC is bound by a computer science (CS) concept called NP-complete, which stipulates that there is a direct connection between the size of a problem and the amount of resources needed to solve it, called the "grand challenge problem". SC helps to surmount NP-complete problems by using inexact methods to give useful but inexact answers to intractable problems. SC became a formal CS area of study in the early 1990s. Earlier computational approaches could only model and precisely analyze relatively simple systems. More complex systems arising in biology, medicine, the humanities, management sciences, and similar fields often remained intractable to HC. It should be pointed out that the simplicity and complexity of systems are relative, and many conventional mathematical models have been both challenging and very productive. SC techniques resemble biological processes more closely than traditional techniques, which are largely based on formal logical systems, such as Boolean logic, or rely heavily on computer-aided numerical analysis techniques (such as finite element analysis). SC techniques are intended to complement HC techniques. Unlike HC schemes, which strive for exactness and full truth, SC techniques exploit the given tolerance of imprecision, partial truth, and uncertainty for a particular problem. Inductive reasoning plays a larger role in SC than in HC. SC and HC can be used together in certain fusion techniques. SC can deal with ambiguous or noisy data, and it is tolerant of imprecision, uncertainty, partial truth, and approximation. In effect, the role model for SC is the human mind. Artificial intelligence and computational intelligence based on SC provide the background for the development of smart management systems and decisions in the case of ill-posed problems.

The present book contains 14 articles accepted for publication among the 40 manuscripts in total that were submitted to the Special Issue of the MDPI Mathematics journal entitled "Fuzzy Logic and Soft Computing – Dedicated to the Centenary of the Birth of Lotfi A. Zadeh (1921-2017)". These articles have been published in Volume 9 (2021) of the journal and cover a wide variety of topics related to fuzzy logic and soft computing. We hope that this book will be useful for those who work in the domains of fuzzy logic and soft computing or for those who want to familiarize themselves with the most advanced knowledge in the field of fuzzy mathematics.

As the Guest Editor of this Special Issue, I am grateful to the authors of the papers for their high-quality contributions, to the reviewers for their valuable comments towards the improvement of the articles submitted, and to the administrative staff of the MDPI publications for their support in completing this project. Special thanks are due to the Managing Editor of the Special Issue, Mr. Claude Zhang, for his excellent collaboration and valuable assistance.

Ioan Dzitac and Sorin Nadaban
Editors

mathematics MDPI

Editorial

Fuzzy Logic and Soft Computing—Dedicated to the Centenary of the Birth of Lotfi A. Zadeh (1921–2017)

Sorin Nădăban

Department of Mathematics and Computer Science, Aurel Vlaicu University of Arad, Elena Drăgoi 2, RO-310330 Arad, Romania; snadaban@gmail.com

1. Introduction

In 1965, Lotfi A. Zadeh published "Fuzzy Sets", his pioneering and controversial paper, which has now reached over 115,000 citations. Zadeh's papers have altogether been cited over 248,000. Starting from the ideas presented in that paper, Zadeh later founded the Fuzzy Logic Theory, which proved to have useful applications from consumer to industrial intelligent products.

In accordance with Zadeh's definition, soft computing (SC) consists of computational techniques in computer science, machine learning, and some engineering disciplines to study, model, and analyze very complex realities, for which more traditional methods have been either unusable or inefficient. SC uses soft techniques, contrasting it with classical artificial intelligence hard computing (HC) techniques, and includes fuzzy logic, neural computing, evolutionary computation, machine learning, and probabilistic reasoning. HC is bound by a computer science (CS) concept called NP-complete, which means that there is a direct connection between the size of a problem and the amount of resources needed to solve it called the "grand challenge problem". SC helps to surmount NP-complete problems by using inexact methods to give useful but inexact answers to intractable problems. SC became a formal CS area of study in the early 1990s. Earlier computational approaches could model and precisely analyze only relatively simple systems. More complex systems arising in biology, medicine, the humanities, management sciences, and similar fields often remained intractable to HC. It should be pointed out that the simplicity and complexity of systems are relative, and many conventional mathematical models have been both challenging and very productive. SC techniques resemble biological processes more closely than traditional techniques, which are largely based on formal logical systems, such as Boolean logic, or rely heavily on computer-aided numerical analysis (such as finite element analysis). SC techniques are intended to complement HC techniques. Unlike HC schemes, which strive for exactness and full truth, SC techniques exploit the given tolerance of imprecision, partial truth, and uncertainty for a particular problem. Inductive reasoning plays a larger role in SC than in HC. SC and HC can be used together in certain fusion techniques. SC can deal with ambiguous or noisy data and it is tolerant of imprecision, uncertainty, partial truth, and approximation. In effect, the role model for SC is the human mind. Artificial intelligence and computational intelligence based on SC provide the background for the development of smart management systems and decisions in the case of ill-posed problems.

2. Contributions

In the following, a brief overview of the published papers is presented.

Finding a suitable definition of fuzzy inner product space have concerned many mathematicians. In [1], first, various approaches are presented for the concept of fuzzy inner product space existing in the specialized literature, and then a new definition is introduced. In fact, the authors modified P. Majumdar and S.K. Samanta's definition of inner product space and proved some new properties of the fuzzy inner product function.

Citation: Nădăban, S. Fuzzy Logic and Soft Computing—Dedicated to the Centenary of the Birth of Lotfi A. Zadeh (1921–2017). *Mathematics* **2022**, *10*, 3216. https://doi.org/10.3390/math10173216

Received: 22 August 2022
Accepted: 2 September 2022
Published: 5 September 2022

Finally, in this paper, it is also proved that this fuzzy inner product generates a fuzzy norm of the type Nădăban-Dzitac.

The paper [2] proposes a new generalization of vector spaces over field, which is called M-hazy vector spaces over M-hazy field. Some fundamental properties of M-hazy field, M-hazy vector spaces, and M-hazy subspaces are studied, and some important results are also proved. Furthermore, the linear transformation of M-hazy vector spaces is studied, and their important results are also proved.

Optimization problems in the fuzzy environment are widely studied in the literature. In paper [3] the authors restrict their attention to mathematical programming problems with coefficients and/or decision variables expressed by fuzzy numbers. This paper identifies the current position and role of the extension principle in solving mathematical programming problems that involve fuzzy numbers in their models, highlighting the indispensability of the extension principle in approaching this class of problems. Finally, some research directions focusing on using the extension principle in all stages of the optimization process are proposed.

The interval range is an important characterization of a fuzzy set. The interval range is also useful for analyses and applications of arithmetic. In paper [4], the authors presented general conclusions on crucial problems related to interval ranges of fuzzy sets.

In paper [5], the authors present a group decision-making solution based on a preference relationship of interval-valued fuzzy soft information. Further, two crisp topological spaces, namely, lower topology and upper topology, are introduced based on the interval-valued fuzzy soft topology. Then, a score function-based ranking system is also defined to design an adjustable multi-steps algorithm. Finally, some illustrative examples are given to compare the effectiveness of the present approach with some existing methods.

In paper [6], the notions of L-fuzzy subalgebra degree and L-subalgebras on an effect algebra are introduced and some characterizations are given. The authors use four kinds of cut sets of L-subsets to characterize the L-fuzzy subalgebra degree. Finally, it is proved that the set of all L-subalgebras on an effect algebra can form an L-convexity, and its L-convex hull formula is given.

The paper [7] is dedicated to Professor Ioan Dzitac (1953–2021). Therefore, his life is briefly presented, as well as a comprehensive overview of his major contributions in the domain of soft computing methods in a fuzzy environment. Finally, some future trends are discussed.

In paper [8], a certain fuzzy class of analytic functions is defined in the open unit disc and some interesting results related to this class are obtained using the concept of fuzzy differential subordination.

In paper [9], the authors introduce the concept of an m-polar fuzzy set (m-PFS) in semigroups. This paper provides some important results related to m-polar fuzzy subsemigroups (m-PFSSs), m-polar fuzzy ideals (m-PFIs), m-polar fuzzy generalized bi-ideals (m-PFGBIs), m-polar fuzzy bi-ideals (m-PFBIs), m-polar fuzzy quasi-ideals (m-PFQIs) and m-polar fuzzy interior ideals (m-PFIIs) in semigroups.

In paper [10], a new implementation of the marks library is presented. Examples in dynamical systems simulation, fault detection and control are also included to exemplify the practical use of the marks.

In paper [11], a fuzzy logic approach is proposed for the decision-making system in management control in small and medium enterprises. The C. Mamdani fuzzy inference system (MFIS) was applied as a decision-making technique to explore the influence of the use of management control tools on the organizational performance of SMEs.

In paper [12], fuzzy differential subordination results are obtained using a new integral operator introduced by the author, using the well-known confluent hypergeometric function, also known as the Kummer hypergeometric function. The new hypergeometric integral operator is defined by choosing particular parameters, having as inspiration the operator studied by Miller, Mocanu and Reade in 1978.

The paper [13] presents the multigranulation roughness of an intuitionistic fuzzy set based on two soft relations over two universes with respect to the aftersets and foresets. Finally, a decision-making algorithm is presented with a suitable example.

The aim of the paper [14] is to present new fuzzy differential subordinations and superordinations for which the fuzzy best dominant and, respectively, fuzzy best subordinant are given. The original theorems proved in the paper generate interesting corollaries for particular choices of functions acting as fuzzy best dominant and fuzzy best subordinant.

Funding: This research received no external funding.

Acknowledgments: As the Guest Editor of this Special Issue, I am grateful to the authors of the papers for their quality contributions, to the reviewers for their valuable comments towards the improvement of the submitted works and to the administrative staff of the MDPI publications for the support to complete this project.

Conflicts of Interest: The author declares no conflict of interest.

References

1. Popa, L.; Sida, L. Fuzzy Inner Product Space: Literature Review and a New Approach. *Mathematics* **2021**, *9*, 765. [CrossRef]
2. Mehmood, F.; Shi, F.-G. M-Hazy Vector Spaces over M-Hazy Field. *Mathematics* **2021**, *9*, 1118. [CrossRef]
3. Stanojević, B.; Stanojević, M.; Nădăban, S. Reinstatement of the Extension Principle in Approaching Mathematical Programming with Fuzzy Numbers. *Mathematics* **2021**, *9*, 1272. [CrossRef]
4. Mao, Q.; Huang, H. Interval Ranges of Fuzzy Sets Induced by Arithmetic Operations Using Gradual Numbers. *Mathematics* **2021**, *9*, 1351. [CrossRef]
5. Ali, M.; Kiliçman, A.; Zahedi Khameneh, A. Application of Induced Preorderings in Score Function-Based Method for Solving Decision-Making with Interval-Valued Fuzzy Soft Information. *Mathematics* **2021**, *9*, 1575. [CrossRef]
6. Dong, Y.-Y.; Shi, F.-G. L-Fuzzy Sub-Effect Algebras. *Mathematics* **2021**, *9*, 1596. [CrossRef]
7. Dzitac, S.; Nădăban, S. Soft Computing for Decision-Making in Fuzzy Environments: A Tribute to Professor Ioan Dzitac. *Mathematics* **2021**, *9*, 1701. [CrossRef]
8. Lupaş, A.A.; Oros, G.I. New Applications of Sălăgean and Ruscheweyh Operators for Obtaining Fuzzy Differential Subordinations. *Mathematics* **2021**, *9*, 2000. [CrossRef]
9. Bashir, S.; Shahzadi, S.; Al-Kenani, A.N.; Shabir, M. Regular and Intra-Regular Semigroups in Terms of m-Polar Fuzzy Environment. *Mathematics* **2021**, *9*, 2031. [CrossRef]
10. Sainz, M.A.; Calm, R.; Jorba, L.; Contreras, I.; Vehi, J. Marks: A New Interval Tool for Uncertainty, Vagueness and Indiscernibility. *Mathematics* **2021**, *9*, 2116. [CrossRef]
11. Nicolas, C.; Müller, J.; Arroyo-Cañada, F.-J. A Fuzzy Inference System for Management Control Tools. *Mathematics* **2021**, *9*, 2145. [CrossRef]
12. Oros, G.I. Fuzzy Differential Subordinations Obtained Using a Hypergeometric Integral Operator. *Mathematics* **2021**, *9*, 2539. [CrossRef]
13. Anwar, M.Z.; Bashir, S.; Shabir, M.; Alharbi, M.G. Multigranulation Roughness of Intuitionistic Fuzzy Sets by Soft Relations and Their Applications in Decision Making. *Mathematics* **2021**, *9*, 2587. [CrossRef]
14. Lupaş, A.A. Applications of the Fractional Calculus in Fuzzy Differential Subordinations and Superordinations. *Mathematics* **2021**, *9*, 2601. [CrossRef]

mathematics

MDPI

Article

Fuzzy Inner Product Space: Literature Review and a New Approach

Lorena Popa and Lavinia Sida *

Department of Mathematics and Computer Science, Aurel Vlaicu University of Arad, Elena Dragoi 2, RO-310330 Arad, Romania; lorena.popa@uav.ro
* Correspondence: lavinia.sida@uav.ro

Abstract: The aim of this paper is to provide a suitable definition for the concept of fuzzy inner product space. In order to achieve this, we firstly focused on various approaches from the already-existent literature. Due to the emergence of various studies on fuzzy inner product spaces, it is necessary to make a comprehensive overview of the published papers on the aforementioned subject in order to facilitate subsequent research. Then we considered another approach to the notion of fuzzy inner product starting from P. Majundar and S.K. Samanta's definition. In fact, we changed their definition and we proved some new properties of the fuzzy inner product function. We also proved that this fuzzy inner product generates a fuzzy norm of the type Nădăban-Dzitac. Finally, some challenges are given.

Keywords: fuzzy Hilbert space; fuzzy inner product; fuzzy norm

MSC: 46A16; 46S40

Citation: Popa, L.; Sida, L. Fuzzy Inner Product Space: Literature Review and a New Approach. *Mathematics* **2021**, *9*, 765. https://doi.org/10.3390/math9070765

Academic Editor: Basil Papadopoulos

Received: 22 February 2021
Accepted: 27 March 2021
Published: 1 April 2021

1. Introduction

The research papers of A.K. Katsaras [1,2] laid the foundations of the fuzzy functional analysis. Moreover, he was the first one who introduced the concept of a fuzzy norm. This concept has amassed great interest among mathematicians. Thus, in 1992, C. Felbin [3] introduced a new idea of a fuzzy norm in a linear space by associating a real fuzzy number to each element of the linear space. In 2003, T. Bag and S.K. Samanta [4] put forward a new concept of a fuzzy norm, which was a fuzzy set on $X \times \mathbb{R}$. New fuzzy norm concepts were later introduced by R. Saadati and S.M. Vaezpour [5], C. Alegre and S.T. Romaguera [6], R. Ameri [7], I. Goleţ [8], A.K. Mirmostafaee [9]. In this paper we use the definition introduced by S. Nădăban and I. Dzitac [10].

Hilbert spaces lay at the core of functional analysis. They frequently and naturally appear in the fields of Mathematics and Physics, being indispensable in the theory of differential equations, quantum mechanics, quantum logic, quantum computing, the Fourier analysis (with applications in the signal theory). Therefore, many mathematicians have focused on finding an adequate definition of the fuzzy inner product space. Although there are many research papers focused on the concept of a fuzzy norm and its diverse applications, there are few papers which study the concept of a fuzzy inner product.

Even though there are few results, we consider that every breakthrough has been an important one. Moreover, we are certain that there exists a correct definition of the fuzzy inner product that when discovered, will not only generate a worldwide consensus on this subject but trigger countless applications in various fields.

The importance of this subject has led us to write this paper and we have tried to further the knowledge on the matter.

Furthermore, we wish to mention the main results of the already existent literature.

R. Biswas in [11] defined the fuzzy inner product of elements in a linear space and two years later J.K. Kohli and R. Kumar altered the Biswas's definition of inner product space [12]. In fact, they showed that the definition of a fuzzy inner product space in terms of the conjugate of a vector is redundant and that those definitions are only restricted to the real linear spaces. They also introduced the fuzzy co-inner product spaces and the fuzzy co-norm functions in their paper.

Two years later, in 1995, Eui-Whan Cho, Young-Key Kim and Chae-Seob Shin introduced and defined in [13] a fuzzy semi-inner-product space and investigated some properties of this fuzzy semi inner product space, those definitions are not restricted to the real linear spaces.

In 2008, P. Majumdar and S.K. Samanta [14] succeeded in taking the first step towards finding a reliable definition of a fuzzy inner product space. From their definition we can identify a serious problem in regard to finding a new reliable definition for the fuzzy inner product. The classical inequality Cauchy-Schwartz cannot be obtained by applying the other axioms and thus had to be introduced itself as an axiom (axiom (FIP2)).

In 2009, M. Goudarzi, S.M. Vaezpour and R. Saadati [15] introduced the concept of intuitionistic fuzzy inner product space. In this context, the Cauchy-Schwartz inequality, the Pythagorean Theorem and some convergence theorems were established.

In the same year, M. Goudarzi and S.M. Vaezpour [16] alter the definition of the fuzzy inner product space and prove several interesting results which take place in each fuzzy inner product space. More specifically, they introduced the notion of a fuzzy Hilbert space and deduce a fuzzy version of Riesz representation theorem.

In 2013, S. Mukherjee and T.Bag [17] amends the definition put forward by M.Goudarzi and S.M. Vaezpour by discarding the (FI-6) condition and by enacting minor changes to the (FI-4) and (FI-5) conditions.

In 2010, A. Hasankhani, A. Nazari and M. Saheli [18] introduced a new concept of a fuzzy Hilbert space. This concept is entirely different from the previous ones as this fuzzy inner product generates a new fuzzy norm of type Felbin.

The disadvantage of this definition is that only linear spaces over $\mathbb{R}$ can be considered. Another disadvantage is the difficulty of working with real fuzzy numbers.

In the subsequent years, many papers addressing this theme were published (see [7,19–24]).

In 2016, M. Saheli and S.Khajepour Gelousalar [25] modified the definition of the fuzzy inner product space and proved some properties of the new fuzzy inner product space.

Also, in 2016, Z. Solimani and B. Daraby [26] slightly altered the definition of a fuzzy scalar product introduced in [18] by changing the (IP2) condition and merging the (IP4) and (IP5).

In 2017, E. Mostofian, M. Azhini and A. Bodaghi [27] presented two new concepts of fuzzy inner product spaces and investigated some of basic properties of these spaces.

This paper is organized as follows—in Section 2 we make a literature review. Such an approach is deemed useful for the readers as it would enable them to better understand the evolution of the fuzzy inner product space concepts. Thus, this section can constitute a starting point for other mathematicians interested in this subject. In Section 3 we introduce a new definition of the fuzzy inner product space starting from P. Majumdar and S.K. Samanta's definition [14]. In fact, we modified the P. Majumdar and S.K. Samanta's definition of inner product space and we introduced and proved some new properties of the fuzzy inner product function. This paper ends up with some conclusions and future works in Section 4.

2. Preliminaries

Definition 1. *[10] Let X be a vector space over a field $\mathbb{K}$ and $*$ be a continuous t-norm. A fuzzy set N in $X \times [0, \infty]$ is called a fuzzy norm on X if it satisfies:*

(N1) $N(x, 0) = 0, (\forall)x \in X;$

(N2) $[N(x,t) = 1, (\forall)t > 0]$ *iff* $x = 0;$

(N3) $N(\lambda x, t) = N\left(x, \frac{t}{|\lambda|}\right), (\forall)x \in X, (\forall)t \geq 0, (\forall)\lambda \in \mathbb{K}^*;$

(N4) $N(x + y, t + s) \geq N(x, t) * N(y, t), (\forall)x, y \in X, (\forall)t, s \geq 0;$

(N5) $(\forall)x \in X, N(x,\cdot)$ *is left continuous and* $\lim_{t\to\infty} N(x,t) = 1$.

The triplet $(X, N, *)$ *will be called fuzzy normed linear space (briefly FNLS).*

Definition 2. [14] *A fuzzy inner product space (FIP-space) is a pair* (X, P), *where X is a linear space over* $\mathbb{C}$ *and P is a fuzzy set in* $X \times X \times \mathbb{C}$ *s.t.*

(FIP1) For $s,t \in \mathbb{C}$, $P(x+y,z,|t|+|s|) \geq \min\{P(x,z,|t|), P(y,z,|s|)\}$;
(FIP2) For $s,t \in \mathbb{C}$, $P(x,y,|st|) \geq \min\{P(x,x,|s|^2), P(y,y,|t|^2)\}$;
(FIP3) For $t \in \mathbb{C}$, $P(x,y,t) = P(y,x,\bar{t})$;
(FIP4) $P(\alpha x,y,t) = P\left(x,y,\frac{t}{|\alpha|}\right), t \in \mathbb{C}, \alpha \in \mathbb{C}^*$;
(FIP5) $P(x,x,t) = 0, (\forall)t \in \mathbb{C} \setminus \mathbb{R}^+$;
(FIP6) $[P(x,x,t) = 1, (\forall)t > 0]$ *iff* $x = 0$;
(FIP7) $P(x,x,\cdot) : \mathbb{R} \to [0,1]$ *is a monotonic non-decreasing function of* $\mathbb{R}$ *and* $\lim_{t\to\infty} P(x,x,t) = 1$.

P will be called the fuzzy inner product on X.

Definition 3. [15] *A fuzzy inner product space (FIP-space) is a triplet* $(X;P;*)$, *where X is a real linear space,* $*$ *is a continuous t-norm and P is a fuzzy set on* $X^2 \times \mathbb{R}$ *satisfying the following conditions for every* $x;y;z \in X$ *and* $t \in \mathbb{R}$.

(FIP1) $P(x,y,0) = 0$;
(FIP2) $P(x,y,t) = P(y,x,t)$;
(FIP3) $P(x,x,t) = H(t), \forall t \in \mathbb{R}$ *iff* $x = 0$, *where* $H(t) = \begin{cases} 1, & if\ t > 0 \\ 0, & if\ t \leq 0 \end{cases}$;
(FIP4) For any real number α, $P(\alpha x,y,t) = \begin{cases} P(x,y,\frac{t}{\alpha}), & if\ \alpha > 0 \\ H(t), & if\ \alpha = 0; \\ 1 - P(x,y,\frac{t}{-\alpha}), & if\ \alpha < 0 \end{cases}$
(FIP5) $\sup_{s+r=t} (P(x,z,s) * P(y,z,r)) = P(x+y,z,t)$;
(FIP6) $P(x,y,\cdot) : \mathbb{R} \to [0,1]$ *is continuous on* $\mathbb{R} \setminus \{0\}$;
(FIP7) $\lim_{t\to\infty} P(x,y,t) = 1$.

Definition 4. [16] *A fuzzy inner product space (FIP - space) is a triplet* $(X, P, *)$, *where X is a real linear space,* $*$ *is a continuous t-norm and P is a fuzzy set in* $X \times X \times \mathbb{R}$ *s.t. the following conditions hold for every* $x,y,z \in X$ *and* $s,t,r \in \mathbb{R}$.

(FI-1) $P(x,x,0) = 0$ *and* $P(x,x,t) > 0$, $(\forall)t > 0$;
(FI-2) $P(x,x,t) \neq H(t)$ *for same* $t \in \mathbb{R}$ *iff* $x \neq 0$;
(FI-3) $P(x,y,t) = P(y,x,t)$;
(FI-4) For any real number α, $P(\alpha x,y,t) = \begin{cases} P(x,y,\frac{t}{\alpha}), & if\ \alpha > 0 \\ H(t), & if\ \alpha = 0; \\ 1 - P(x,y,\frac{t}{-\alpha}), & if\ \alpha < 0 \end{cases}$
(FI-5) $\sup_{s+r=t} (P(x,z,s) * P(y,z,r)) = P(x+y,z,t)$;
(FI-6) $P(x,y,\cdot) : \mathbb{R} \to [0,1]$ *is continuous on* $\mathbb{R} \setminus \{0\}$;
(FI-7) $\lim_{t\to\infty} P(x,y,t) = 1$.

Definition 5. [14] *Let X be a linear space over* $\mathbb{R}$. *A fuzzy set P in* $X \times X \times \mathbb{R}$ *is called fuzzy real inner product on X if* $(\forall)x,y,z \in X$ *and* $t \in \mathbb{R}$, *the following conditions hold:*

(FI-1) $P(x,x,0) = 0$, $(\forall)t < 0$;
(FI-2) $[P(x,x,t) = 1, (\forall)t > 0]$ *iff* $x = 0$;
(FI-3) $P(x,y,t) = P(y,x,t)$;

(FI-4) $P(\alpha x, y, t) = \begin{cases} P(x, y, \frac{t}{\alpha}), & if\ \alpha > 0 \\ H(t), & if\ \alpha = 0; \\ 1 - P(x, y, \frac{t}{\alpha}), & if\ \alpha < 0 \end{cases}$

(FI-5) $P(x+y, z, t+s) \geq \min\{P(x,z,t), P(y,z,s)\}$;

(FI-6) $\lim_{t \to \infty} P(x,y,t) = 1$.

The pair (X, P) is called fuzzy real inner space.

In order to present the definition of A. Hasankhani, A. Nazari and M. Saheli, we firstly need to define some concepts.

Definition 6. [28] *A fuzzy set in $\mathbb{R}$, namely a mapping $x : \mathbb{R} \to [0,1]$, with the following properties:*

(1) *x is convex, that is, $x(t) \geq \min\{x(s), x(r)\}$ for $s \leq t \leq r$;*

(2) *x is normal, that is, $(\exists) t_0 \in \mathbb{R} : x(t_0) = 1$;*

(3) *x is upper semicontinuous, that is, $(\forall) t \in \mathbb{R}, (\forall) \alpha \in [0,1] : x(t) < \alpha, (\exists) \delta > 0$ s.t.$|s - t| < \delta \Rightarrow x(s) < \alpha$*

is called fuzzy real number. We denote by $\mathfrak{F}(\mathbb{R})$ the set of all fuzzy real numbers.

Definition 7. [29] *The arithmetic operation $+, -, \cdot, /$ on $\mathfrak{F}(\mathbb{R})$ are defined by:*

$$(x+y)(t) = \bigvee_{s \in \mathbb{R}} \min\{x(s), y(t-s)\},\ (\forall) t \in \mathbb{R};$$
$$(x-y)(t) = \bigvee_{s \in \mathbb{R}} \min\{x(s), y(s-t)\},\ (\forall) t \in \mathbb{R};$$
$$(xy)(t) = \bigvee_{s \in \mathbb{R}^*} \min\{x(s), y(t/s)\},\ (\forall) t \in \mathbb{R};$$
$$(x/y)(t) = \bigvee_{s \in \mathbb{R}} \min\{x(ts), y(s)\},\ (\forall) t \in \mathbb{R}.$$

Remark 1. *Let $x \in \mathfrak{F}(\mathbb{R})$ and $\alpha \in (0,1]$. The α-level sets $[x]_\alpha = \{t \in \mathbb{R} : x(t) \geq \alpha\}$ are closed intervals $[x_\alpha^-, x_\alpha^+]$.*

Definition 8. [18] *Let X be a linear space over $\mathbb{R}$. A fuzzy inner product on X is a mapping $< \cdot, \cdot >: X \times X \to \mathfrak{F}(\mathbb{R})$ s.t. $(\forall) x, y, z \in X, (\forall) r \in \mathbb{R}$, we have:*

(IP1) $< x+y, z > = < x, z > \oplus < y, z >$;

(IP2) $< rx, y > = \tilde{r} < x, y >$, *where* $\tilde{r} = \begin{cases} 1, & if\ t = r \\ 0, & if\ t \neq r \end{cases}$;

(IP3) $< x, y > = < y, x >$;

(IP4) $< x, x > \geq 0$;

(IP5) $\inf_{\alpha \in (0,1]} < x, x >_\alpha^- = 0$ *if* $x \neq 0$;

(IP6) $< x, x > = \tilde{0}$ *iff* $x = 0$.

The pair $(X, \langle \cdot, \cdot \rangle)$ is called fuzzy inner product space.

Definition 9. [25] *A fuzzy inner product space is a triplet $(X, P, *)$, where X is a fuzzy set in $X \times X \times \mathbb{R}$ satisfying the following conditions for every $x, y, z \in X$ and $t, s \in \mathbb{R}$:*

(FI1) $P(x, y, 0) = 0$;

(FI2) $P(x, y, t) = P(y, x, t)$;

(FI3) $[P(x, x, t) = 1, (\forall) t > 0]$ *iff* $x = 0$;

(FI4) $(\forall) \alpha \in \mathbb{R}, t \neq 0, \quad P(\alpha x, y, t) = \begin{cases} P(x, y, \frac{t}{\alpha}), & if\ \alpha > 0 \\ H(t), & if\ \alpha = 0; \\ 1 - P(x, y, \frac{t}{\alpha}), & if\ \alpha < 0 \end{cases}$

(FI5) $P(x, z, t) * P(y, z, s) \leq P(x+y, z, t+s), (\forall) t, s > 0$;

(FI6) $\lim_{t \to \infty} P(x, y, t) = 1$.

Definition 10. [26] *Let X be a linear space over* $\mathbb{R}$. *A fuzzy inner product on X is a mapping* $< \cdot, \cdot >: X \times X \to \mathfrak{F}^*(\mathbb{R})$, *where* $\mathfrak{F}^*(\mathbb{R}) = \{\eta \in \mathfrak{F}(\mathbb{R}) : \eta(t) = 0 \text{ if } t < 0\}$, *with the following properties* $(\forall)x, y, z \in X, (\forall)r \in \mathbb{R}$:

(FIP1) $< x + y, z > = < x, z > \oplus < y, z >$;
(FIP2) $< rx, y > = | \tilde{r} | < x, y >$;
(FIP3) $< x, y > = < y, x >$;
(FIP4) $x \neq 0 \Rightarrow < x, x > (t) = 0, (\forall)t < 0$;
(FIP5) $< x, x > = \tilde{0}$ *iff* $x = 0$.

The pair $(X, < \cdot, \cdot >)$ *is called fuzzy inner product space.*

3. A New Approach for Fuzzy Inner Product Space

We will denote by $\mathbb{C}$ the space of complex numbers and we will denote by $\mathbb{R}_+^*$ the set of all strict positive real numbers.

Definition 11. *Let H be a linear space over* $\mathbb{C}$. *A fuzzy set P in* $H \times H \times \mathbb{C}$ *is called a fuzzy inner product on H if it satisfies:*

(FIP1) $P(x, x, v) = 0, (\forall)x \in H, (\forall)v \in \mathbb{C} \setminus \mathbb{R}_+^*$;
(FIP2) $P(x, x, t) = 1, (\forall)t \in \mathbb{R}_+^*$ *if and only if* $x = 0$;
(FIP3) $P(\alpha x, y, v) = P\left(x, y, \frac{v}{|\alpha|}\right), (\forall)x, y \in H, (\forall)v \in \mathbb{C}, (\forall)\alpha \in \mathbb{C}^*$;
(FIP4) $P(x, y, v) = P(y, x, \overline{v}), (\forall)x, y \in H, (\forall)v \in \mathbb{C}$;
(FIP5) $P(x + y, z, v + w) \geq \min\{P(x, z, v), P(y, z, w)\}, (\forall)x, y, z \in H, (\forall)v, w \in \mathbb{C}$;
(FIP6) $P(x, x, \cdot) : \mathbb{R}_+ \to [0, 1], (\forall)x \in H$ *is left continuous and* $\lim_{t \to \infty} P(x, x, t) = 1$;
(FIP7) $P(x, y, st) \geq \min\{P(x, x, s^2), P(y, y, t^2)\}, (\forall)x, y \in H, (\forall)s, t \in \mathbb{R}_+^*$.

The pair (H, P) *will be called fuzzy inner product space.*

Example 1. *Let H be a linear space over* $\mathbb{C}$ *and* $< \cdot, \cdot >: H \times H \to \mathbb{C}$ *be an inner product. Then* $P : H \times H \times \mathbb{C} \to [0, 1]$,

$$P(x, y, s) = \begin{cases} \frac{s}{s + |<x, y>|}, & \text{if } s \in \mathbb{R}_+^* \\ 0, & \text{if } s \in \mathbb{C} \setminus \mathbb{R}_+^* \end{cases}$$

is a fuzzy inner product on H.

Let verify now the conditions from the definition.

(FIP1) $P(x, x, v) = 0, (\forall)x \in H, (\forall)v \in \mathbb{C} \setminus \mathbb{R}_+^*$ is is obvious from definition of P.
(FIP2) $P(x, x, t) = 1, (\forall)v \in \mathbb{R}_+^* \Leftrightarrow t + |< x, x >| = t, (\forall)t > 0 \Leftrightarrow |< x, x >| = 0 \Leftrightarrow x = 0$.
(FIP3) $P(\alpha x, y, v) = P\left(x, y, \frac{v}{|\alpha|}\right), (\forall)x, y \in H, (\forall)v \in \mathbb{C}, (\forall)\alpha \in \mathbb{C}$ is obvious for $v \in \mathbb{C} \setminus \mathbb{R}_+^*$.
If $v \in \mathbb{R}_+^*$, than

$$P(\alpha x, y, v) = \frac{v}{v + |< \alpha x, y >|} = \frac{v}{v + |\alpha| \cdot |< x, y >|} = \frac{\frac{v}{|\alpha|}}{\frac{v}{|\alpha|} + |< x, y >|} = P\left(x, y, \frac{v}{|\alpha|}\right).$$

(FIP4) $P(x, y, v) = P(y, x, \overline{v}), (\forall)x, y \in H, (\forall)v \in \mathbb{C}$ is obvious for $v \in \mathbb{C} \setminus \mathbb{R}_+^*$.
If $v \in \mathbb{R}_+^*$, then $v = \overline{v}$ and

$$P(x, y, v) = \frac{v}{v + |< x, y >|} = \frac{\overline{v}}{\overline{v} + |< y, x >|} = P(y, x, \overline{v})$$

(FIP5) $P(x + y, z, v + w) \geq \min\{P(x, z, v), P(y, z, w)\}, (\forall)x, y, z \in H, (\forall)v, w \in \mathbb{C}$.
If at least one of v and w is from $\mathbb{C} \setminus \mathbb{R}_+^*$, then the result is obvious.

If $v, w \in \mathbb{R}_+^*$, let us assume without loss of generality that $P(x,z,v) \leq P(y,z,w)$. Then

$$\frac{v}{v+ |< x,z >|} \leq \frac{w}{w+ |< y,z >|} \Rightarrow$$
$$\Rightarrow \frac{v+ |< x,z >|}{v} \geq \frac{w+ |< y,z >|}{w} \Rightarrow$$
$$\Rightarrow 1 + \frac{|< x,z >|}{v} \geq 1 + \frac{|< y,z >|}{w} \Rightarrow$$
$$\Rightarrow \frac{|< x,z >|}{v} \geq \frac{|< y,z >|}{w} \Rightarrow$$
$$\Rightarrow \frac{w}{v} |< x,z >| \geq |< y,z >| \Rightarrow$$
$$\Rightarrow |< x,z >| + \frac{w}{v} |< x,z >| \geq |< x,z >| + |< y,z >| \Rightarrow$$
$$\Rightarrow \frac{v+w}{v} |< x,z >| \geq |< x+y,z >| \Rightarrow$$
$$\Rightarrow \frac{|< x,z >|}{v} \geq \frac{|< x+y,z >|}{v+w} \Rightarrow$$
$$\Rightarrow \frac{|< x,z >|}{v} + 1 \geq \frac{|< x+y,z >|}{v+w} + 1 \Rightarrow$$
$$\Rightarrow \frac{v+ |< x,z >|}{v} \geq \frac{(v+w)+ |< x+y,z >|}{v+w} \Rightarrow$$
$$\Rightarrow \frac{v}{v+ |< x,z >|} \leq \frac{v+w}{(v+w)+ |< x+y,z >|} \Rightarrow$$
$$\Rightarrow P(x,z,v) \leq P(x+y,z,v+w)$$
$$\Rightarrow P(x+y,z,v+w) \geq \min\{P(x,z,v), P(y,z,w)\}, (\forall)x,y,z \in H, (\forall)v,w \in \mathbb{C}.$$

(FIP6) $P(x,x,\cdot) : \mathbb{R}_+ \to [0,1], (\forall)x \in H$ is left continuous function and $\lim_{t\to\infty} P(x,x,t) = 1$.

$$\lim_{t\to\infty} P(x,x,t) = \lim_{t\to\infty} \frac{t}{t+ |< x,x >|} = \lim_{t\to\infty} \frac{t}{t(1 + \frac{|<x,x>|}{t})} = 1.$$

$F(x,x,\cdot)$ is left continuous in $t > 0$ follows from definition.

(FIP7) $P(x,y,st) \geq \min\{P(x,x,s^2), P(y,y,t^2)\}, (\forall)x,y \in H, (\forall)s,t \in \mathbb{R}_+^*$. If at least one of s and t is from $\mathbb{C} \setminus \mathbb{R}_+^*$, then the result is obvious.
If $s,t \in \mathbb{R}_+^*$, let us assume without loss of generality that $P(x,x,s^2) \leq P(y,y,t^2)$. Then

$$\frac{s^2}{s^2+ |< x,x >|} \leq \frac{t^2}{t^2+ |< y,y >|} \Leftrightarrow$$
$$t^2 |< x,x >| \geq s^2 |< y,y >| .$$

Thus by Cauchy–Schwartz inequality we obtain

$$s |< x,y >| \leq \sqrt{|< x,x >|} \cdot s\sqrt{|< y,y >|} \leq \sqrt{|< x,x >|} \cdot t\sqrt{|< x,x >|} = t |< x,x >| \Rightarrow$$
$$\Rightarrow s^2 |< x,y >| \leq st |< x,x >| \Rightarrow$$
$$\Rightarrow s^3t + s^2 |< x,y >| \leq s^3t + st |< x,x >| \Rightarrow$$
$$\Rightarrow s^2(st+ |< x,y >|) \leq st(s^2+ |< x,x >|) \Rightarrow$$
$$\frac{s^2}{s^2+ |< x,x >|} \leq \frac{st}{st+ |< x,y >|} \Rightarrow$$
$$\Rightarrow P(x,x,s^2) \leq P(x,y,st) \Rightarrow$$
$$\Rightarrow P(x,y,st) \geq \min\left\{P(x,x,s^2), P(y,y,t^2)\right\}, (\forall)x,y \in H, (\forall)s,t \in \mathbb{R}_+^*.$$

Proposition 1. *For $x, y \in H$, $v \in \mathbb{C}$ and $\alpha \in \mathbb{C}$ we have*

$$P(x, \alpha y, v) = P\left(x, y, \frac{v}{|\alpha|}\right).$$

Proof. From (FIP3) and (FIP4) it follows $P(x, \alpha y, v) = P(\alpha y, x, \overline{v}) = P\left(y, x, \frac{\overline{v}}{|\alpha|}\right) = P\left(x, y, \frac{\overline{\overline{v}}}{|\alpha|}\right) = P\left(x, y, \frac{v}{|\alpha|}\right)$. □

Proposition 2. *For $x \in H$, $v \in \mathbb{R}_+^*$ we have*

$$P(x, 0, v) = 1.$$

Proof. From (FIP3) and (FIP6) it follows

$$P(x, 0, v) = P(x, 0, 2nv) = P(x, x - x, nv + nv) \geq \min\{P(x, x, nv), P(x, x, nv)\} =$$

$$P(x, x, nv) \overset{n \to \infty}{\longrightarrow} 1$$

So $P(x, 0, v) = 1$. □

Proposition 3. *For $y \in H$, $v \in \mathbb{R}_+^*$ we have*

$$P(0, y, v) = 1.$$

Proposition 4. *$P(x, y, \cdot) : \mathbb{R}_+ \to [0, 1]$ is a monotonic non-decreasing function on $\mathbb{R}_+$, $(\forall) x, y \in H$.*

Proof. Let $s, t \in \mathbb{R}_+$, $s \leq t$. Then $(\exists) p$ such that $t = s + p$ and

$$P(x, y, t) = P(x + 0, y, s + p) \geq \min\{P(x, y, s), P(0, y, p)\} = P(x, y, s).$$

Hence $P(x, y, s) \leq P(x, y, t)$ for $s \leq t$. □

Corollary 1. $P(x, y, st) \geq \min\{P(x, y, s^2), P(x, y, t^2)\}$, $(\forall) x, y \in H$, $(\forall) s, t \in \mathbb{R}_+^*$.

Proof. Let $s, t \in \mathbb{R}_+$, $s \leq t$. Then

$$P(x, y, s^2) \leq P(x, y, st) \leq P(x, y, t^2).$$

Hence $P(x, y, st) \geq \min\{P(x, y, s^2), P(x, y, t^2)\}$.
Let now $s, t \in \mathbb{R}_+$, $t \leq s$. Then

$$P(x, y, t^2) \leq P(x, y, st) \leq P(x, y, s^2).$$

Hence $P(x, y, st) \geq \min\{P(x, y, s^2), P(x, y, t^2)\}$. □

Proposition 5. $P(x, y, v) \geq \min\{P(x, y - z, v), P(x, y + z, v)\}$, $(\forall) x, y, z \in H$, $(\forall) v \in \mathbb{C}$.

Proof.

$$P(x, y, v) = P(x, 2y, 2v) = P(x, y + z + y - z, v + v) \geq \min\{P(x, y + z, v), P(x, y - z, v)\}.$$

Hence $P(x, y, v) \geq \min\{P(x, y - z, v), P(x, y + z, v)\}$. □

Theorem 1. *If (H, P) be a fuzzy inner product space, then $N : X \times [0, \infty) \to [0, 1]$ defined by*

$$N(x, t) = P(x, x, t^2)$$

is a fuzzy norm on X.

Proof. ***(N1)*** $N(x,0) = P(x,x,0) = 0, (\forall)x \in H$ from **(FIP1)**;

(N2) $[N(x,t) = 1, (\forall)t > 0] \Leftrightarrow [P(x,x,t^2) = 1, (\forall)t > 0] \Leftrightarrow x = 0$ from **(FIP2)**;

(N3) $N(\lambda x,t) = P(\lambda x, \lambda x, t^2) = P\Big(x, \lambda x, \frac{t^2}{|\lambda|}\Big) = P\Big(\lambda x, x, \frac{t^2}{|\lambda|}\Big) = P\Big(\lambda x, x, \frac{t^2}{|\lambda|}\Big) = P\Big(x, x, \frac{t^2}{|\lambda|^2}\Big) == N\Big(x, \frac{t}{|\lambda|}\Big), (\forall)t \geq 0, (\forall)\lambda \in \mathbb{K}^*$;

(N4) $N(x+t, t+s) \geq \min\{N(x,t), N(y,s)\}, (\forall)x,y \in H, (\forall)t,s \geq 0$.

If $t = 0$ or $s = 0$ the previous inequality is obvious. We asume that $t,s > 0$.

$$\begin{aligned} N(x+y,t+s) &= P\Big(x+y, x+y, (t+s)^2\Big) = \\ &= P(x+y, x+y, t^2+s^2+ts+ts) \geq \\ &\geq P(x, x+y, t^2+ts) \wedge P(y, x+y, s^2+ts) \geq \\ &\geq P(x,x,t^2) \wedge P(x,y,ts) \wedge P(y,x,ts) \wedge P(y,y,s^2) = \\ &= P(x,x,t^2) \wedge P(y,y,s^2) = \\ &= \min\{N(x,t), N(y,s)\}; \end{aligned}$$

(N5) From **(FIP6)** it result that $N(x,\cdot)$ is left continuous and $\lim_{t\to\infty} N(x,t) = 1$.

□

4. Conclusions and Future Works

In this paper, we wrote a literature review regarding the diverse approaches of fuzzy inner product space concept, but we also introduced a new approach.

We have thus laid the ground for further research on the problems within the fuzzy Hilbert space theory, searching for analogies in this fuzzy context for the Pythagorean theorem, for the parallelogram law, as well as for other orthogonality problems.

The following step would be to study the linear and bounded operators in a fuzzy Hilbert space. Recently, there have been many important results concerning the linear and bounded operators in a fuzzy Banach space (see [30–34]), fact which motivates us even further to try to achieve this goal.

This research will be followed by other papers in which we will firstly define the concept of an adjoint of a linear and bounded operator on a fuzzy Hilbert space. This concept will allow us to study important classes of operators such as self-adjoint operators, normal operators and unitary operators. We will then follow up with the spectral theory and we will also construct a analytic functional calculus.

Last, but not least, we will study the aforementioned orthogonality in the fuzzy Hilbert space. Thus, this will enable us to observe the properties of the self-adjoint projections and to undertake directly decompositions of the fuzzy Hilbert space.

This paper summarized the current research status of the fuzzy inner product spaces and can thus facilitate researchers in writing their future papers on this topic.

Author Contributions: Conceptualization, L.P.; Data curation, L.S.; Formal analysis, L.S. Funding acquisition, L.P.; Resources, L.S.; Writing–review and editing, L.P. Both authors have read and agreed to the published version of the manuscript.

Funding: This research received no funding.

Institutional Review Board Statement: Not applicable.

Informed Consent Statement: Not applicable.

Data Availability Statement: Not applicable.

Acknowledgments: The authors are extremely grateful to the editor and reviewers, for their very carefully reading of the paper and for their valuable comments and suggestions which have been useful to increase the scientific quality and presentation of the paper. The authors wish to record their sincere gratitude to S. Nădăban for his help in the preparation of this paper.

Conflicts of Interest: The authors declare no conflict of interest.

References

1. Katsaras, A.K. Fuzzy topological vector spaces I. *Fuzzy Sets Syst.* **1981**, *6*, 85–95. [CrossRef]
2. Katsaras, A.K. Fuzzy topological vector spaces II. *Fuzzy Sets Syst.* **1984**, *12*, 143–154. [CrossRef]
3. Felbin, C. Finite dimensional fuzzy normed linear space. *Fuzzy Sets Syst.* **1992**, *48*, 239–248. [CrossRef]
4. Bag, T.; Samanta, S.K. Finite dimensional fuzzy normed linear spaces. *J. Fuzzy Math.* **2003**, *11*, 687–705.
5. Saadati, R.; Vaezpour S.M. Some results on fuzzy Banach spaces. *J. Appl. Math. Comput.* **2005**, *17*, 475–484. [CrossRef]
6. Alegre, C.; Romaguera, S.T. Characterizations of fuzzy metrizable topological vector spaces and their asymmetric generalization in terms of fuzzy (quasi-) norms. *Fuzzy Sets Syst.* **2010**, *161*, 2181–2192. [CrossRef]
7. Ameri, R. Fuzzy inner product and fuzzy norm of hyperspaces. *Iran. J. Fuzzy Syst.* **2014**, *11*, 125–135.
8. Goleţ, I. On generalized fuzzy normed spaces and coincidence point theorems. *Fuzzy Sets Syst.* **2010**, *161*, 1138–1144. [CrossRef]
9. Mirmostafaee, A.K. Perturbation of generalized derivations in fuzzy Menger normed algebras. *Fuzzy Sets Syst.* **2012**, *195*, 109–117. [CrossRef]
10. Nădăban, S.; Dzitac I. Atomic decompositions of fuzzy normed linear spaces for wavelet applications. *Inform. Vilnius* **2014**, *25*, 643–662. [CrossRef]
11. Biswas, R. Fuzzy inner product spaces and fuzzy norm functions. *Inf. Sci.* **1991**, *53*, 185–190. [CrossRef]
12. Kohli, J.K.; Kumar, R. On fuzzy inner product spaces and fuzzy co-inner product spaces. *Fuzzy Sets Syst.* **1993**, *53*, 227–232. [CrossRef]
13. Cho, E.-I.; Kim, Y.-K.; Shin, C.-S. Fuzzy semi-inner product space. *J. Korea Soc. Math. Educ. Ser. B Pure Appl. Math.* **1995**, *2*, 163–171.
14. Majumdar, P.; Samanta, S.K. On fuzzy inner product spaces. *J. Fuzzy Math.* **2008**, *16*, 377–392.
15. Goudarzi, M.; Vaezpour, S.M.; Saadati, R. On the intuitionistic fuzzy inner product spaces. *Chaos Solitons Fractals* **2009**, *41*, 1105–1112. [CrossRef]
16. Goudarzi, M.; Vaezpour, M.S. On the definition of fuzzy Hilbert spaces and its application. *J. Nonlinear Sci. Appl.* **2009**, *2*, 46–59. [CrossRef]
17. Mukherjee, S.; Bag, T. Fuzzy real inner product spaces. *Ann. Fuzzy Math. Inform.* **2013**, *6*, 377–389.
18. Hasankhani, A.; Nazari, A.; Saheli, M. Some properties of fuzzy Hilbert spaces and norm of operators. *Iran. J. Fuzzy Syst.* **2010**, *7*, 129–157.
19. Al-Mayahi, N.F.; Radhi, I.H. On fuzzy co-pre-Hilbert spaces. *J. Kufa Math. Comput.* **2013**, *1*, 1–6.
20. Das, S.; Samanta, S.K. Operators on soft inner product spaces. *Fuzzy Inf. Eng.* **2014**, *6*, 435–450. [CrossRef]
21. Dey, A.; Pal, M. Properties of fuzzy inner product spaces. *Int. J. Fuzzy Log. Syst.* **2014**, *4*, 23–39. [CrossRef]
22. Gebray, G.; Reddy, B.K. Fuzzy metric on fuzzy linear spaces. *Int. J. Sci. Res.* **2014**, *3*, 2286–2288.
23. Mukherjee, S.; Bag, T. Some fixed point results in fuzzy inner product spaces. *Int. J. Math. Sci. Comput.* **2015**, *5*, 44–48.
24. Saheli, M. A comparative study of fuzzy inner product spaces. *Iran. J. Fuzzy Syst.* **2015**, *12*, 75–93.
25. Saheli, M.; Khajepour Gelousalar, S. Fuzzy inner product spaces. *Fuzzy Sets Syst.* **2016**, *303*, 149–162. [CrossRef]
26. Solimani, Z.; Daraby, B. A note on fuzzy inner product spaces. In Proceedings of the Extended Abstracts of the 4th Seminar on Functional Analysis and Its Applications, Ferdowsi University of Mashhad, Mashhad, Iran, 2–3 March 2016.
27. Mostofian, E.; Azhini, M.; Bodaghi, A. Fuzzy inner product spaces and fuzzy ortogonality. *Tbilisi Math. J.* **2017**, *10*, 157–171. [CrossRef]
28. Dzitac, I. The fuzzification of classical structures: A general view. *Int. J. Comput. Commun. Control.* **2015**, *10*, 772–788. [CrossRef]
29. Mizumoto, M.; Tanaka, J. Some properties of fuzzy numbers. In *Advances in Fuzzy Set Theory and Applications*; Gupta, M.M., Ragade, R.K., Yager, R.R., Eds.; Publishing House: New York, NY, USA; North-Holland, The Netherlands, 1979; pp. 153–164.
30. Bag, T.; Samanta, S.K. Fuzzy bounded linear operators. *Fuzzy Sets Syst.* **2005**, *151*, 513–547. [CrossRef]
31. Bag, T.; Samanta, S.K. Fuzzy bounded linear operators in Felbin's type fuzzy normed linear spaces. *Fuzzy Sets Syst.* **2008**, *159*, 685–707. [CrossRef]
32. Bînzar, T.; Pater, F.; Nădăban S. A Study of Boundedness in Fuzzy Normed Linear Spaces. *Symmetry* **2019**, *11*, 923. [CrossRef]
33. Bînzar, T.; Pater, F.; Nădăban S. Fuzzy bounded operators with application to Radon transform. *Chaos Solitons Fractals* **2020**, *141*, 110359. [CrossRef]
34. Kim, J.M.; Lee, K.Y. Approximation Properties in Felbin Fuzzy Normed Spaces. *Mathematics* **2019**, *7*, 1003. [CrossRef]

mathematics

MDPI

Article

M-Hazy Vector Spaces over M-Hazy Field

Faisal Mehmood and Fu-Gui Shi *

Beijing Key Laboratory on MCAACI, School of Mathematics and Statistics, Beijing Institute of Technology, Beijing 102488, China; 3820170030@bit.edu.cn or faisal.mehmood007@gmail.com
* Correspondence: fuguishi@bit.edu.cn

Abstract: The generalization of binary operation in the classical algebra to fuzzy binary operation is an important development in the field of fuzzy algebra. The paper proposes a new generalization of vector spaces over field, which is called M-hazy vector spaces over M-hazy field. Some fundamental properties of M-hazy field, M-hazy vector spaces, and M-hazy subspaces are studied, and some important results are also proved. Furthermore, the linear transformation of M-hazy vector spaces is studied and their important results are also proved. Finally, it is shown that M-fuzzifying convex spaces are induced by an M-hazy subspace of M-hazy vector space.

Keywords: M-hazy group; M-hazy ring; M-hazy field; M-hazy vector space; M-hazy subspace; M-fuzzifying convex space

Citation: Mehmood, F.; Shi, F.-G. M-Hazy Vector Spaces over M-Hazy Field. *Mathematics* **2021**, *9*, 1118. https://doi.org/10.3390/math9101118

Academic Editors: Sorin Nadaban and Ioan Dzitac

Received: 4 April 2021
Accepted: 12 May 2021
Published: 14 May 2021

1. Introduction

In 1971, Rosenfeld [1] published an innovative paper on fuzzy subgroups. This article introduced the new field of abstract algebra and the new field of fuzzy mathematics. Many scientists and researchers worked in this field and obtained fruitful research. Liu [2,3] gave an important generalization in the field of fuzzy algebra by introducing fuzzy subrings of a ring and fuzzy ideals. Demirci [4] firstly introduced the fuzzification of binary operation to group structure through fuzzy equality [5] and introduced "vague groups." After this work, many researchers used this concept and extended it to several other useful directions such as [6–10]. In Demirci's approach, the characteristic of the degree between the fuzzy binary operation is not used, and the identity and inverse element of an element are also not unique. Liu and Shi [11] proposed a new approach to fuzzify the group structure by characterizing the degree of fuzzy binary operation, which is called M-hazy groups. It is important to mention that M-hazy associative law has been defined in order to obtain M-hazy groups. Mehmood et al. [12] extended this concept to the ring structure and gave a new method to the fuzzification of rings, which is defined by M-hazy rings. It is also worth mentioning that an M-hazy distributive law has been proposed so as to define M-hazy rings. Furthermore, Mehmood et al. [13] also provided the homomorphism theorems of M-hazy rings with its induced fuzzifying convexities. Liu and Shi [14] proposed M-hazy lattices. Fan et al. [15] introduced an M-hazy Γ-semigroup.

Vector space has been the most widely studied and used in linear algebra theory. A vector space is a set of elements with a binary addition operator and a multiplication operator that has closure under these two operations over a field, all while satisfying a set of axioms. Vector spaces are the realm of linear combinations, also known as superpositions, weighted sums, and sums with coefficients. Such sums occur throughout mathematics, both pure and applied, including statistics, science, engineering, and economics. The key word is "linear". Even when studying nonlinear phenomena, it is often useful to approximate with a simpler linear model. You can say that vector spaces are one of the great organizing tools of mathematics, helping reveal a structural similarity in a wide variety of topics found in such different contexts that they may seem completely different. Suppose you stand in front of a house. It is rather old but beautifully constructed of masonry that exhibits excellent craftsmanship. You point at a brick in one of the lower layers and ask "What is the need for this brick?" Short answer: it helps the structure. Long answer: it can be missed in the sense that the building won't fall apart if you take it away, but it will damage it in various ways that will become clear if you live there for a couple of years. This house is a metaphor for mathematics. A vector space is a lot more than just a brick. It is one of the fundamental notions, and so is part of the foundation. The most fundamental notion is "Set," and a vector space is one notch higher, a set with a specific structure. Can you do without it? Not if you want to do any serious mathematics. You can refine the structure to get topological vector space, metric vector space, complete vector space, normed vector space, and inner product vector space, each a refinement of the former. The beauty of this is that a refinement inherits all properties of its ancestor, so you are saved a lot of groundwork and can explore additional properties. The need for refinement is usually triggered by questions from physics or engineering. A nice example is Fourier analysis that fits smoothly in a Hilbert space structure.

Since Katsaras and Liu [16] presented the notion of fuzzy vector spaces, many scientists and researchers explored its properties and obtained fruitful research such as [17–21]. Mordeson [22] defined bases of fuzzy vector spaces. Shi and Huang [23] defined fuzzy bases and fuzzy dimension of fuzzy vector spaces. Nanda [24,25] introduced fuzzy fields and linear spaces. Malik and Mordeson [26] defined fuzzy subfield of a field. Fang and Yan [27] introduced the notion of L-fuzzy topological vector spaces. Zhang and Xu [28,29] presented the concept of topological L-fuzzifying neighborhood structures. Furthermore, Yan and Wu [30] extended this concept by introducing fuzzifying topological vector spaces on completely distributive lattices. Wen et al. [31] gave the degree to which an L-subset of a vector space is an L-convex set.

In the past few years, theory of convexity has emerged more and more important study for exceptional problems in many fields of applied mathematics. Since the 1950s, convexity theory has developed into several related theories. Van de Vel [32] conducted an inventive investigation. His work was praised as excellence. An interesting question about the application part of convex theory attempts to include determining the computational complexity of convexity, pattern recognition problems, optimization, etc. Fuzzy set theory is an emerging discipline in different fields of abstract algebra such as topology and convexity theory, etc. Rosa [33] firstly presented the notion of fuzzy convex spaces. Shi and Xiu [34] proposed a new technique for the fuzzification of convex structures, which is described as M-fuzzifying convex structures. In this technique, each subset of a set X has a certain degree of convexity. Furthermore, Shi and Xiu [35] gave the generalization of L-convex structure and M-fuzzifying convex structure by introducing (L, M)-fuzzy convex structure. In their approach, every L-fuzzy subset can be considered as an L-convex set to some extent. With repeated progress in the area of convexity theory, fuzzy convex structures have become a major research area such as [36–52]. Pang and Xiu [53] introduced the notion of lattice-valued interval operators and described their connection between L-fuzzifying convex structures. Liu and Shi [54] proposed M-fuzzifying median algebra, which is obtained through fuzzy binary operation. This study provided the characteristics of M-fuzzifying median algebra and M-fuzzifying convex spaces. This work provided motivation to extend it on more algebraic structures like groups, rings, lattices, and vector spaces. Liu and Shi [11] introduced M-hazy groups by using the M-hazy binary operation. Mehmood et al. [12,13] extended this idea by defining M-hazy rings and obtained its induced fuzzifying convexities. By getting the motivations of these new proposed concepts through M-hazy operations, we proposed a new generalization of vector spaces over field based on M-hazy binary operation, which is denoted as M-hazy vector spaces over M-hazy field.

The paper is organized as follows: Section 2 consists of fundamental notions about completely residuated lattices, field and vector spaces, M-hazy groups, M-hazy rings, and M-fuzzifying convex spaces. In Section 3, the concept of M-hazy vector space is defined and obtained its fundamental properties. In Section 4, the concept of M-hazy subspaces is introduced, and it has been shown that all the M-hazy subspaces of M-hazy vector space form a convex structure. In Section 5, the linear transformation of M-hazy vector spaces is introduced. Finally, M-fuzzifying convex spaces are induced by M-hazy subspaces of M-hazy vector spaces. Section 6 concludes the paper.

2. Preliminaries

This section contains the fundamental definitions about completely residuated lattices, fields, vector spaces, vector subspaces, M-hazy groups, M-hazy rings, and M-fuzzifying convexity.

All through this paper, $(M, \vee, \wedge, \diamond, \rightarrow, \perp, \top)$ represents a completely residuated lattice, and $\leq$ denotes the partial order of M. Assume P is a nonempty set and $K \subseteq M$, then $\bigvee K$ denotes the least upper bound of K and $\bigwedge K$ the greatest lower bound of K. 2^P (resp., M^P) denotes the collection of all subsets (resp., M-subsets) on P. A family $\{A_i \mid i \in \omega\}$ is up-directed provided for each $A_1, A_2 \in \{A_i \mid i \in \omega\}$ that there exists a third element $A_3 \in \{A_i \mid i \in \omega\}$ such that $A_1 \subseteq A_3$ and $A_2 \subseteq A_3$ are denoted by: $\{A_i \mid i \in \omega\} \overset{dir}{\subseteq} 2^P$.

Definition 1 ([55]). *Assume that $\diamond : M \times M \longrightarrow M$ is a function. $\diamond$ is defined to be a triangular norm (for short, t-norm) on M, if the following conditions holds:*

(1) $u \diamond v = v \diamond u$,
(2) $(u \diamond v) \diamond w = u \diamond (v \diamond w)$,
(3) $u \leq w, v \leq x$ *implies* $u \diamond v \leq w \diamond x$,
(4) $u \diamond 1 = u$ *for all* $u \in M$.

Definition 2 ([55]). *Assume that $\rightarrow: M \times M \longrightarrow M$ is a function, and $\diamond$ is a t-norm in M. Then, $\rightarrow$ is defined to be the residuum of $\diamond$, if, for all $u, v, w \in M$,*

$$u \leq v \rightarrow w \Leftrightarrow u \diamond v \leq w.$$

Definition 3 ([56]). *Assume that $(M, \vee, \wedge, \perp, \top)$ is a bounded lattice, where $\perp$ represents the least element, $\top$ the greatest element, $\diamond$ is a t-norm on M, and $\rightarrow$ denotes the residuum of $\diamond$. Then, $(M, \vee, \wedge, \diamond, \rightarrow, \perp, \top)$ is said to be a residuated lattice.*

A residuated lattice is defined to be a completely residuated lattice if the primary lattice is complete. In addition, we define $u \leftrightarrow v = (u \rightarrow v) \wedge (v \rightarrow u)$. The proposition below shows properties of the implication operation.

Proposition 1 ([57,58]). *Assume $(M, \vee, \wedge, \diamond, \rightarrow, \perp, \top)$ is a completely residuated lattice. Then, for every $u, v, w \in M$, $\{u_i\}_{i \in I}, \{v_i\}_{i \in I} \subseteq M$, the below statements are valid:*

(1) $u \rightarrow v = \bigvee\{w \in M \mid u \diamond w \leq v\}$.
(2) $v \leq u \rightarrow v$, $\top \rightarrow u = u$.
(3) $u \rightarrow (\bigwedge\limits_{i \in I} v_i) = \bigwedge\limits_{i \in I} (u \rightarrow v_i)$.
(4) $(\bigvee\limits_{i \in I} u_i) \rightarrow v = \bigwedge\limits_{i \in I} (u_i \rightarrow v)$.

Definition 4 ([11]). *Assume that $* : P \times P \longrightarrow M^P$ is a function; then, $*$ is defined to be an M-hazy operation on P, if the conditions given below hold:*

(MH1) $\forall u, v \in P$, *we have* $\bigvee\limits_{p \in P} (u * v)(p) \neq \perp$.

(MH2) $\forall u, v, p, q \in P$, $(u * v)(p) \diamond (u * v)(q) \neq \perp \Rightarrow p = q$.

Definition 5 ([11]). *Assume $* : P \times P \longrightarrow M^P$ is an M-hazy operation on a nonempty set P. Then, $(P, *)$ is defined to be an M-hazy group (in short, MHG) if the following conditions hold:*

(MG1) $\forall u, v, w, p, q \in P$, $(u * v)(p) \diamond (v * w)(q) \leq \bigwedge_{r \in P}((p * w)(r) \leftrightarrow (u * q)(r))$, *i.e., the M-hazy associative law holds.*

(MG2) *An element $o \in P$ is said to be the left identity element of P, if $o * u = u_\top$ for all $u \in P$.*

(MG3) *An element $v \in P$ is said to be the left inverse of u, if for each $u \in P$, $v * u = o_\top$, and is denoted by u^{-1}.*

$(P, *)$ is defined to be an abelian MHG if the following condition holds:

(MG4) *$u * v = v * u$ for all $u, v \in P$.*

Definition 6 ([12]). *Assume $+ : R \times R \longrightarrow M^R$ and $\bullet : R \times R \longrightarrow M^R$ are the M-hazy addition operation and M-hazy multiplication operation on R, respectively. Then, $(R, +, \bullet)$ is defined to be an M-hazy ring (in short, MHR) if the below conditions hold:*

(MHR1) *$(R, +)$ is an abelian MHG.*

(MHR2) *$(R, \bullet)$ is an M-hazy semigroup.*

(MHR3) $\forall u, v, w, p, q, r \in R, (u \bullet v)(p) \diamond (v + w)(q) \diamond (u \bullet w)(r) \leq \bigwedge_{s \in R}((u \bullet q)(s) \leftrightarrow (p + r)(s))$.

We now give the definition of M-fuzzifying convex space, and we refer to Vel [32] for all of the background on the convexity theory that may be required.

Definition 7 ([34]). *A function $\mathscr{S} : 2^P \to M$ is said to be an M-fuzzifying convexity on a nonempty set P if the below conditions hold:*

(1) $\mathscr{S}(\emptyset) = \mathscr{S}(P) = \top$;

(2) *If $\{D_i | i \in \Omega\} \subseteq 2^P$ is nonempty, then $\mathscr{S}(\bigcap_{i \in \Omega} D_i) \geq \bigwedge_{i \in \Omega} \mathscr{S}(D_i)$;*

(3) *If $\{D_i | i \in \Omega\} \overset{dir}{\subseteq} 2^P$, then $\mathscr{S}(\bigcup_{i \in \Omega} D_i) \geq \bigwedge_{i \in \Omega} \mathscr{S}(D_i)$.*

Then, $(P, \mathscr{S})$ is said to be an M-fuzzifying convex space.

A function $\pi : (P, \mathscr{S}_P) \longrightarrow (Q, \mathscr{S}_Q)$ is defined as M-fuzzifying convexity preserving (M-CP, in short) given that $\mathscr{S}_P(\pi^{\leftarrow}(B)) \geq \mathscr{S}_Q(B)$ for each $B \in 2^Q$; π is called M-fuzzifying convex-to-convex (M-CC, in short) provided that $\mathscr{S}_Q(\pi^{\rightarrow}(A)) \geq \mathscr{S}_P(A)$ for each $A \in 2^P$.

Definition 8 ([22]). *A field is a set F with two operations, called addition and multiplication, which satisfy the following conditions:*

(F1) $\forall u, v \in F, u + v = v + u$,

(F2) $\forall u, v, w \in F, (u + v) + w = u + (v + w)$,

(F3) *F contains an element 0 such that $0 + u = u, \forall u \in F$,*

(F4) *For each $u \in F$, there is an element $-u \in F$ such that $u + (-u) = 0$,*

(F5) $\forall u, v \in F \implies u \cdot v \in F$,

(F6) $\forall u, v \in F, u \cdot v = v \cdot u$,

(F7) $\forall u, v, w \in F, (u \cdot v) \cdot w = u \cdot (v \cdot w)$,

(F8) *F contains an element $1 \neq 0$ and $\forall u \in F$ such that $1 \cdot u = u$,*

(F9) *For each $0 \neq u \in F$, there is an element $u^{-1} \in F$ such that $u \cdot u^{-1} = 1$,*

(F10) $\forall u, v, w \in F, u \cdot (v + w) = u \cdot w + v \cdot w$.

Definition 9 ([22]). *A vector space is a nonempty set V over a field F, whose objects are called vectors equipped with two operations, called addition and scalar multiplication: for any two vectors u, v in V and a scalar a in F defined by the mappings $+ : V \times V \longrightarrow V$ and $\cdot : F \times V \longrightarrow V$, the following conditions are satisfied:*

(V1) $\forall u, v, w \in V, (u+v)+w = u+(v+w)$,

(V2) *There is a vector 0, called the zero vector, such that* $u+0=u$,

(V3) *For any vector u, there is a vector* $-u$ *such that* $u+(-u)=0$,

(V4) $\forall u, v \in V, u+v = v+u$,

(V5) $\forall u, v \in V, \forall a \in F, a\cdot(u+v) = a\cdot u + a\cdot v$,

(V6) $\forall a, b \in F, \forall u \in V, (a+b)\cdot u = a\cdot u + b\cdot u$,

(V7) $\forall a, b \in F, \forall u \in V, a\cdot(b\cdot u) = (a\cdot b)\cdot u$,

(V8) $\forall 1 \in F, \forall u \in V, 1\cdot u = u$.

3. *M*-Hazy Vector Spaces

In this section, we introduce the concept of M-hazy vector spaces over the M-hazy field.

We first introduce the concept of M-hazy field and give its properties, which are necessary to present the concept of M-hazy vector space.

Definition 10. *Assume* $+ : F\times F \longrightarrow M^F$ *and* $\bullet : F\times F \longrightarrow M^F$ *are the M-hazy addition operation and M-hazy multiplication operation on F, respectively. Then, the 3-tuple* $(F, +, \bullet)$ *is defined to be an M-hazy field (in short, MHF) if the below conditions hold:*

(MHF1) $(F, +)$ *is an abelian MHG.*

(MHF2) $(F, \bullet)$ *is an abelian MHG.*

(MHF3) $\forall u, v, w, p, q, r \in F, (u\bullet v)(p) \diamond (v+w)(q) \diamond (u\bullet w)(r) \leq \bigwedge_{s\in F}((u\bullet q)(s) \leftrightarrow (p+r)(s))$.

Proposition 2. *Assume that* $(F, +, \bullet)$ *is an M-hazy field, and o and e are the additive and multiplicative identity elements of F, respectively. Then,* $\forall u \in F$;

(1) $o+u = u+o = u_\top$.

(2) $e\bullet u = u\bullet e = u_\top$.

Proof. The proof is similar to the proof of Proposition 3.8 in [11] so it is omitted. □

Proposition 3. *In an M-hazy field* $(F, +, \bullet)$*, the left additive inverse* $-u$ *of u is also the right additive inverse of u in* $(F, +, \bullet)$*. In addition, the left multiplicative inverse* u^{-1} *of u is also the right multiplicative inverse of u in* $(F, +, \bullet)$*. That is, the following conditions hold:*

(1) $(-u)+u = u+(-u) = o_\top$.

(2) $u^{-1}\bullet u = u\bullet u^{-1} = e_\top$.

Proof. The proof is similar to the proof of Proposition 3.7 in [11] so it is omitted. □

Example 1. *Assume that* $F = \{o, e, u, v\}$ *is a set and assume* $(M, \diamond) = ([0,1], \wedge)$*. The mappings* $+ : F\times \times F \longrightarrow [0,1]^F$ *and* $\bullet : F\times F \longrightarrow [0,1]^F$ *are defined by the following tables:*

(a) Values of the $[0,1]$-hazy operation $+$.

$x+y$ \ y ; x	o	e	u	v
o	o_1	e_1	u_1	v_1
e	e_1	o_1	$v_{0.2}$	$u_{0.2}$
u	u_1	$v_{0.2}$	o_1	e_1
v	v_1	$u_{0.2}$	e_1	o_1

(b) Values of the $[0,1]$-hazy operation $\bullet$.

$x\bullet y$ \ y ; x	o	e	u	v
o	o_1	o_1	o_1	o_1
e	o_1	e_1	$u_{0.3}$	$v_{0.4}$
u	o_1	$u_{0.3}$	$v_{0.4}$	e_1
v	o_1	$v_{0.4}$	e_1	$u_{0.3}$

It is easy to verify (MHF1), (MHF2), and (MHF3) analogous to Example 3.3 in [12].

Proposition 4. *Assume that* $(F, +, \bullet)$ *is an M-hazy field; then, the following equations hold:*

(1) $(u+v)(w) = ((-u)+w)(v) = (w+(-v))(u) = (v+(-w))(-u)$
$= ((-w)+u)(-v) = ((-v)+(-u))(-w).$

(2) $(u \bullet v)(w) = (u^{-1} \bullet w)(v) = (w \bullet v^{-1})(u) = (v \bullet w^{-1})(u^{-1}) = (w^{-1} \bullet u)(v^{-1})$
$= (v^{-1} \bullet u^{-1})(w^{-1}).$

Proof. The proof is similar to the proof of Proposition 3.11 and Corollary 3.12 in [11] so it is omitted. □

We now present the concept of M-hazy vector space.

Definition 11. *Assume* $(F, +, \bullet)$ *is an M-hazy field and* $(V, \oplus)$ *is an abelian M-hazy group. We define an M-hazy vector space over F as a quadruple* $(V, \oplus, \circ, F)$*, where* $\circ$ *is a mapping* $\circ : F \times V \longrightarrow M^V$ *such that the following conditions hold:*

(MHV1) $\forall u, v, p, q, r \in V$ *and* $a \in F$, $(a \circ u)(p) \diamond (u \oplus v)(q) \diamond (a \circ v)(r) \leq \bigwedge_{s \in V} ((a \circ q)(s) \leftrightarrow (p \oplus r)(s))$.

(MHV2) $\forall u, p, q \in V$ *and* $a, b \in F$, $(a \circ u)(p) \diamond (a+b)(q) \diamond (b \circ u)(r) \leq \bigwedge_{s \in V} ((q \circ u)(s) \leftrightarrow (p \oplus r)(s))$.

(MHV3) $\forall u, q \in V$*, and* $\forall a, b, c \in F$, $(a \bullet b)(c) \diamond (b \circ u)(q) \leq \bigwedge_{r \in V} ((c \circ u)(r) \leftrightarrow (a \circ q)(r))$.

(MHV4) $\forall u \in V$ *and* $e \in F$, $e \circ u = e_\top$.

Proposition 5. *Assume* $\oplus : V \times V \longrightarrow M^V$ *and* $\circ : F \times V \longrightarrow M^V$ *are the M-hazy operations under addition and under scalar multiplication on V, respectively; then, the following statements are equivalent for all* $u, v \in V$*, and* $\forall a \in F$.

(MHV1) $\forall u, v, p, q, r \in V$*, and* $\forall a \in F$,

$$(a \circ u)(p) \diamond (u \oplus v)(q) \diamond (a \circ v)(r) \leq \bigwedge_{s \in V} ((a \circ q)(s) \leftrightarrow (p \oplus r)(s)).$$

(MHV1′) $\forall u, v, p, q, r, s \in V$*, and* $\forall a \in F$,

$$(a \circ u)(p) \diamond (u \oplus v)(q) \diamond (a \circ v)(r) \diamond (a \circ q)(s) \leq (p \oplus r)(s)$$

and

$$(a \circ u)(p) \diamond (u \oplus v)(q) \diamond (a \circ v)(r) \diamond (p \oplus r)(s) \leq (a \circ q)(s).$$

(MHV1″)

$$If\, a \circ u = p_\lambda, u \oplus v = q_\mu, a \circ v = r_\nu,$$

then $(a \circ q) \diamond \lambda \diamond \mu \diamond \nu \leq (p \oplus r)$ *and* $(p \oplus r) \diamond \lambda \diamond \mu \diamond \nu \leq (a \circ q)$.

(MHV1‴) *If* $a \circ u = p_\lambda$, $u \oplus v = q_\mu$, $a \circ v = r_\nu$, $a \circ q = t_\alpha$, $p \oplus r = u_\beta$, *then*

$$t = u, \lambda \diamond \mu \diamond \nu \diamond \alpha \leq \beta \,and\, \lambda \diamond \mu \diamond \nu \diamond \beta \leq \alpha.$$

Proof. (MHV1) ⇒ (MHV1′) The proof is simple so it is omitted.
(MHV1′) ⇒ (MHV1″) $\forall a \in F$ and $\forall q \in V$, we have

$$\begin{aligned} & ((a \circ q) \diamond \lambda \diamond \mu \diamond \nu)(s) \\ = & (a \circ q)(s) \diamond \lambda \diamond \mu \diamond \nu \\ = & (a \circ q)(s) \diamond (a \circ u)(p) \diamond (u \oplus v)(q) \diamond (a \circ v)(r) \\ \leq & (p \oplus r)(s), \end{aligned}$$

that is, $(a \circ q) \diamond \lambda \diamond \mu \diamond \nu \leq p \oplus r$. A similar argument shows the other inequality.

(MHV1$''$) $\Rightarrow$ (MHV1$'''$) We need to only verify $t = u$. According to (MHV1$''$), we have $t_\alpha \diamond \lambda \diamond \mu \diamond \nu \leq u_\beta$, that is, $t_{\alpha \diamond \lambda \diamond \mu \diamond \nu} \leq u_\beta$, whence by (MH2), $t = u$.

(MHV1$'''$) $\Rightarrow$ (MHV1) $\forall p, q, r, s \in P$ and $\forall a \in F$, we have

$$\begin{aligned} & (a \circ u)(p) \diamond (u \oplus v)(q) \diamond (a \circ v)(r) \\ = \; & \lambda \diamond \mu \diamond \nu \\ \leq \; & (\alpha \to \beta) \wedge (\beta \to \alpha) \\ = \; & \alpha \leftrightarrow \beta = (a \circ q)(t) \leftrightarrow (p \oplus r)(u) \end{aligned}$$

and, by (MH2), we can complete the proof. □

Example 2. *(1) The Euclidean space R^n is an M-hazy vector space under the addition and scalar multiplication.*

(2) The set P_n of all polynomials of degree less than or equal to n is an M-hazy vector space under the addition and scalar multiplication of polynomials.

(3) The set $M(m, n)$ of all $m \times n$ matrices is an M-hazy vector space under the addition and scalar multiplication of matrices.

In the following discussion, we assume that the operation $\diamond$ in the completely residuated lattice M is $\wedge$; that is, the lattice valued environment M is degenerated to complete Heyting algebra. We also assume that the smallest element $\perp$ is prime in M.

Theorem 1. *Assume that $(V, \oplus, \circ, F)$ is an M-hazy vector space over an M-hazy field $(F, +, \bullet)$. Then, $\forall u \in V$ and $\forall a \in F$:*

(1) $(a \circ o)(o) \neq \perp$.

(2) $(o \circ u)(o) \neq \perp$.

(3) If $(a \circ u)(o) \neq \perp$, then $a = o$ or $u = o$.

Proof. We only prove (1). Assume $(a \circ o)(o) = \perp$. By (MHV1) $\forall a \in F$ and $o \in V$, we have

$$(a \circ o)(p) \diamond (o \oplus o)(o) \diamond (a \circ o)(r) \leq \bigwedge_{s \in V} ((a \circ o)(s) \leftrightarrow (p \oplus r)(s)).$$

When $r = s = p$, we have,

$$(a \circ o)(p) \diamond (a \circ o)(p) \diamond (a \circ o)(p) \leq (p \oplus p)(p)).$$

If $p \neq o$, then $(p \oplus p)(p) = (p \oplus (-p))(p) = \perp$ by the condition (1) of Proposition 4, which is a contradiction. Hence, $(a \circ o)(o) \neq \perp$. □

4. *M*-Hazy Subspaces

In this section, we introduce the concept of M-hazy subspaces of M-hazy vector space.

Definition 12. *Assume that $(V, \oplus, \circ, F)$ is an M-hazy vector space over an M-hazy field $(F, +, \bullet)$. A nonempty subset W of V is called an M-hazy subspace of V if W itself is an M-hazy vector space over F.*

Theorem 2. *Assume W is a nonempty subset of an M-hazy vector space $(V, \oplus, \circ, F)$ over an M-hazy field $(F, +, \bullet)$; then, $(W, \oplus, \circ, F)$ is an M-hazy subspace of $(V, \oplus, \circ, F)$ over $(F, +, \bullet)$ if and only if the following conditions hold:*

(1) $\forall u, v \in W$, we have $\bigvee_{p \in W} (u \oplus v)(p) \neq \perp$,

(2) $\forall u \in W$ and $\forall a \in F$, we have $\bigvee_{p \in W} (a \circ u)(p) \neq \perp$,

(3) $\forall u \in W$, we have $-u \in W$.

Proof. The proof is simple and omitted. □

Theorem 3. *Assume W is a nonempty subset of an M-hazy vector space $(V, \oplus, \circ, F)$ over an M-hazy field $(F, +, \bullet)$; then, $(W, \oplus, \circ, F)$ is an M-hazy subspace of $(V, \oplus, \circ, F)$ over $(F, +, \bullet)$ if and only if the following conditions hold:*

(1) $\forall u, v \in W$, we have $\bigvee_{p \in W} (u \oplus (-v))(p) \neq \perp$,

(2) $\forall u \in W$ and $\forall a \in F$, we have $\bigvee_{p \in W} (a \circ u)(p) \neq \perp$.

Proof. The proof is similar to the proof of Theorem 4.4 in [12] so it is omitted. □

Theorem 4. *Assume W is a nonempty subset of an M-hazy vector space $(V, \oplus, \circ, F)$ over an M-hazy field $(F, +, \bullet)$; then, $(W, \oplus, \circ, F)$ is an M-hazy subspace of $(V, \oplus, \circ, F)$ over $(F, +, \bullet)$ if and only if $\forall u, v, p, q, r \in W$ and $\forall a, b \in F$, we have*

$$\bigvee_{r \in W} ((a \circ u) \oplus (b \circ v))(r) \neq \perp.$$

Proof. Assume that W is an M-hazy subspace of V over an M-hazy field F. Suppose that

$$\bigvee_{r \in W} ((a \circ u) \oplus (b \circ v))(r) = \perp.$$

On the other hand, $\bigvee_{p \in W} (a \circ u)(p) \neq \perp$ and $\bigvee_{q \in W} (b \circ v)(q) \neq \perp$ by Theorem 3. Hence, $\bigvee_{r \in W} (p \oplus q)(r) \neq \perp$, which is a contradiction. Hence,

$$\bigvee_{r \in W} ((a \circ u) \oplus (b \circ v))(r) \neq \perp.$$

Conversely, suppose $\bigvee_{r \in W} ((a \circ u) \oplus (b \circ v))(r) \neq \perp$. Since W is a nonempty subset of V and we know that (MH1) and (MH2) holds in V, thus it holds in W. Hence, $\forall u, v \in W$ and $e \in F$, we have $\bigvee_{u \in W} (e \circ u)(u) \neq \perp$ and $\bigvee_{v \in W} (e \circ v)(v) \neq \perp$. Hence, $\bigvee_{r \in W} (u \oplus v)(r) \neq \perp$. In addition, $\forall b \in F$ and $o \in W$, we have $(b \circ o)(o) \neq \perp$ by statement (1) of Theorem 1. Since $\forall a \in F$ and $\forall u \in W$, we have $\bigvee_{p \in W} (a \circ u)(p) \neq \perp$; thus, $\bigvee_{r \in W} (p \oplus o)(r) \neq \perp$. Hence, $\bigvee_{r \in W} (a \circ u)(r) \neq \perp$ by (MG2). Now, we already know that (MG3) holds in V, so it holds in W. Hence, $\forall u \in W$, we have $-u \in W$. Hence, W is an M-hazy subspace of V over F. □

Proposition 6. *The intersection of a family of M-hazy subspace of an M-hazy vector space $(V, \oplus, \circ, F)$ over an M-hazy field $(F, +, \bullet)$ is an M-hazy subspace of $(V, +, \circ, F)$.*

Proof. Assume Λ is an index set and W_i is an M-hazy subspace of $(V, \oplus, \circ, F)$. Assume $K = \bigcap_{i \in \Lambda} W_i$.

(1) Since $o \in W_i$ for each $i \in \Lambda$ and K is a nonempty subset of V, we have $o \in K$.

(2) For every $u, v \in K, a \in F$ and for every $i \in \Lambda$, we obtain $u, v \in W_i$. Since W is an M-hazy subspace of V, we obtain $\bigvee_{p \in W_i} (u \oplus (-v))(p) \neq \perp$ and $\bigvee_{p \in W_i} (a \circ u)(p) \neq \perp$ by Theorem 3. Since $\bigvee_{p \in W_i} (a \circ u)(p) \neq \perp$. This implies that there exists $x_{au} \in W_i$ such that $(a \circ u)(x_{au}) \neq \perp$. This implies, for all $i \in \Lambda$, $x_{au} \in W_i$. Thus, we can obtain $x_{au} \in \bigcap_{i = \Lambda} W_i$.

Hence, $\bigvee_{p\in \bigcap_{i\in\Lambda} W_i} (a \circ u)(p) \geq (a \circ u)(x_{au}) \neq \bot$. Similarly, $\bigvee_{p\in \bigcap_{i\in\Lambda} W_i} (u \oplus (-v))(p) \neq \bot$. Hence, $K = \bigcap_{i=\Lambda} W_i$ is an M-hazy subspace of $(V, \oplus, \circ, F)$. □

Proposition 7. *The union of a nonempty up-directed family of M-hazy subspace of M-hazy vector space $(V, \oplus, \circ, F)$ over an M-hazy field $(F, +, \bullet)$ is an M-hazy subspace of $(V, \oplus, \circ, F)$. In particular, the union of a nonempty chain of M-hazy subspace of M-hazy vector space $(V, \oplus, \circ, F)$ is an M-hazy subspace of $(V, \oplus, \circ, F)$.*

Proof. Assume Λ is an index set and W_i is an M-hazy subspace of $(V, \oplus, \circ, F)$, where $\{W_i \mid i \in \Lambda\}$ is an up-directed subfamily of 2^V. Let $N = \bigcup_{i\in\Lambda} W_i$.

(1) Clearly, N is a nonempty subset of V.

(2) For every $u, v \in N$ and $a \in F$, there exists $i, j \in \Lambda$ such that $u \in W_i$ and $v \in W_j$. Since N is an up-directed family, then there exists $m \in \Lambda$ such that $W_i \subseteq W_m$; this implies that $u, v \in W_m$. As $(W, \oplus, \circ, F)$ is an M-hazy subspace of $(V, \oplus, \circ, F)$, we obtain $\bigvee_{p\in W_m} (u \oplus (-v))(p) \neq \bot$ and $\bigvee_{p\in W_m} (a \circ u)(p) \neq \bot$ by Theorem 3. Hence, $\bigvee_{p\in N} (u \oplus (-v))(p) \neq \bot$ and $\bigvee_{p\in N} (a \circ u)(p) \neq \bot$. Hence, $\bigcup_{i\in\Lambda} W_i$ is an M-hazy subspace of $(V, \oplus, \circ, F)$. □

Based on the above results, we can draw an important and interesting conclusion; that is, we have the following result.

Proposition 8. *All of the M-hazy subspaces of M-hazy vector space and the empty set form a convex structure.*

5. Linear Transformation of M-Hazy Vector Spaces

In this section, we introduce the linear transformation of M-hazy vector spaces. We have also shown that M-fuzzifying convex spaces are induced by M-hazy subspace of M-hazy vector space.

Definition 13. *Assume that $(V, \oplus, \circ, F)$ and $(W, \boxplus, \odot, F)$ are two M-hazy vector spaces over an M-hazy field $(F, +, \bullet)$. Then, the mapping $\tau : V \longrightarrow W$ is called a linear transformation if the following conditions hold:*

(1) $\forall u, v \in V, \tau_M^{\rightarrow}(u \oplus v) = (\tau(u) \boxplus \tau(v))$,

(2) $\forall u \in V, \forall a \in F, \tau_M^{\rightarrow}(a \circ u) = (a \odot \tau(u))$.

Definition 14. *Assume V and W are two M-hazy vector spaces over an M-hazy field F, $\tau : V \longrightarrow W$ is a linear transformation, and o' is the additive identity element of W. Then, the kernel of τ, $\mathrm{Ker}\tau$ is determined by*

$$\mathrm{Ker}\tau = \tau^{\leftarrow}(\{o'\}) = \{p \in V \mid \tau(p) = o'\}.$$

Example 3. *(1)* *Assume that $(V, \oplus, \circ, F)$ is an M-hazy vector space over an M-hazy field $(F, +, \bullet)$, the set $\{o\}$ and the whole M-hazy vector space V are M-hazy subspaces of V; they are called the trivial M-hazy subspaces of V.*

(2) *Assume R^n and R^m are the Euclidean spaces and $\tau : R^n \longrightarrow R^m$ is a linear transformation. The image*

$$\tau^{\rightarrow}(p) = \{\tau(p) : p \in R^n\}$$

of τ is an M-hazy subspace of R^m, and the inverse image

$$\tau^{\leftarrow}(\{o'\}) = \{p \in R^n \mid \tau(p) = o'\}$$

is an M-hazy subspace of R^m.

Proposition 9. *Assume the mapping $\tau : V \longrightarrow W$ is a linear transformation, $(V, \oplus, \circ, F)$ and $(W, \boxplus, \odot, F)$ are two M-hazy vector spaces over an M-hazy field $(F, +, \bullet)$. Then, the following statements are valid:*

(1) *If J is an M-hazy subspace of V, then $\tau^{\rightarrow}(J)$ is an M-hazy subspace of W.*
(2) *If K is an M-hazy subspace of W, then $\tau^{\leftarrow}(K)$ is an M-hazy subspace of V containing* $\ker\tau$.

Proof. (1) For all $u, v \in J$, we have $\bigvee_{p\in J}(u \oplus (-v))(p) \neq \perp$. Then,

$$\begin{aligned}
& \bigvee_{\tau(p)\in\tau^{\rightarrow}(J)} (\tau(u) \boxplus (-\tau(v)))(\tau(p)) \\
= & \bigvee_{\tau(p)\in\tau^{\rightarrow}(J)} (\tau(u) \boxplus \tau(-v))(\tau(p)) \\
= & \bigvee_{\tau(p)\in\tau^{\rightarrow}(J)} \tau_M^{\rightarrow}(u \oplus (-v))(\tau(p)) \\
= & \bigvee_{\tau(p)\in\tau^{\rightarrow}(J)} \bigvee_{\tau(x)=\tau(p)} (u \oplus (-v))(x) \\
= & \bigvee_{\tau(p)\in\tau^{\rightarrow}(J)} (u \oplus (-v))(p) \\
\geq & \bigvee_{p\in J} (u \oplus (-v))(p) \\
\neq & \perp.
\end{aligned}$$

Similarly,

$$\bigvee_{\tau(p)\in\tau^{\rightarrow}(J)} (a \odot \tau(u))(\tau(p)) \geq \bigvee_{p\in J} (a \circ u)(p) \neq \perp.$$

Then, by Theorem 3, it follows that $\tau^{\rightarrow}(J)$ is an M-hazy subspace of W.

(2) For all $u, v \in \tau^{\leftarrow}(K)$, we have $\tau(u), \tau(v) \in K$. We find that $\bigvee_{\tau(p)\in K} (\tau(u) \boxplus (-\tau(v)))(\tau(p)) \neq \perp$ and $\bigvee_{\tau(p)\in K} (a \odot \tau(u))(\tau(p)) \neq \perp$, since K is an M-hazy subspace of W. Furthermore,

$$\begin{aligned}
& \bigvee_{\tau(p)\in K} (\tau(u) \boxplus (-\tau(v)))(\tau(p)) \\
= & \bigvee_{\tau(p)\in K} (\tau(u) \boxplus \tau(-v))(\tau(p)) \\
= & \bigvee_{\tau(p)\in K} \tau_M^{\rightarrow}(u \oplus (-v))(\tau(p)) \\
= & \bigvee_{\tau(p)\in K} \bigvee_{\tau(x)=\tau(p)} (u \oplus (-v))(x) \\
= & \bigvee_{\tau(p)\in K} (u \oplus (-v))(p) \\
= & \bigvee_{p\in\tau^{\leftarrow}(K)} (u \oplus (-v))(p) \\
& \neq \perp.
\end{aligned}$$

Similarly,

$$\bigvee_{\tau(p)\in K} (a \odot \tau(u))(\tau(p)) = \bigvee_{p\in\tau^{\leftarrow}(K)} (a \circ u)(p) \neq \perp.$$

Consequently, $\tau^{\leftarrow}(K)$ is an M-hazy subspace of V.

Now, assume $p \in \ker\tau$. Since K is an M-hazy subspace of W, then $\tau(p) = o' \in K$, and so $p \in \tau^{\leftarrow}(K)$. Hence, $\ker\tau \subseteq \tau^{\leftarrow}(K)$. □

Proposition 10. *Assume V and W are two M-hazy vector spaces over an M-hazy field F and $\tau : V \longrightarrow W$ is a linear transformation. Then,* $\mathrm{Ker}\tau$ *is an M-hazy subspace of V.*

Proof. It is easy to see that $\{o'\}$ is an M-hazy subspace of W. Then, by Proposition 9, we have that $\mathrm{Ker}\tau$ is an M-hazy subspace of V. □

The theorems below give an approach to induce M-fuzzifying convex spaces using M-hazy subspace of M-hazy vector space.

Theorem 5. *Assume $(V, \oplus, \circ, F)$ is an M-hazy vector space over an M-hazy field $(F, +, \bullet)$ and define $\mathscr{S} : 2^V \to M$ as follows:*

$$\forall A \in 2^V,\ \mathscr{S}(A) = \bigwedge_{p \in V} \Big(\big(\big(\bigvee_{u,v \in A} (u \oplus (-v))(p) \big) \to A(p) \big) \wedge \big(\big(\bigvee_{a \in F, u \in A} (a \circ u)(p) \big) \to A(p) \big) \Big).$$

Then, $(V, \mathscr{S})$ is an M-fuzzifying convex space.

Proof. The proof is similar to the proof of Theorem 7 in [13] so it is omitted. □

Theorem 6. *Assume that the mapping $\tau : V \longrightarrow W$ is a linear transformation, $(V, \oplus, \circ, F)$ and $(W, \boxplus, \odot, F)$ are two M-hazy vector spaces over an M-hazy field $(F, +, \bullet)$; then, $\tau : (V, \mathscr{S}_V) \longrightarrow (W, \mathscr{S}_W)$ is an M-CP mapping.*

Proof. The proof is similar to the proof of Theorem 8 in [13] so it is omitted. □

Theorem 7. *Assume the mapping $\tau : V \longrightarrow W$ is a linear transformation, $(V, \oplus, \circ, F)$ and $(W, \boxplus, \odot, F)$ are two M-hazy vector spaces over an M-hazy field $(F, +, \bullet)$; then, $\tau : (V, \mathscr{S}_V) \longrightarrow (W, \mathscr{S}_W)$ is an M-CC mapping.*

Proof. Since the mapping $\tau : V \longrightarrow W$ is a linear transformation if the following conditions hold:

(1) $\forall u, v \in V, \tau_M^{\rightarrow}(u \oplus v) = (\tau(u) \boxplus \tau(v))$,
(2) $\forall u \in V, \forall a \in F, \tau_M^{\rightarrow}(a \circ u) = (a \odot \tau(u))$.

Then, for all $A \in 2^V$, we have

$$\begin{aligned}
&\mathscr{S}_W(\tau^{\rightarrow}(A)) = \bigwedge_{p \in W} \Big(\big(\big(\bigvee_{\tau(u),\tau(v) \in \tau^{\rightarrow}(A)} (\tau(u) \boxplus \tau(-v))(\tau(p)) \big) \to (\tau^{\rightarrow}(A))(\tau(p)) \big) \\
&\wedge \big(\big(\bigvee_{a \in F, \tau(u) \in \tau^{\rightarrow}(A)} (a \odot \tau(u))(\tau(p)) \big) \to (\tau^{\rightarrow}(A))(\tau(p)) \big) \Big) \\
&\geq \bigwedge_{p \in W} \Big(\big(\big(\bigvee_{\tau(u),\tau(v) \in \tau^{\rightarrow}(A)} (\tau_M^{\rightarrow} \cdot \tau_M^{\leftarrow}(\tau(u) \boxplus \tau(-v)))(\tau(p)) \big) \to (\tau^{\rightarrow}(A))(\tau(p)) \big) \\
&\wedge \big(\big(\bigvee_{a \in F, \tau(u) \in \tau^{\rightarrow}(A)} (\tau_M^{\rightarrow} \cdot \tau_M^{\leftarrow}(a \odot \tau(u)))(\tau(p)) \big) \to (\tau^{\rightarrow}(A))(\tau(p)) \big) \Big) \\
&= \bigwedge_{p \in W} \Big(\big(\big(\bigvee_{\tau(u),\tau(v) \in \tau^{\rightarrow}(A)} (\tau_M^{\rightarrow}(u \oplus (-v)))(\tau(p)) \big) \to (\tau^{\rightarrow}(A))(\tau(p)) \big) \\
&\wedge \big(\big(\bigvee_{a \in F, \tau(u) \in \tau^{\rightarrow}(A)} (\tau_M^{\rightarrow}(a \circ u))(\tau(p)) \big) \to (\tau^{\rightarrow}(A))(\tau(p)) \big) \Big) \\
&= \bigwedge_{x \in V} \Big(\big(\big(\bigvee_{\tau(p) \in \tau_M^{\rightarrow}(A)} \bigvee_{\tau(x) = \tau(p)} (u \oplus (-v))(x) \big) \to (A)(x) \big) \\
&\wedge \big(\big(\bigvee_{a \in F, \tau(p) \in \tau_M^{\rightarrow}(A)} \bigvee_{\tau(x) = \tau(p)} ((a \circ u))(x) \big) \to (A)(x) \big) \Big) \\
&\geq \bigwedge_{p \in V} \Big(\big(\big(\bigvee_{u, -v \in A} (u \oplus (-v))(p) \big) \to (A)(p) \big) \\
&\wedge \big(\big(\bigvee_{a \in F, u \in A} (a \odot u)(p) \big) \to (A)(p) \big) \Big) \\
&= \mathscr{S}_V(A).
\end{aligned}$$

This implies that $\tau : (V, \mathscr{S}_V) \longrightarrow (W, \mathscr{S}_W)$ is an M-CC mapping. □

6. Conclusions

Liu and Shi [11] introduced M-hazy groups by using the M-hazy binary operation. Mehmood et al. [12,13] extended this idea by defining M-hazy rings and obtained its induced fuzzifying convexities. By getting the motivations of these new proposed concepts through M-hazy operations, we proposed a new generalization of vector spaces over field based on M-hazy binary operation, which is denoted as M-hazy vector spaces over M-hazy field. In addition, by using the completely residuated lattice valed logic, some important properties of M-hazy field, M-hazy vector space, and M-hazy subspace are introduced. Based on these properties, it is shown that an M-hazy subspace of M-hazy vector space forms a convex structure. In addition, the linear transformation of M-hazy vector space is defined and proves its important results. The convexity of the M fuzzy set on the set P is the M fuzzy set on the power set with certain properties. Therefore, to a certain extent, each subset of P can be regarded as a convex set. Finally, considering the importance of this fact, a method is given that uses the M-hazy subspace of M-hazy vector space to induce the M-fuzzifying convex space.

The possible directions for the future work could be M-hazy modules, bases, dimensions of M-hazy vector spaces, M-hazy topological vector spaces, and other fuzzy algebra.

Author Contributions: Conceptualization, F.M.; formal analysis, F.M. and F.-G.S.; funding acquisition, F.-G.S.; investigation, F.M. and F.-G.S.; methodology, F.M. and F.-G.S.; project administration, F.M. and F.-G.S.; supervision, F.-G.S.; validation, F.M. and F.-G.S.; visualization, F.-G.S.; writing—original draft, F.M.; writing—review and editing, F.M. and F.-G.S. Both authors have read and agreed to the published version of the manuscript.

Funding: The project is funded by the National Natural Science Foundation of China (11871097).

Institutional Review Board Statement: Not applicable.

Informed Consent Statement: Not applicable.

Data Availability Statement: Not applicable.

Acknowledgments: The authors would like to thank the editors and the anonymous reviewers for their fruitful comments and suggestions, which led to a number of improvements of the paper.

Conflicts of Interest: The authors declare no conflict of interest.

References

1. Rosenfeld, A. Fuzzy groups. *J. Math. Anal. Appl.* **1971**, *35*, 512–517. [CrossRef]
2. Liu, W.J. Fuzzy invariant subgroups and fuzzy ideals. *Fuzzy Set. Syst.* **1982**, *8*, 133–139. [CrossRef]
3. Liu, W.J. Operations on fuzzy ideals. *Fuzzy Set. Syst.* **1983**, *11*, 31–41.
4. Demirci, M. Vague groups. *J. Math. Anal. Appl.* **1999**, *230*, 142–156. [CrossRef]
5. Demirci, M. Fuzzy functions and their fundamental properties. *Fuzzy Set. Syst.* **1999**, *106*, 239–246. [CrossRef]
6. Aktas, H.; Cagman, N. A type of fuzzy ring. *Arch. Math. Log.* **2007**, *46*, 165–177. [CrossRef]
7. Crist, R.; Mordeson, J.N. Vague Rings. *New Math. Nat. Comput.* **2005**, *1*, 213–228. [CrossRef]
8. Demirci, M. Smooth groups. *Fuzzy Set. Syst.* **2001**, *117*, 431–437. [CrossRef]
9. Sezer, S. Some properties of vague rings. *Int. J. Algebra* **2010**, *4*, 751–760.
10. Sezer, S. Vague rings and vague ideals. *Iran. J. Fuzzy Syst.* **2011**, *8*, 145–157.
11. Liu, Q.; Shi, F.-G. A new approach to the fuzzification of groups. *J. Intell. Fuzzy Syst.* **2019**, *37*, 6429–6442. [CrossRef]
12. Mehmood, F.; Shi, F.-G.; Hayat, K. A new approach to the fuzzification of rings. *J. Nonlinear Convex Anal.* **2020**, *21*, 2637–2646.
13. Mehmood, F.; Shi, F.-G.; Hayat, K.; Yang, X.-P. The homomorphism theorems of M-hazy rings and their induced fuzzifying convexities. *Mathematics* **2020**, *8*, 411. [CrossRef]
14. Liu, Q.; Shi, F.-G. M-hazy lattices and its induced fuzzifying convexities. *J. Intell. Fuzzy Syst.* **2019**, *37*, 2419–2433. [CrossRef]
15. Fan, C.-Z.; Shi, F.-G.; Mehmood, F. M-hazy Γ-semigroup. *J. Nonlinear Convex Anal.* **2020**, *21*, 2659–2669.
16. Katsaras, A.K.; Liu, D.B. Fuzzy vector spaces and fuzzy topological vector spaces. *J. Math. Anal. Appl.* **1977**, *58*, 135–146. [CrossRef]
17. Abdulkhalikov, K.S. The dual of a fuzzy subspace. *Fuzzy Set. Syst.* **1996**, *82*, 375–381. [CrossRef]
18. Lowen, R. Convex fuzzy sets. *Fuzzy Set. Syst.* **1980**, *3*, 291–310. [CrossRef]
19. Lubczonok, G.; Murali, V. On flags and fuzzy subspaces of vector spaces. *Fuzzy Set. Syst.* **2002**, *125*, 201–207. [CrossRef]
20. Lubczonok, P. Fuzzy vector spaces. *Fuzzy Set. Syst.* **1990**, *38*, 329–343. [CrossRef]

21. Malik D.S.; Mordeson, J.N. Fuzzy vector spaces. *Inform. Sci.* **1991**, *55*, 271–281. [CrossRef]
22. Mordeson, J.N. Bases of fuzzy vector spaces. *Inform. Sci.* **1993**, *67*, 87–92. [CrossRef]
23. Shi F.-G.; Huang, C.E. Fuzzy bases and the fuzzy dimension of fuzzy vector spaces. *Math. Commun.* **2010**, *15*, 303–310.
24. Nanda, S. Fuzzy fields and linear spaces. *Fuzzy Set. Syst.* **1986**, *19*, 89–94.
25. Nanda, S. Fuzzy algebras over fuzzy fields. *Fuzzy Set. Syst.* **1990**, *37*, 99–103. [CrossRef]
26. Malik D.S.; Mordeson, J.N. Fuzzy subfields. *Fuzzy Set. Syst.* **1990**, *37*, 383–388. [CrossRef]
27. Fang, J.-X.; Yan, C.-H. L-fuzzy topological vector spaces. *J. Fuzzy Math.* **1997**, *5*, 133–-144.
28. Zhang, D.; Xu, L. Categories isomorphic to FNS. *Fuzzy Set. Syst.* **1999**, *104*, 373–-380. [CrossRef]
29. Zhang, D. On the reflectivity and coreflectivity of L-fuzzifying topological spaces in L-topological spaces. *Acta Math. Sin. (Engl. Ser.)* **2002**, *18*, 55–-68. [CrossRef]
30. Yan, C.-H.; Wu, C.-X. Fuzzifying topological vector spaces on completely distributive lattices. *Int. J. Gen. Syst.* **2007**, *36*, 513–525. [CrossRef]
31. Wen, Y.-F.; Zhong, Y.; Shi, F.-G. L-fuzzy convexity induced by L-convex degree on vector spaces. *J. Intell. Fuzzy Syst.* **2017**, *33*, 4031–4041. [CrossRef]
32. Van de Vel, M. *Theory of Convex Structures*; North Holland: New York, NY, USA, 1993.
33. Rosa, M.V. On fuzzy topology fuzzy convexity spaces and fuzzy local convexity. *Fuzzy Set. Syst.* **1994**, *62*, 97–100. [CrossRef]
34. Shi F.-G.; Xiu, Z.Y. A new approach to the fuzzification of convex structures. *J. Appl. Math.* **2014**, *2014*, 1–12. [CrossRef]
35. Shi, F.-G.; Xiu, Z.-Y. (L,M)-fuzzy convex structures. *J. Nonlinear Sci. Appl.* **2017**, *10*, 3655–3669. [CrossRef]
36. Dong, Y.Y.; Li, J. Fuzzy convex structure and prime fuzzy ideal space on residuated lattices. *J. Nonlinear Convex Anal.* **2020**, *21*, 2725–2735.
37. Li, J.; Shi, F.-G. L-fuzzy convexity induced by L-convex fuzzy sublattice degree. *Iran. J. Fuzzy Syst.* **2017**, *14*, 83–102.
38. Li, E.Q.; Shi, F.-G. Some properties of M-fuzzifying convexities induced by M-orders. *Fuzzy Set. Syst.* **2018**, *350*, 41–54. [CrossRef]
39. Li, Q.-H.; Huang, H.-L.; Xiu, Z.-Y. Degrees of special mappings in the theory of L-convex spaces. *J. Intell. Fuzzy Syst.* **2019**, *37*, 2265–2274. [CrossRef]
40. Pang, B.; Shi, F.-G. Strong inclusion orders between L-subsets and its applications in L-convex spaces. *Quaest. Math.* **2018**, *41*, 1021–1043. [CrossRef]
41. Pang, B. Convergence structures in M-fuzzifying convex spaces. *Quaest. Math.* **2019**, *43*, 1–21. [CrossRef]
42. Pang, B. Hull operators and interval operators in (L, M)-fuzzy convex spaces. *Fuzzy Set. Syst.* **2019**, *405*, 106–127. [CrossRef]
43. Pang, B. L-fuzzifying convex structures as L-convex structures. *J. Nonlinear Convex Anal.* **2020**, *21*, 2831–2841.
44. Pang, B. Bases and subbases in (L,M)-fuzzy convex spaces. *Comput. Appl. Math.* **2020**, *39*, 41. [CrossRef]
45. Wang, B.; Li, Q.; Xiu, Z.-Y. A categorical approach to abstract convex spaces and interval spaces. *Open Math. J.* **2019**, *17*, 374–384. [CrossRef]
46. Wang, K.; Shi, F.-G. M-fuzzifying topological convex spaces. *Iran. J. Fuzzy Syst.* **2018**, *15*, 159–174.
47. Wang, K.; Shi, F.-G. Many-valued convex structures induced by fuzzy inclusion orders. *J. Intell. Fuzzy Syst.* **2019**, *36*, 3373–3383. [CrossRef]
48. Wang, L.; Pang, B. Coreflectivities of (L, M)-fuzzy convex structures and (L, M)-fuzzy cotopologies in (L, M)-fuzzy closure systems. *J. Intell. Fuzzy Syst.* **2019**, *37*, 3751–3761. [CrossRef]
49. Wang, L.; Wu, X.-Y.; Xiu, Z.-Y. A degree approach to relationship among fuzzy convex structures, fuzzy closure systems and fuzzy Alexandrov topologies. *Open Math. J.* **2019**, *17*, 913–928. [CrossRef]
50. Wu, X.Y.; Li, E.Q.; Bai, S.Z. Geometric properties of M-fuzzifying convex structures. *J. Intell. Fuzzy Syst.* **2017**, *32*, 4273–4284. [CrossRef]
51. Xiu, Z.Y.; Pang, B. M-fuzzifying cotopological spaces and M-fuzzifying convex spaces as M-fuzzifying closure spaces. *J. Intell. Fuzzy Syst.* **2017**, *33*, 613–620. [CrossRef]
52. Zhong, Y.; Shi, F.-G. Characterizations of (L, M)-Fuzzy topological degrees. *Iran. J. Fuzzy Syst.* **2018**, *15*, 129–149.
53. Pang, B.; Xiu, Z.Y. Lattice-valued interval operators and its induced lattice-valued convex structures. *IEEE Trans. Fuzzy Syst.* **2018**, *26*, 1525–1534. [CrossRef]
54. Liu, Q.; Shi, F.-G. M-fuzzifying median algebras and its induced convexities. *J. Intell. Fuzzy Syst.* **2019**, *36*, 1927–1935. [CrossRef]
55. Zhang, W.; Wang, J.; Liu, W.; Fang, J. *An Introduction to Fuzzy Mathematics*; Xi'an Jiaotong University Press: Xi'an, China, 1991.
56. Ward M.; Dilworth, R.P. Residuated lattices. *Trans. Am. Math. Soc.* **1939**, *45*, 335–354. [CrossRef]
57. Blount, K.; Tsinakis, C. The structure of residuated lattices. *Int. J. Algebra Comput.* **2003**, *13*, 437–461. [CrossRef]
58. Turunen, E. *Mathematics Behind Fuzzy Logic*; Physica: Heidelberg, Germany, 1999.

Σ mathematics

Review

Reinstatement of the Extension Principle in Approaching Mathematical Programming with Fuzzy Numbers

Bogdana Stanojević [1], Milan Stanojević [2] and Sorin Nădăban [3,*]

1 Mathematical Institute of the Serbian Academy of Sciences and Arts, Kneza Mihaila 36, 11000 Belgrade, Serbia; bgdnpop@mi.sanu.ac.rs
2 Faculty of Organizational Sciences, University of Belgrade, Jove Ilića 154, 11000 Belgrade, Serbia; milan.stanojevic@fon.bg.ac.rs
3 Department of Mathematics and Computer Science, Aurel Vlaicu University of Arad, Elena Drăgoi 2, RO-310330 Arad, Romania
* Correspondence: snadaban@gmail.com

Abstract: Optimization problems in the fuzzy environment are widely studied in the literature. We restrict our attention to mathematical programming problems with coefficients and/or decision variables expressed by fuzzy numbers. Since the review of the recent literature on mathematical programming in the fuzzy environment shows that the extension principle is widely present through the fuzzy arithmetic but much less involved in the foundations of the solution concepts, we believe that efforts to rehabilitate the idea of following the extension principle when deriving relevant fuzzy descriptions to optimal solutions are highly needed. This paper identifies the current position and role of the extension principle in solving mathematical programming problems that involve fuzzy numbers in their models, highlighting the indispensability of the extension principle in approaching this class of problems. After presenting the basic ideas in fuzzy optimization, underlying the advantages and disadvantages of different solution approaches, we review the main methodologies yielding solutions that elude the extension principle, and then compare them to those that follow it. We also suggest research directions focusing on using the extension principle in all stages of the optimization process.

Keywords: fuzzy numbers; extension principle; mathematical programming

MSC: 03E72; 90C70

Citation: Stanojević, B.; Stanojević, M.; Nădăban, S. Reinstatement of the Extension Principle in Approaching Mathematical Programming with Fuzzy Numbers. *Mathematics* **2021**, *9*, 1272. https://doi.org/10.3390/math9111272

Academic Editor: Francisco Javier Cabrerizo-Lorite

Received: 7 May 2021
Accepted: 28 May 2021
Published: 1 June 2021

1. Introduction

Zadeh's fuzzy set theory [1] is an accurate mathematical tool that is able to model the uncertainty widely present in real-life problems. It has a wide range of applications in various scientific fields from medicine, engineering, and computer science to artificial intelligence. Dubois [2] emphasized that one of the roles of the fuzzy set theory is to facilitate a joint functionality of the numerical and qualitative approaches in decision-making.

Dzitac et al. [3] recently presented several important aspects of Zadeh's fuzzy logic theory that were proved to have useful applications. A discussion on the need of fuzzy logic and a nonstandard perspective on it was given in [4]. Wu and Xu [5] presented a wide range of applications of the fuzzy logic in decision making that proved fuzzy logic's ability in handling uncertain linguistic information. Shi [6] introduced several results from fuzzy group's theory that could represent a good foundation when the multivalued computer systems will be redeveloped in the future. Nădăban [7] presented a concise and unitary general view on the algebraic connections between classic, fuzzy, and quantum logics.

In this study, we restrict our attention to fuzzy mathematical programming. Zimmerman [8,9] emphasized the role of the fuzzy set theory in mathematical programming, introducing a solution approach to multiple objective optimization problems based on aggregation of fuzzy goals and fuzzy constraints. Verdegay [10] emphasized that the fuzzy linear programming is one of the most studied topics in the theory of the fuzzy sets and systems. We focus especially on optimization problems that involve fuzzy numbers as coefficients and/or variables aiming to rehabilitate the position of Zadeh's extension principle [1] in approaching such problems.

When a solution approach to fuzzy optimization problems strictly follows the extension principle, the ranking of the involved fuzzy quantities is avoided. From our perspective, this fact is a real advantage since there are many ranking functions defined in the literature (Abbasbandy [11] mentioned more than thirty); each of them might generate a solution approach to certain classes of optimization problems, and any comparison of their effectiveness is almost impossible.

After a brief presentation of the basic notation and terminology related to fuzzy sets and mathematical programming given in Section 2, we include in Section 3 a discussion on the indispensability of the extension principle in solving mathematical programming problems with fuzzy numbers. In Section 4, we survey the main methodologies that address full fuzzy optimization problems and analyze the effects of neglecting the extension principle in some of their optimization steps. In Section 5, we suggest research directions focusing on using the extension principle in all stages of the optimization process. Our concluding remarks are presented in Section 6.

2. Notation and Terminology

2.1. Fuzzy Sets

Zadeh [1] introduced the concept of fuzzy set $\widetilde{A}$ over the universe X as a collection of pairs $(x, \mu_{\widetilde{A}}(x))$ such that the first component of each pair $x \in X$ is an element of the universe, while the second element $\mu_{\widetilde{A}}(x) \in [0,1]$ is its corresponding membership degree. Function $\mu_{\widetilde{A}} : X \to [0,1]$ is called the membership function of the fuzzy set $\widetilde{A}$.

Atanassov [12] introduced the intuitionistic fuzzy sets as a generalization to the fuzzy sets. An intuitionistic fuzzy set $\widetilde{A}^I$ of a universe X is a set of triples

$$(x, \mu_{\widetilde{A}^I}(x), \nu_{\widetilde{A}^I}(x)) \tag{1}$$

such that $x \in X$, $\mu_{\widetilde{A}^I}(x), \nu_{\widetilde{A}^I}(x) \in [0,1]$, and $0 \leq \mu_{\widetilde{A}^I}(x) + \nu_{\widetilde{A}^I}(x) \leq 1$. The membership function of $\widetilde{A}^I$ is $\mu_{\widetilde{A}^I} : X \to [0,1]$, and the nonmembership function of $\widetilde{A}^I$ is $\nu_{\widetilde{A}^I} : X \to [0,1]$ of $\widetilde{A}^I$ in X. For each $x \in X$ the value $\mu_{\widetilde{A}^I}(x)$, called the membership degree, the value $\nu_{\widetilde{A}^I}(x)$ is called the nonmembership degree, and the value

$$h(x) = 1 - \mu_{\widetilde{A}^I}(x) - \nu_{\widetilde{A}^I}(x) \tag{2}$$

is called the degree of hesitancy of x in $\widetilde{A}^I$. Deng [13] proposed a new way to measure the information volume of fuzzy and intuitionistic fuzzy membership functions.

2.1.1. Fuzzy Numbers

Fuzzy numbers (FNs) are special cases of fuzzy sets. A fuzzy set $\widetilde{A}$ of the universe of real numbers R is called a fuzzy number if and only if: (i) it is fuzzy normal and fuzzy convex; (ii) the membership function $\mu_{\widetilde{A}}$ is upper semicontinuous; and (iii) its support, i.e., $\{x \in R | \mu_{\widetilde{A}}(x) > 0\}$ is bounded. Similarly, an intuitionistic fuzzy set of R is an intuitionistic fuzzy number (IFN) if and only if its membership function fulfills all conditions in the definition of a fuzzy number; the nonmembership function is fuzzy concave and lower semicontinuous; and its support $\{x \in R | \nu_{\widetilde{A}}(x) < 1\}$ is bounded.

An LR flat fuzzy number is defined using two reference functions for the left and right sides of the fuzzy number, respectively. The reference functions L and R are both defined on the interval $[0, \infty)$, take values from the interval $[0,1]$, and have two essential characteristics: (i) $L(0) = R(0) = 1$; and (ii) both L and R are nonincreasing on $[0, \infty)$. We refer the reader to the book of Dubois and Prade [14] for more details.

In what follows, we are interested in triangular and trapezoidal fuzzy and intuitionistic fuzzy numbers. The graph of the nonzero piece of the membership function of a triangular fuzzy number (TFN) forms a triangle with the abscissa, and is generally expressed by its components, as a triple (a_1, a_2, a_3), $a_1 \leq a_2 \leq a_3$. The interval (a_1, a_3) is the support of the fuzzy set and the component a_2 is the value with the maximal amplitude. Similarly, the graph of the nonzero piece of the membership function of a trapezoidal fuzzy number (TrFN) forms a trapezoid with the abscissa, and is generally expressed by its components, as a quadruple (a_1, a_2, a_3, a_4), $a_1 \leq a_2 \leq a_3 \leq a_4$. The interval (a_1, a_4) is the support of the fuzzy set and all values in the interval $[a_1, a_4]$ have the maximal amplitude.

A triangular intuitionistic fuzzy number (TIFN) is generally denoted by

$$\widetilde{A}^I = (a_1, a_2, a_3; a_1', a_2', a_3'). \tag{3}$$

Its first three components are related to the membership function (that is identical to a membership function of a triangular fuzzy number), and last three related to the nonmembership function. Similarly, a trapezoidal intuitionistic fuzzy number (TrIFN) is generally described by

$$\widetilde{A}^I = (a_1, a_2, a_3, a_4; a_1', a_2', a_3', a_4'), \tag{4}$$

first four components being related to the membership function that is in fact a membership function of a trapezoidal fuzzy number.

2.1.2. The Extension Principle

Bellman and Zadeh [15] introduced the concepts of fuzzy decisions and fuzzy constraints, and proposed a principle to aggregate them. The fuzzy arithmetic was developed with the help of the extension principle mentioned by Zadeh from the beginning in [1].

According to this principle, the fuzzy set $\widetilde{B}$ of the universe Y that is the result of evaluating the function f at the fuzzy sets $\widetilde{A}_1, \widetilde{A}_2, \ldots, \widetilde{A}_r$ over their universes $X_1, X_2, \ldots, X_r$ is defined through its membership function as

$$\mu_{\widetilde{B}}(y) = \begin{cases} \sup\limits_{(x_1,\ldots,x_r)\in f^{-1}(y)} \left(\min\left\{\mu_{\widetilde{A}_1}(x_1), \ldots, \mu_{\widetilde{A}_r}(x_r)\right\}\right), & f^{-1}(y) \neq \varnothing \\ 0, & \text{otherwise.} \end{cases} \tag{5}$$

See also Zimmerman [16] for more details.

Fuzzy addition and subtraction of triangular and trapezoidal fuzzy numbers both yield triangular and trapezoidal numbers, respectively. Strictly following the extension principle, neither fuzzy multiplication nor division of triangular/trapezoidal fuzzy numbers yield triangular/trapezoidal numbers. This issue is generally overcome using a relative innocent approximation that replaces the exact results by those triangular/trapezoidal numbers that keep the extreme values (i.e., the endpoints of their support, and the values with maximal amplitude) the same.

Diniz et al. [17] discussed the optimization of a fuzzy-valued function using Zadeh's extension principle. The objective function was a Zadeh's extension of a function with respect to a parameter and an independent variable. Kupka [18] introduced some results on the approximation of Zadeh's extension of a given function, and studied the quality of the approximation with respect to the choice of the metric on the space of the fuzzy sets.

2.2. Mathematical Programming Models

2.2.1. The General Crisp Model

Model (6) generally defines a crisp mathematical programming problem that consists in maximizing the objective function f that depends on the coefficients c, and the decision variables x over the feasible set $X(b)$ that depends on the coefficients b.

$$\begin{array}{ll} \max & f(x, c), \\ \text{s.t.} & x \in X(b), \end{array} \tag{6}$$

Under additional assumptions (imposed on the objective function and/or the constraints), Model (6) becomes convex (if the objective function is convex over the feasible set that is also convex); linear (if both the objective function and constraint system are defined by linear expressions, i.e., $f(x,c) = c^T x$, and $X(A,b) = \{x | Ax \leq b, x \geq 0\}$, with A, b, and c being matrices of certain dimensions); linear fractional (when the objective function is a ratio of linear functions); mixed integer (if some decision variables take integer and/or binary values); multiple objective (if the values of the objective function are vectors), etc. Specific models have specific solution methods. The basic references here are [19] for linear programming, [20] for multiple objective programming, [21] for linear fractional programming, and [22] for integer programming.

2.2.2. Fuzzified Models

Model (7) generally defines the class of fuzzy mathematical programming problems that use the fuzzy coefficients $\widetilde{b}$ and $\widetilde{c}$ to describe the uncertainty of the real system on which the feasible set and objective function depend, respectively.

$$\begin{array}{ll} \max & f(x,\widetilde{c}), \\ \text{s.t.} & x \in X(\widetilde{b}), \end{array} \tag{7}$$

Problems belonging to this class are known as mathematical programming problems with fuzzy coefficients. Depending on the type of the constraints that define the feasible set $X(\widetilde{b})$ a solution approach to (7) provides either crisp optimal values (real numbers) or "optimal" fuzzy sets values for the decision variables. In the second case, the fuzziness of the decision variables cannot be concluded from the model; sometimes, it is ignored, especially when the focus is exclusively put on the objective function's fuzzy set value, and sometimes, it is derived with respect to the endpoints of the support of the fuzzy set of optimal objective values.

Whenever the values of the decision variables are fuzzy sets the model should be formalized as

$$\begin{array}{ll} \max & f(\widetilde{x},\widetilde{c}), \\ \text{s.t.} & \widetilde{x} \in X(\widetilde{b}), \end{array} \tag{8}$$

in order to explicitly show the fuzziness of x-es. In fact, Model (8) describes the so-called full fuzzy (FF) mathematical programming problems. Again, particular forms of the objective function and/or constraints in Model (8) contribute to classifying it as full fuzzy linear programming model, full fuzzy multiple objective model, full fuzzy fractional model, etc. Similar concepts and models exist in an intuitionistic fuzzy (IF) environment.

We will use the following abbreviations related to mathematical programming: LP for linear programming; LFP for linear fractional programming; MO for multiple objective; TP for transportation problem. More elaborate abbreviations can be obtained by concatenating some of the basic abbreviations. For instance FIF-LP stands for full intuitionistic fuzzy linear programming.

3. The Necessity of the Extension Principle

The extension principle was created to generalize the crisp mathematical concepts to concepts compatible with the fuzzy set theory. Neglecting the extension principle when dealing with optimization problems whose models use fuzzy numbers either move the derived fuzzy solutions out of their proper bounds or damage their shapes.

The solution approaches to Model (8) a priori impose certain shapes for the decision variables (see FIS in Figure 1), mainly the same shape as the shapes used for the coefficients, generally fuzzy numbers. Further on, they formally use the fuzzy arithmetic theory to evaluate the expressions of the objective function and constraints with respect to the fuzzy coefficients and variables, and then propose different methods relying on fuzzy numbers optimization and ranking.

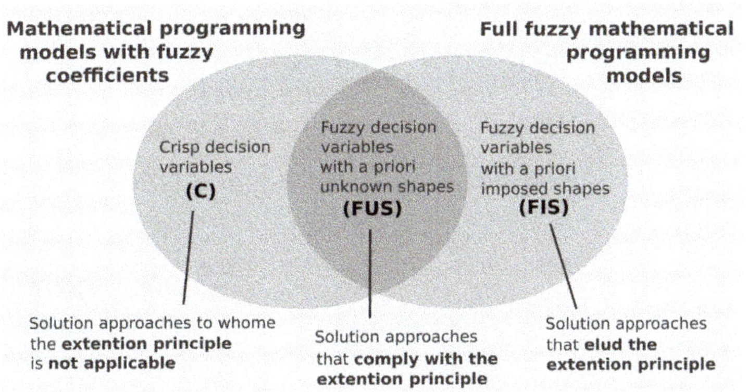

Figure 1. Decision variables shapes with respect to the two types of fuzzy mathematical programming and how they comply with the extension principle.

Figure 2 visually presents the current position on which the extension principle is mainly involved in the solution approaches to mathematical programming with fuzzy numbers; and the desired position on which the extension principle should be generally reinstated.

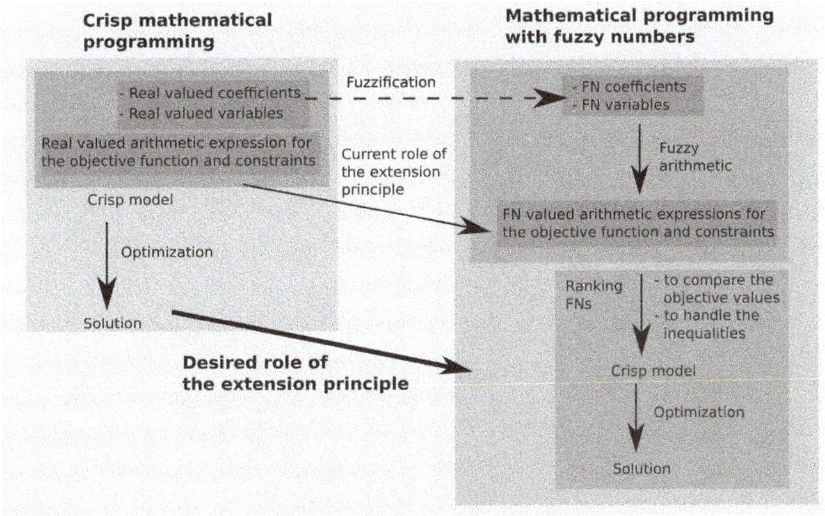

Figure 2. Current versus desired roles of the extension principle in mathematical programming with fuzzy numbers.

Comparing Models (7) and (8) and the solution approaches found in the literature, it is noticeable that problems with fuzzy coefficients and their corresponding full fuzzy problems share some hidden characteristics visible through certain solution approaches. To derive solution values to Model (7), one must use crisp decision variables (see C in Figure 1) through the computation, and can easily ignore the ranges of their values whenever the main focus is put on the objective function, i.e., on finding the fuzzy set that in the best way describes the possible values of the objective function. In some cases, paying more attention to the values of the decision variables, one can see that they describe fuzzy sets that easily fit in Model (8). Consequently, the border between the class of problems with fuzzy coefficients and the class of full fuzzy problems cannot be clearly stated, and solution approaches to problems with fuzzy coefficients might be extremely useful in solving full fuzzy problems, especially when they strictly comply with the extension principle. Figure 1 visually presents how the shapes of the decision variables in Models (7) and (8) comply with the extension principle.

By the literature review given in the next section, we aim to present the existing approaches to both classes of fuzzy mathematical programming problems, grouping them by the models they solve and their methodological particularities, and also conclude about the negative effects of eluding the extension principle in the crucial stages of the optimization process.

4. Literature Review from the Extension-Principle-Based Perspective

LP models both crisp and fuzzy are widely studied due to their simplicity. Ghanbari et al. [23] surveyed the literature on fuzzy LP problems discussing both the mathematical models and the solutions. Fractional programming models are the next most used models since they offer a relatively simple generalization of linear programming toward hard nonlinearity. Stanojević et al. [24] reviewed the literature on LFP problems with fuzzy numbers. Transportation problems in both forms linear and fractional are also widely studied in the literature due to their particularities that bring consistent simplifications to the general solution approaches.

The main facts that make a difference among solution approaches to full fuzzy optimization problems are related to the way of ranking the fuzzy numbers when optimizing the objective function and/or evaluating the fuzzy constraints.

We organize our literature review in three subsections that cover the three classes of problems mentioned above and restrict our attention to those papers that are relevant to our discussion, which is focused on how the presence of the extension principle in formulating the solution concept contributes to the relevance of the derived fuzzy set solutions.

4.1. Fuzzy Transportation Problems

A full fuzzy linear TP consists of minimizing the fuzzy cost of transferring fuzzy amounts of goods from sources to destinations with fuzzy demands. Model (9)

$$\begin{array}{ll} \min & \sum_{i=1}^{m}\sum_{j=1}^{n}\widetilde{c}_{ij}\widetilde{x}_{ij} \\ \text{s.t.} & \\ & \sum_{j=1}^{n}\widetilde{x}_{ij}=\widetilde{a}_i, \quad i=\overline{1,m}, \\ & \sum_{i=1}^{m}\widetilde{x}_{ij}=\widetilde{b}_j, \quad j=\overline{1,n}, \\ & \widetilde{x}_{ij}\geq\widetilde{0} \qquad i=\overline{1,m}, j=\overline{1,n}, \end{array} \tag{9}$$

summarizes the essential details, assuming that there are m sources with the amount of available goods expressed by the fuzzy quantities $\widetilde{a}_i$, $i=\overline{1,m}$ and n destinations with the fuzzy demands $\widetilde{b}_j$, $j=\overline{1,n}$. The decision variables $\widetilde{x}_{ij}$, $i=\overline{1,m}$, $j=\overline{1,n}$, represent the amount of goods transported from the source i to the destination j, with a transport cost per unit equal to $\widetilde{c}_{ij}$.

Liu and Kao [25] and Liu [26] proposed solution approaches to linear and linear fractional TPs with fuzzy coefficients, respectively, that fit to Model (7). Their solution concept is based on defining the fuzzy set of the optimal solution values of the objective function through a membership function that complies to the extension principle. Equation (10) shows their membership function adapted to the notation used in Model (7).

$$\mu_{\widetilde{f}}(z)=\begin{cases} \max\limits_{(b,c)|z=\max\limits_{x\in X(b)}f(x,c)}\left(\mu_{(\widetilde{b},\widetilde{c})}(b,c)\right), & \exists b,c|z=\max\limits_{x\in X(b)}f(x,c), \\ 0, & \text{otherwise}, \end{cases} \tag{10}$$

where $\mu_{(\widetilde{b},\widetilde{c})}(b,c) = \min\{\mu_{\widetilde{b}}(b), \mu_{\widetilde{c}}(c)\}$, and $\mu_{\widetilde{b}}(b)$ and $\mu_{\widetilde{c}}(c)$ represent the membership functions of the coefficients $\widetilde{b}$ and $\widetilde{c}$, respectively.

Neither Liu and Kao [25] nor Liu [26] reported the shapes of the fuzzy sets of the optimal values of the decision variables. Later on, based on Liu and Kao's solution concept [25], the optimal values of the decision variables were empirically disclosed by a Monte Carlo simulation in [27] to be trapezoidal-like FNs whenever the problem's coefficients were TrFNs. In that way, it was proved that the problems solved by Liu and Kao [25] belong to the class of problems that can be modeled using fuzzy decision variables with a priori unknown shapes (see FUS in Figure 1). Stanojević and Stanojević [28] introduced an extension-principle-based formulation for the fuzzy set value of each decision variable and proposed mathematical models able to derive numerical approximations to the membership functions of the fuzzy sets representing the best values for the decision variables. Formulation (11) represents their definition for each decision variable $\widetilde{x}_i$, $i = \overline{1,n}$ adapted to the notation used in Model (7).

$$\mu_{\widetilde{x}_i}(t) = \begin{cases} \max\limits_{(b,c)|t = \arg_i \max\limits_{y \in X(b)} f(y,b)} \left(\mu_{(\widetilde{b},\widetilde{c})}(b,c)\right), & \exists b,c | t = \arg_i \max\limits_{y \in X(b)} f(y,b), \\ 0, & \text{otherwise,} \end{cases} \tag{11}$$

where $\arg_i \max\limits_{y \in X(b)} f(y,b)$ represents the optimal value of the crisp scalar decision variable y_i obtained when maximizing $f(y,b)$ over the feasible set $X(b)$, and $\mu_{(\widetilde{b},\widetilde{c})}(b,c)$ has the same meaning as in the definition of the membership function of optimal solution values (10). The membership function of the fuzzy set $\widetilde{x}$ that collects the crisp values of the vector x of the decision variables, i.e.,

$$\mu_{\widetilde{x}}(t) = \begin{cases} \max\limits_{(b,c)|t = \arg \max\limits_{y \in X(b)} f(y,b)} \left(\mu_{(\widetilde{b},\widetilde{c})}(b,c)\right), & \exists b,c | t = \arg \max\limits_{y \in X(b)} f(y,b), \\ 0, & \text{otherwise,} \end{cases} \tag{12}$$

was first given in [29] for the general linear case. The graph of the membership function given in (11) represents in fact the projection of the graph of the membership function (12) on the plane $(Ox_i, O\alpha)$, $i = \overline{1,n}$, where the axes Ox_i and $O\alpha$ contain the values of the variable x_i and the membership degrees α, respectively.

The first step in solving a general crisp TP is reserved to a simple balancing procedure. In the case of solving fuzzy TPs, the situation is quite different: some approaches can be applied only to balanced problems; several approaches focus only on balancing a general fuzzy TP; and also some approaches can be applied to unbalanced problems. Unfortunately, the solution provided by an approach applied to an unbalanced problem is far from being similar to the solution derived after a balancing procedure is applied. Mishra and Kumar [30] proposed a balancing procedure for FIF-TP that is an adaptation of Kumar and Kaur's method [31] to balance FF-TPs. Any solution approach to fuzzy TP based on a solution concept that is in accordance to the extension principle does not need a fuzzy balancing procedure, since crisp TPs are solved and their optimal solutions build the final fuzzy set solution to the original fuzzy problem. The methodology proposed in Stanojević and Stanojević [28] to solve FIF-TPs discussed this issue in detail.

Kumar and Kaur [31], after balancing the fuzzy transportation problem, proposed two approaches to solved it. In both approaches, they first applied algebraic operators on the fuzzy numbers to derive the fuzzy number values of the objective function and left-hand sides of the constraints; and then focused on the optimization of the objective function. In the first method, Kumar and Kaur applied a ranking function on the TrFN representing the objective function, obtained a crisp expression of the components of the involved TrFNs, and optimized it subject to a conjunctive constraint system. In the second method, they solved four crisp TPs, one for each component of the TrFN representing the objective function. None of these methods respects the extension principle.

Singh and Yadav [32] proposed several algorithms to determine a basic feasible solution to a balanced FIF-TP. All their algorithms are analogous to the well known methods developed for crisp TPs. This simple analogy also discards any possibility to follow the extension principle.

Stanojević and Stanojević [28] improved Liu and Kao's approach [25] by proposing a mathematical model with disjunctive constraint system and then extended it to solving TPs with TrIFN as parameters. They compare their results with the results reported by Kumar and Hussain [33], Ebrahimnejad and Verdegay [34] (who formulated their solution approach based on an accuracy function used for ordering the TrIFNs), and Mahmoodirad et al. [35]. All of these papers addressed FIF-TPs. The experiments showed that, due to its compliance with the extension principle, the method proposed in [28] found a fuzzy set solution with a wider support and smaller optimal values for the objective function that had to be minimized than the methods introduced in [33–35].

Further investigations are necessary to conclude about Mahajan and Gupta's approach [36] for solving a FIF-MO-TP problem. Using an accuracy function on each objective, they first reduced the problem to a crisp MO-TP, and developed an algorithm to solve it. To handle the intuitionistic fuzzy constraints, they used linear, exponential, and hyperbolic membership functions. Unifying the arithmetic and optimization, and formulating an extension-principle-based approach to solve the same problem, possibly better results can be derived.

Table 1 reports results found in the literature for three numerical examples of full fuzzy TPs. The first example uses trapezoidal fuzzy numbers, whereas the next two of them consider intuitionistic fuzzy numbers for all coefficients and decision variables. The results derived with a full respect to the extension principle correspond to the references written in bold. All full fuzzy TPs were minimization problems, and all optimal fuzzy numbers that comply to the extension principle are clearly smaller than those that elude it. For the first example, the values with maximal amplitude are the same in both references, but the second reference reports smaller minimal values with nonzero membership function. For the second and third examples, one can notice that smaller minimal values with nonzero membership function were obtained by the solution approach that strictly followed the extension principle. Figure 3 shows graphically the results reported in Table 1 for the second example.

Table 1. Comparative numerical results for full fuzzy and full intuitionistic fuzzy TPs.

Ex.	Fuzzy Type	Ref.	$\widetilde{z}_{\min}$
1	trapezoidal	[25]	(2100, 2900, 3500, 5800)
		[28]	(1500, 2900, 3500, 5800)
2	trapezoidal intuitionistic	[34]	(3300, 5800, 9100, 13,200; 2350, 4450, 11,050, 15,550)
		[28]	(1400, 3760, 8750, 13,500; 700, 2280, 9550, 16,000)
3	triangular intuitionistic	[33]	(137, 292, 502; 12, 292, 961)
		[34]	(63, 313, 773; 2, 313, 1726)
		[35]	(63, 310, 757; 2, 310, 1806)
		[28]	(32, 305.4, 765; 0, 305.4, 1697)

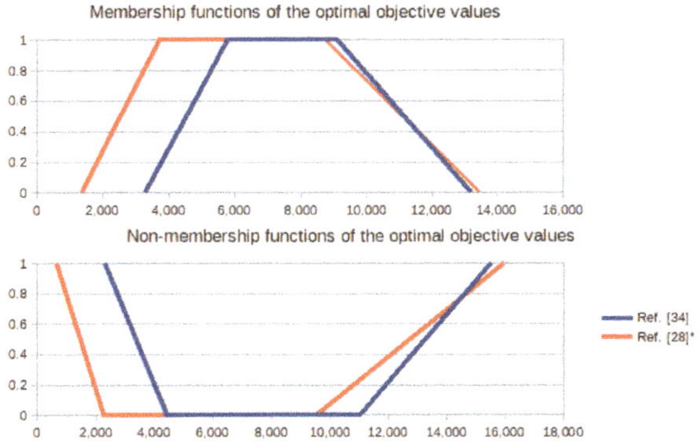

Figure 3. Graphic representation of the results reported in Table 1, example 2. Symbol * marks the reference that introduced the approach that comply to the extension principle.

4.2. Fuzzy Linear Programming Problems

A full fuzzy LP problem is generally described by Model (13)

$$\begin{array}{ll} \max & \tilde{c}^T \tilde{x} \\ \text{s.t.} & \tilde{A}\tilde{x} \leq \tilde{b}, \\ & \tilde{x} \geq 0. \end{array} \tag{13}$$

Several variants of Model (13) are sometimes addressed in the literature—for instance, those that minimize the objective function and/or use equality constraints to define the feasible set, but they are not different in nature, and their approaches can be easily adapted to solve Model (13) as well.

Perez-Canedo et al. [37] surveyed the literature on fuzzy linear programming, paying attention to those papers that used a lexicographic method to rank the fuzzy numbers. Their up-to-date review shows that fuzzy LP relying on lexicographic methods is an active research area with a wide range of applications in practice.

We briefly survey several papers on fuzzy linear programming, including notes from the extension-principle-based perspective.

Hosseinzadeh Lotfi et al. [38] used lexicographic method and fuzzy approximate solutions to FF-LP problems with TFNs. They transformed all TFNs coefficients in symmetric TFNs and assumed that all decision variables were symmetric TFNs. They also used a special ranking function on fuzzy numbers, and obtained a crisp MO-LP to solve. They did not involve the extension principle in the optimization step, and derived only one efficient solution to the crisp MO-LP.

Khan et al. [39] introduced a simplex-like technique for solving FF-LP problems. They involved a ranking function in the Gaussian elimination process and developed a flexible easy and reasonable solution algorithm. The use of any ranking function supposes defuzzification before optimization that is not in the desired accordance to the extension principle.

To transform FF-LP problems into crisp LPs, Kumar et al. [40] used a ranking function in order to compare the FN values of the objective function and a component-wise comparison of the left and right hand sides of the constraints. They finally solved one single objective linear programming problem deriving optimal values for all components of all decision variables. The use of two distinct methods to compare fuzzy quantities embedded in the same model is against the essence of the extension principle.

Ezzati et al. [41] addressed FF-LP problems by applying first fuzzy arithmetic to the TFN values of the decision variables and coefficients in order to derive the TFN value of the objective function; then, they constructed a three-objective crisp problem and solved it by a lexicographic method. Stanojević and Stanojević [29] proposed empirical solutions to FF-LP problems based on a Monte Carlo simulation that fully comply with the extension principle. They analyzed one of Ezzati et al.'s examples and conclude that Ezzati et al.'s methodology fails to follow the extension principle.

Noticing a shortcoming in the solution approach introduced in [41], Bhardwaj and Kumar [42] corrected and improved that approach by proposing a proper transformation able to replace the fuzzy inequality constraints by equivalent equality constraints. The effort put to rewrite the inequalities in a convenient form is unnecessary whenever the extension principle is applied.

The subsequent papers have come to our attention and require further examination. Das et al. [43] introduced a new method to solve FF-LP problems with TrFNs. Their method was based on solving a mathematical model derived from the MO-LP problem and lexicographic ordering method. They solved real-life problems as production planning and diet problem to illustrate the applicability of their approach.

Pérez-Cañedo and Concepción-Morales [44] derived a unique optimal fuzzy value to a FF-LP problem with inequality constraints containing unrestricted LR flat fuzzy coefficients and decision variables. In [45], they ranked LR-type IFNs using a lexicographic criterion, and introduced a method to derive solutions to FIF-LP problems with unique optimal values.

Khalifa [46] addressed FF-LP problems with coefficients and variables expressed by LR fuzzy numbers. He transformed the original fuzzy problem into a three-objective crisp LP problem, and solved it using a classic weighted sum method. The derived crisp solution was next used to construct the fuzzy solution to the original problem.

Khalili Goudarzi et al. [47] proposed a solution approach to FF-MILP problems. First, they formulate a crisp three-objective problem, and find the positive and negative ideal solutions to each objective. Then, they determined a linear membership function for each objective, and constructed a new achievement function defined as a convex combination of the lower bound for satisfaction degree of all objectives, and the weighted sum of the satisfaction degrees of all objectives. In this way, their approach aimed to ensure a balanced compromise solution.

Hamadameen and Hassan [48] introduced a compromise solution approach to FF-MO-LP problems based on a revised simplex method and a Gaussian elimination method.

Table 2 presents the numerical results derived in the literature to a maximization full fuzzy LP problem with triangular fuzzy coefficients and decision variables. Analyzing the reported results one can notice the improvement brought by the approach that complies to the extension principle: the optimal objective function value with maximal amplitude is significantly greater for the third reference than for the other two references. A significant difference can be seen observing the optimal values of the decision variables: the supports of the fuzzy numbers derived by the approach described in the third reference are much wider than those derived by the approaches introduced in the first two references. All values belonging to the supports of the reported fuzzy numbers are feasible values with certain nonzero membership degrees. The narrow supports seen for the first two references show that many relevant feasible values of the decision variables were ignored by the approaches that eluded the extension principle in the optimization step. Figure 4 gives a graphic representation of the results reported in Table 2 for the optimal objective values.

Table 2. Comparative numerical results for a full fuzzy LP problem.

	[41]	[40]	[29]
$\tilde{z}_{\max}$	$(304.58, 509.79, 704.37)$	$(301.83, 503.23, 724.15)$	$(279.37, 579.32, 985.13)$
$\tilde{x}_1^{\max}$	$(17.27, 17.27, 17.27)$	$(15.28, 15.28, 15.28)$	$(0, 38.28, 61.24)$
$\tilde{x}_2^{\max}$	$(2.16, 2.16, 2.16)$	$(2.40, 2.40, 9.10)$	$(0; 1.32, 53.08)$
$\tilde{x}_3^{\max}$	$(4.64, 9.97, 16.36)$	$(6.00, 11.25, 11.25)$	$(0, 1.28, 51.54)$
$\tilde{x}_4^{\max}$	$(6.36, 6.36, 6.36)$	$(6.49, 6.49, 9.49)$	$(0, 1.14, 45.60)$

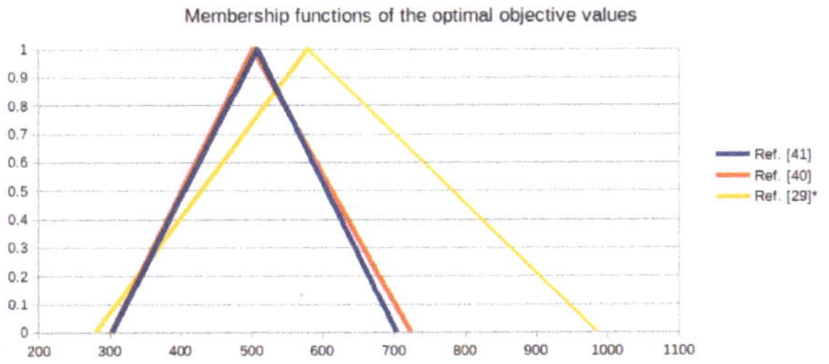

Figure 4. Graphic representation of optimal objective values reported in Table 2. Symbol * marks the reference that introduced the approach that comply to the extension principle.

4.3. Fuzzy Linear Fractional Programming Problems

A full fuzzy LFP problem differs from a full fuzzy LP problem only through its objective function. However, the nonlinearity of the objective function brings certain complications to the solution approaches. Model (14)

$$\begin{array}{ll} \max & \dfrac{\tilde{c}^T\tilde{x}+\tilde{c}_0}{\tilde{d}^T\tilde{x}+\tilde{d}_0} \\ \text{s.t.} & \tilde{A}\tilde{x} \leq \tilde{b}, \\ & \tilde{x} \geq 0, \end{array} \tag{14}$$

has a linear fractional objective function that has to be maximized subject to linear constraints. Due to the presence of the denominator, an additional condition assuming its strictly positiveness over the feasible set has to be imposed. This condition does not reduce the generality of the problem, since the case of an objective function with strictly negative denominator can be equivalently transformed to a case complying to the initially imposed condition.

Except the methodology for solving a linear fractional transportation problem proposed by Liu [26], no other attempts were made in the literature to approach the fractional programming problems with an exclusive focus on the extension principle. All of the results reported in the papers that are briefly described below might be amended by an approach that fully comply to the extension principle. The analogical treatment of the FF-LFP problems inspired from the approaches to FF-LP problems is the main fact that underpins this conclusion.

Pop and Stancu-Minasian [49] transformed a FF-LFP problem into a crisp MO-LFP problem, and applied Buckley and Feuring's approach [50] to derive the final solution. Stanojević and Stancu-Minasian [51] evaluated the fuzzy inequalities of a FF-LFP problem, and using a generalized form of Charnes-Cooper transformation linearized the original problem. Zadeh's principle was used exclusively for the fuzzy arithmetic between coefficients and variables. Das et al. [52] developed an algorithm for solving the FF-LFP problem. They used a generalized form of Charnes–Cooper transformation and obtained a crisp MO-LP model solved by a lexicographic method. For their numerical experiments, they used real-life problems, and for comparison, they used the approaches introduced in [49,51]. Any linearization of a fractional expression in fuzzy environment is not conformed with the extension principle if the same fuzzy quantity appears in both numerator and denominator of the ratio.

Stanojević et al. [53] evaluated the approximation error that arises when the exact membership function of a ratio of TFNs is replaced by the membership function of a TFN, and its applications to decision-making. They mentioned for the first time the importance of using crisp decision variables in the optimization models even when fuzzy solutions were desired, which was a step toward unification of full fuzzy models with no a priori shapes for the decision variables and models with fuzzy coefficients and constraints that dictate the fuzziness of the decision variables.

Chinnadurai and Muthukumar [54] proposed a numerical approach to solving LFP problems in a fuzzy environment. They obtained (α, r)-acceptable optimal values by solving crisp bi-objective LFP problems. Their procedure was improved by Ebrahimnejad et al. [55], who modified the optimization model, and ensured the non-negativity of the fuzzy valued decision variables. A new extension principle-based solution concept able to embed the (α, r)-acceptable optimality is desired in order to assure further improvements.

Kaur and Kumar [56] discussed the shortcoming of papers [49,51] that resides in an improper interpretation of the fuzzy constraints of FF-LFP models. To avoid the constraint interpretation, they used Yager's ranking approach [57]. An effective way to remove the constraints interpretation from a solution approach is to transfer them in crisp environment, as proved in [28] for unbalanced FF-TP problems.

Arya et al. [58] are the first authors that addressed FF-MO-LFP problems. Their approach analogically follows Chakraborty and Gupta's method [59] designed to solve MO-LFP problems via fuzzy goals, and it is essentially based on a generalized Charnes–Cooper transformation whose arithmetic is in discordance with the extension principle.

Loganathan and Ganesan [60] transformed the original FF-LFP problem into a FF-LP problem and then replaced all coefficients and variables by their parametric forms derived from the α-cuts of triangular fuzzy numbers. They solved the parametric problem using a simplex-like algorithm, and derived parametric solutions. The fuzzy solutions were further derived numerically for different values of the parameter $\alpha \in [0,1]$. Because of the linearization step, the paper by Loganathan and Ganesan [60] might be affected by not complying to the extension principle.

4.4. Summary

A summary of the surveyed papers can be found in Table 3. In this table, the references are grouped with respect to the class of problems they address. The references marked with a superscript ^{r} are review papers, those written in bold introduce solution approaches that strictly comply to the extension principle, and those marked with a superscript * need further investigations in order to conclude whether the presented methodologies strictly followed the extension principle or not. The rest of them elude the extension principle within the optimization step.

Table 3. A summary of the surveyed papers (the symbol ^{r} marks the review papers, the symbol * marks the papers that need further investigations, and the references written in bold strictly comply to the extension principle).

Problem	References
Fuzzy TP	[30], [31], [32], [33], [34], [35], **[26]**, **[25]**, **[27]**, **[28]**, [36]*
Fuzzy LP	[50], [37], [38], [39], [40], [41], **[29]**, [42], [43]*, [44]*, [45]*, [46]*, [47]*, [48]*, [23]^r
Fuzzy LFP	[49], [51], [53]^r, [54], [55], [56], [58], [60], [24]^r

We restricted our attention to about thirty papers from the literature that proposed solution approaches to full fuzzy mathematical programming problems, i.e., problems with both coefficients and variables assumed to be fuzzy quantities. This study is not meant to be an extensive survey of the above mentioned topic, but we hope that it succeeds to provide a hint on the extent to which the extension principle is currently ignored in the literature. We focused on basic methodologies for optimizing fuzzy number valued objective functions rather than on case studies dealing with fuzzy optimization problems involving fuzzy goals and/or fuzzy decisions.

5. Research Directions

Our main reason for writing this review paper is our belief that the extension principle, which is one of the main principles in fuzzy sets theory, is neglected within mathematical programming with fuzzy numbers.

As can be seen from the literature review presented in previous section, there are various solving methods that comply with the extension principle when using fuzzy arithmetic but globally neglect it, since they rely on the ranking of fuzzy numbers when optimizing the objective function and interpreting the fuzzy constraints.

In the recent literature, a Monte Carlo simulation algorithm [29] was proposed to derive extension-principle-based fuzzy set solutions to linear programming problems with fuzzy numbers. Deriving empirically such fuzzy set solutions to a fuzzy problem has the same complexity as solving the crisp variant of that problem. Therefore, whenever a crisp problem is generalized to a problem with fuzzy numbers, the Monte Carlo simulation can be employed to disclose the shapes of the fuzzy set solutions. This method works for a wide range of problems; its application is not limited to the optimization problems. After disclosing the shapes of the fuzzy sets solutions, the next step is to develop algorithms able to derive numerically the membership functions of those fuzzy sets solutions.

Due to its simplicity, the class of linear programming problems, including the subclass of transportation problems, were already solved numerically using nonlinear models (see [28,29]). Two important research directions arise and deserve attention: (i) generalize the methodology to solve other classes of problems (for instance, problems from the field of data envelopment analysis), and (ii) find replacements for the nonlinear models in order to simplify the numerical approaches. Moreover, everything already concluded for mathematical optimization problems with fuzzy and/or intuitionistic fuzzy numbers is worth generalizing to another level of uncertainty, e.g., to Pythagorean [61] and Fermatean fuzzy numbers [62].

An argument that supports the relevance of the above mentioned research directions is illustrated in Tables 1 and 2. They disclose the nature of the differences among solutions that follow the extension principle and those that do not comply to it when solving transportation problems and general linear programming problems, respectively. Further on, due to the similarities of the existing solving approaches to full fuzzy LFP problems to those for LP problems, similar improvements are expected when solving full fuzzy LFP problems applying an extension principle-based approach. Such approaches have been lacking in the literature thus far.

Finally, inducing the same logic to a wider class of problems either relying or not on fuzzy LP but using fuzzy quantities within parameters, a similar increase in the accuracy and reliability of the derived solutions is expected when a methodology that strictly follows the extension principle is applied.

6. Conclusions

In this paper, we surveyed the literature on mathematical programming with fuzzy numbers from a perspective that emphasizes the necessity of using the extension principle within the whole optimization process, not only within the fuzzy arithmetic evaluation.

Since our review showed that the extension principle was widely present through the fuzzy arithmetic, but much less involved in the foundations of the solution concepts, we advanced the idea that the rehabilitation of the extension principle when deriving relevant fuzzy descriptions to optimal solutions is highly needed. Our opinion was supported by several recent studies showing that better and more effective solutions can be obtained when the algebraic aggregation of the fuzzy quantities and optimization are unified and together conduce to a solution concept that complies to the extension principle.

We discussed the similarities between optimization problems with fuzzy coefficients and full fuzzy optimization problems, and concluded about the unification of their subclass of optimization problems with fuzzy coefficients that assume fuzzy decision variables but with a priori unknown shapes. The solution approaches to this subclass of optimization problems with fuzzy numbers are expected to follow the extension principle, and a posteriori disclose the shapes of the decision variables.

We believe that this study advances relevant ideas to initiate the revitalization of the extension principle in mathematical optimization with fuzzy numbers. Due to its critical view on the basis of the methodology used to solve certain classes of fuzzy optimization problems, and since its implementation in various fields is able to provide more accurate and reliable optimal solutions, we expect that our conclusions will have an impact on solving practical industrial problems. As it has been said, the impact should be effective on the methodology level, and could be seen as a subtle attempt to filling the gap between the academic researches with their theoretical results and the relevant solutions needed in practice.

Our future research will be focused on deriving analytic solutions to full fuzzy optimization problems that are in accordance to the extension principle, and also simple enough to be able to replace the existing solution approaches. We will follow the research directions proposed in the previous section in order to enlarge the class of problems approachable via the extension principle.

Author Contributions: Conceptualization and methodology, B.S. and M.S.; writing—original draft and visualization, B.S.; writing—review and editing, validation, S.N.; supervision, M.S. All authors have read and agreed to the published version of the manuscript.

Funding: This research received no external funding.

Institutional Review Board Statement: Not applicable.

Informed Consent Statement: Not applicable.

Data Availability Statement: Not applicable.

Acknowledgments: The authors wish to record their sincere gratitude to professor I. Dzitac for his generous help in the preparation of this paper, and also for many valuable and constructive suggestions that guided them for years. This work was partly supported by the Serbian Ministry of Education, Science and Technological Development through Mathematical Institute of the Serbian Academy of Sciences and Arts and Faculty of Organizational Sciences of the University of Belgrade.

Conflicts of Interest: The authors declare no conflict of interest.

Abbreviations

The following abbreviations are used in this manuscript:

LP	Linear Programming
LFP	Linear Fractional Programming
TP	Transportation Problem
MO	Multiple Objective
FN	Fuzzy Number
LR-FN	Left-Right flat Fuzzy Number
IFN	Intuitionistic Fuzzy Number
TFN	Triangular Fuzzy Number
TIFN	Triangular Intuitionistic Fuzzy Number
TrFN	Trapezoidal Fuzzy Number
TrIFN	Trapezoidal Intuitionistic Fuzzy Number
FF	Full Fuzzy
FIF	Full Intuitionistic Fuzzy

References

1. Zadeh, L. Fuzzy sets. *Inf. Control* **1965**, *8*, 338–353. [CrossRef]
2. Dubois, D. The role of fuzzy sets in decision sciences: Old techniques and new directions. *Fuzzy Sets Syst.* **2011**, *184*, 3–28. [CrossRef]
3. Dzitac, I.; Filip, F.; Manolescu, M. Fuzzy Logic Is Not Fuzzy: World-renowned Computer Scientist Lotfi A. Zadeh. *Int. J. Comput. Commun. Control* **2017**, *12*, 748–789. [CrossRef]
4. Zadeh, L.A. Is there a need for fuzzy logic? *Inf. Sci.* **2008**, *178*, 2751–2779. [CrossRef]
5. Wu, H.; Xu, Z. Fuzzy Logic in Decision Support: Methods, Applications and Future Trends. *Int. J. Comput. Commun. Control* **2021**, *16*, 4044. [CrossRef]
6. Shi, Y. My Early Researches on Fuzzy Set and Fuzzy Logic. *Int. J. Comput. Commun. Control* **2021**, *16*, 4090. [CrossRef]

7. Nădăban, S. From Classical Logic to Fuzzy Logic and Quantum Logic: A General View. *Int. J. Comput. Commun. Control* **2021**, *16*, 4125. [CrossRef]
8. Zimmermann, H.J. Fuzzy programming and linear programming with several objective functions. *Fuzzy Sets Syst.* **1978**, *1*, 45–55. [CrossRef]
9. Zimmermann, H.J. Applications of fuzzy set theory to mathematical programming. *Inf. Sci.* **1985**, *36*, 29–58. [CrossRef]
10. Verdegay, J. Progress on Fuzzy Mathematical Programming: A personal perspective. *Fuzzy Sets Syst.* **2015**, *281*, 219–226. [CrossRef]
11. Abbasbandy, S. Ranking of fuzzy numbers, some recent and new formulas. In Proceedings of the IFSA-EUSFLAT 2009, Lisbon, Portugal, 20–24 July 2009.
12. Atanassov, K. Intuitionistic fuzzy sets. *Fuzzy Sets Syst.* **1986**, *20*, 87–96. [CrossRef]
13. Deng, J.; Deng, Y. Information Volume of Fuzzy Membership Function. *Int. J. Comput. Commun. Control* **2021**, *16*, 4106. [CrossRef]
14. Dubois, D.; Prade, H. *Fuzzy Sets and Systems: Theory and Applications*; Academic Press: Cambridge, MA, USA, 1980.
15. Bellman, R.; Zadeh, L. Decision-Making in a Fuzzy Environment. *Manag. Sci.* **1970**, *17*, B-141–B-164. [CrossRef]
16. Zimmermann, H.J., The Extension Principle and Applications. In *Fuzzy Set Theory—And Its Applications*; Springer: Dordrecht, The Netherlands, 1996; pp. 53–67.
17. Diniz, M.; Gomes, L.; Bassanezi, R. Optimization of fuzzy-valued functions using Zadeh's extension principle. *Fuzzy Sets Syst.* **2021**, *404*, 23–37. [CrossRef]
18. Kupka, J. On approximations of Zadeh's extension principle. *Fuzzy Sets Syst.* **2016**, *283*, 26–39. [CrossRef]
19. Dantzig, G. *Linear Programming and Extensions*; RAND Corporation: Santa Monica, CA, USA, 1963. [CrossRef]
20. Ehrgott, M. *Multicriteria Optimization*; Springer: Secaucus, NJ, USA, 2005.
21. Stancu-Minasian, I.M. *Fractional Programming: Theory, Methods and Applications*; Kluwer Academic Publishers: New York, NY, USA, 1997.
22. Wolsey, L.; Nemhauser, G. *Integer and Combinatorial Optimization*; John Wiley & Sons, Inc.: Hoboken, NJ, USA, 1988. [CrossRef]
23. Ghanbari, R.; Ghorbani-Moghadam, K.; De Baets, B. Fuzzy linear programming problems: Models and solutions. *Soft Comput.* **2019**, *24*, 1433–7479. [CrossRef]
24. Stanojevic, B.; Dzitac, S.; Dzitac, I. Fuzzy Numbers and Fractional Programming in Making Decisions. *Int. J. Inf. Technol. Decis. Mak.* **2020**, *19*, 1123–1147. [CrossRef]
25. Liu, S.T.; Kao, C. Solving fuzzy transportation problems based on extension principle. *Eur. J. Oper. Res.* **2004**, *153*, 661–674. [CrossRef]
26. Liu, S.T. Fractional transportation problem with fuzzy parameters. *Soft Comput.* **2015**, *20*, 3629–3636. [CrossRef]
27. Stanojević, B.; Stanojević, M. Solution value envelope to full fuzzy transportation problems. In Proceedings of the SymOrg 2020, online, 7–9 September 2020.
28. Stanojević, B.; Stanojević, M. Approximate membership function shapes of solutions to intuitionistic fuzzy transportation problems. *Int. J. Comput. Commun. Control* **2021**, *16*, 4057. [CrossRef]
29. Stanojević, B.; Stanojević, M., Empirical versus analytical solutions to full fuzzy linear programming. In *Intelligent Methods for Computing, Communications and Control. ICCC2020*; Dzitac, I.,Dzitac, S., Filip, F., Kacprzyk, J., Manolescu, M., Oros, H., Eds.; Springer: Berlin/Heidelberg, Germany, 2020; Volume 1243. [CrossRef]
30. Mishra, A.; Kumar, A. JMD method for transforming an unbalanced fully intuitionistic fuzzy transportation problem into a balanced fully intuitionistic fuzzy transportation problem. *Soft Comput.* **2020**, *24*, 15639–15654. [CrossRef]
31. Kumar, A.; Kaur, A. Methods for solving unbalanced fuzzy transportation problems. *Int. J. Oper. Res.* **2012**, *12*, 287–316. [CrossRef]
32. Singh, S.; Yadav, S. A novel approach for solving fully intuitionistic fuzzy transportation problem. *Int. J. Oper. Res.* **2016**, *26*, 460–472. [CrossRef]
33. Kumar, P.; Hussain, R. Computationally simple approach for solving fully intuitionistic fuzzy real life transportation problems. *Int. J. Syst. Assur. Eng. Manag.* **2016**, *7*, 90–101. [CrossRef]
34. Ebrahimnejad, A.; Verdegay, J. A new approach for solving fully intuitionistic fuzzy transportation problems. *Fuzzy Optim. Decis. Making* **2018**, *17*, 447–474. [CrossRef]
35. Mahmoodirad, A.; Allahviranloo, T.; Niroomand, S. A new effective solution method for fully intuitionistic fuzzy transportation problem. *Soft Comput.* **2019**, *23*, 4521–4530. [CrossRef]
36. Mahajan, S.; Gupta, S. On fully intuitionistic fuzzy multiobjective transportation problems using different membership functions. *Ann. Oper. Res.* **2019**, *296*, 211–241. [CrossRef]
37. Pérez-Cañedo, B.; Verdegay, J.; Concepción-Morales, E.; Rosete, A. Lexicographic Methods for Fuzzy Linear Programming. *Mathematics* **2020**, *8*, 1540. [CrossRef]
38. Hosseinzadeh Lotfi, F.; Allahviranloo, T.; Alimardani Jondabeh, M.; Alizadeh, L. Solving a full fuzzy linear programming using lexicography method and fuzzy approximate solution. *Appl. Math. Model.* **2009**, *33*, 3151–3156. [CrossRef]
39. Khan, I.; Ahmad, T.; Maan, N. A Simplified Novel Technique for Solving Fully Fuzzy Linear Programming Problems. *J. Optim. Theory Appl.* **2013**, 159, *536–546*. [CrossRef]
40. Kumar, A.; Kaur, J.; Singh, P. A new method for solving fully fuzzy linear programming problems. *Appl. Math. Model.* **2011**, *35*, 817–823.. [CrossRef]

41. Ezzati, R.; Khorram, V.; Enayati, R. A new algorithm to solve fully fuzzy linear programming problems using the MOLP problem. *Appl. Math. Model.* **2015**, *39*, 3183–3193. [CrossRef]
42. Bhardwaj, B.; Kumar, A. A note on 'A new algorithm to solve fully fuzzy linear programming problems using the MOLP problem'. *Appl. Math. Model.* **2015**, *39*, 5982–5985. [CrossRef]
43. Das, S.; Mandal, T.; Edalatpanah, S. A mathematical model for solving fully fuzzy linear programming problem with trapezoidal fuzzy numbers. *Appl. Intell.* **2017**, *46*, 509–519. [CrossRef]
44. Pérez-Cañedo, B.; Concepción-Morales, E. A method to find the unique optimal fuzzy value of fully fuzzy linear programming problems with inequality constraints having unrestricted L-R fuzzy parameters and decision variables. *Expert Syst. Appl.* **2019**, *123*, 256–269. [CrossRef]
45. Pérez-Cañedo, B.; Concepción-Morales, E. On LR-type fully intuitionistic fuzzy linear programming with inequality constraints: Solutions with unique optimal values. *Expert Syst. Appl.* **2019**, *128*, 246–255. [CrossRef]
46. Khalifa, H. Utilizing a new approach for solving fully fuzzy linear programmingproblems. *Croat. Oper. Res. Rev.* **2019**, *10*, 337–344. [CrossRef]
47. Khalili Goudarzi, F.; Nasseri, S.; Thagi-Nezhad, N. A new interactive approach for solving fully fuzzy mixed integer linear programming. *Yugoslav J. Oper. Res.* **2020**, *30*, 71–89. [CrossRef]
48. Hamadameen, A.; Hassan, N. A Compromise Solution for the Fully Fuzzy Multiobjective Linear Programming Problems. *IEEE Access* **2018**, *6*, 43696–43711. [CrossRef]
49. Pop, B.; Stancu-Minasian, I. A method of solving fully fuzzified linear fractional programming problems. *J. Appl. Math. Comput.* **2008**, *27*, 227–242. [CrossRef]
50. Buckley, J.J.; Feuring, T. Evolutionary algorithm solution to fuzzy problems: Fuzzy linear programming. *Fuzzy Sets Syst.* **2000**, *109*, 35–53. [CrossRef]
51. Stanojević, B.; Stancu-Minasian, I.M. Evaluating fuzzy inequalities and solving fully fuzzified linear fractional program. *Yugoslav J. Oper. Res.* **2012**, *22*, 41–50. [CrossRef]
52. Das, S.; Mandal, T.; Edalatpanah, S. A new approach for solving fully fuzzy linear fractional programming problems using the multi-objective linear programming. *RAIRO-Oper. Res.* **2017**, *51*, 285–297. [CrossRef]
53. Stanojević, B.; Dziţac, I.; Dziţac, S. On the ratio of fuzzy numbers - exact membership function computation and applications to decision making. *Technol. Econ. Dev. Econ.* **2015**, *21*, 815–832. [CrossRef]
54. Chinnadurai, V.; Muthukumar, S. Solving the linear fractional programming problem in a fuzzy environment: Numerical approach. *Appl. Math. Model.* **2016**, *40*, 6148–6164. [CrossRef]
55. Ebrahimnejad, A.; Ghomi, S.; Mirhosseini-Alizamini, S. A revisit of numerical approach for solving linear fractional programming problem in a fuzzy environment. *Appl. Math. Model.* **2018**, *57*, 459–473. [CrossRef]
56. Kaur, J.; Kumar, A. A novel method for solving fully fuzzy linear fractional programming problems. *J. Intell. Fuzzy Syst.* **2017**, *33*, 1983–1990. [CrossRef]
57. Yager, R.R. A procedure for ordering fuzzy subsets of the unit interval. *Inf. Sci.* **1981**, *24*, 143–161. [CrossRef]
58. Arya, R.; Singh, P.; Kumari, S.; Obaidat, M. An approach for solving fully fuzzy multi-objective linear fractional optimization problems. *Soft Comput.* **2019**, *24*, 9105–9119. [CrossRef]
59. Chakraborty, M.; Gupta, S. Fuzzy mathematical programming for multi objective linear fractional programming problem. *Fuzzy Sets Syst.* **2002**, *125*, 335–342. [CrossRef]
60. Loganathan, T.; Ganesan, K. A solution approach to fully fuzzy linear fractional programming problems. *J. Phys. Conf. Ser.* **2019**, *1377*, 012040. [CrossRef]
61. Yager, R. Pythagorean fuzzy subsets. In Proceedings of the 2013 Joint IFSA World Congress and NAFIPS Annual Meeting (IFSA/NAFIPS), Edmonton, AB, Canada, 24–28 June 2013; pp. 57–61. [CrossRef]
62. Senapati, T.; Yager, R. Fermatean fuzzy sets. *J. Ambient. Intell. Humaniz. Comput.* **2019**, *11*, 663–674. [CrossRef]

mathematics

MDPI

Article

Interval Ranges of Fuzzy Sets Induced by Arithmetic Operations Using Gradual Numbers

Qingsong Mao [1] and Huan Huang [2,*]

1 Teachers College, Jimei University, Xiamen 361021, China; pinem@163.com or 200261000022@jmu.edu.cn
2 Department of Mathematics, Jimei University, Xiamen 361021, China
* Correspondence: hhuangjy@126.com or 200261000004@jmu.edu.cn

Abstract: Wu introduced the interval range of fuzzy sets. Based on this, he defined a kind of arithmetic of fuzzy sets using a gradual number and gradual sets. From the point of view of soft computing, this definition provides a new way of handling the arithmetic operations of fuzzy sets. The interval range is an important characterization of a fuzzy set. The interval range is also useful for analyses and applications of arithmetic. In this paper, we present general conclusions on crucial problems related to interval ranges of fuzzy sets induced by this arithmetic. These conclusions indicate that the corresponding conclusions in previous works should be modified: firstly, we give properties of the arithmetic and the composites of finite arithmetic. Then, we discuss the relationship between the domain of a gradual set and the range of its induced fuzzy set, and the relationship between the domain of a gradual set and the interval range of its induced fuzzy set. Based on the above results, we present the relationship between the intersection of the interval ranges of a group of fuzzy sets and the interval ranges of their resulting fuzzy sets obtained by compositions of finite arithmetic. Furthermore, we construct examples to show that even under conditions stronger than in previous work, there are still various possibilities in the relationship between the intersection of interval ranges of a group of fuzzy sets and the ranges of their resulted fuzzy sets, and there are still various possibilities in the relationship between the intersection of the interval ranges of a group of fuzzy sets and the interval ranges of their resulting fuzzy sets.

Keywords: interval range; arithmetic; gradual numbers; gradual sets

Citation: Mao, Q.; Huang, H. Interval Ranges of Fuzzy Sets Induced by Arithmetic Operations Using Gradual Numbers. *Mathematics* **2021**, *9*, 1351. https://doi.org/10.3390/math9121351

Academic Editors: Sorin Nadaban and Ioan Dzitac

Received: 13 May 2021
Accepted: 9 June 2021
Published: 11 June 2021

1. Introduction

Since Zadeh [1] put forward the concept of a fuzzy set in 1965, the fuzzy set theory has been widely used to handle uncertainties [2–4]. In 2008, new concepts of "gradual set" and "gradual number" were introduced in fuzzy set theory [5,6]. The gradual set and the gradual number are effective tools to study fuzzy sets, and have considerable applications in soft computing and related areas [7–10].

Fortin, Dubois and Fargier [5] introduced the concept of a gradual number and proposed the idea of representing a 1-dimensional fuzzy number as a crisp interval of gradual numbers. Dubois and Prade [6] introduced the concept of a gradual set and put forward the idea of inducing a fuzzy set by a gradual set or even an assignment function.

Fuzzy arithmetic are fundamental and essential in fuzzy set theory and widely used in soft computing. Fuzzy arithmetic and studies related to fuzzy arithmetic have attracted much attention [8,11–16]. For the need of theory and practical applications, various kinds of fuzzy arithmetic are proposed and studied. It is well known that Zadeh's extension principle [17] can be used to induce the arithmetic operations of fuzzy sets.

Interval arithmetic play important roles in many kinds of fuzzy arithmetic. Recently, Wu [15] proposed a new way to handle the arithmetic operations of fuzzy sets which does not directly use the interval arithmetic. In this paper, we discuss the relationship between the arithmetic defined by Wu and the interval arithmetic (see Section 3).

Wu [15] introduced the concept of the interval range of a fuzzy set. Based on the use of an interval range, he introduced the arithmetic operations of fuzzy sets in $\mathbb{R}$ by using gradual sets and gradual numbers.

According to Wu's definition, the resulted fuzzy sets of arithmetic operations of two fuzzy sets are defined as the induced fuzzy sets of the gradual sets on the intersection of the interval ranges of these two fuzzy sets. These gradual sets were generated from the arithmetic of gradual numbers. Here, we mention that in extreme cases, the gradual sets will be replaced by an assignment function with its value at each element being the empty set. Wu's definition provides a new way of handling the arithmetic operations of fuzzy sets.

The interval range is an important characterization of a fuzzy set. The interval range is also useful for the analyses and applications of this arithmetic. Therefore, the discussions on the interval ranges are essential and important in the study of arithmetic operations given by Wu [15].

Wu [15] claimed that, under certain conditions, the domain of a gradual set is equal to the range of its induced fuzzy set.

On the basis of this, Wu [15] claimed that the intersection of the interval ranges of two fuzzy sets is equal to the interval ranges of the resulted fuzzy sets obtained by the arithmetic operations of these two fuzzy sets.

Furthermore, Wu [15] claimed that, under certain conditions, the intersection of the interval ranges of a group of fuzzy sets, the interval ranges and the ranges of their resulted fuzzy sets obtained by compositions of finite arithmetic operations are equal.

In this paper, we pointed out that the above conclusions in [15] should be modified. Based on investigations into the properties of the arithmetic and the composites of finite arithmetic, we further give general conclusions on the above relationships in [15]. Examples are constructed to illustrate various possibilities in the relationships.

Here, we should mention that our results are complementary to [15], not in competition with [15].

The remainder of this paper is organized as follows. Section 2 recalls some basic concepts about fuzzy sets, gradual sets and gradual numbers. Section 3 reviews the arithmetic of fuzzy sets in $\mathbb{R}$ introduced by Wu [15]. Furthermore, we give some properties of the arithmetic and the composites of finite arithmetic. In Sections 4–6, we point out that some conclusions on interval ranges and ranges in [15] should be modified, and have presented general conclusions. Section 4 discusses the domain of a gradual set and the interval range and range of its induced fuzzy set. Section 5 considers the intersection of the interval ranges of a group of fuzzy sets and the interval ranges of their resulted fuzzy sets obtained by compositions of finite arithmetic. Section 6 investigates the intersection of the interval ranges of a group of fuzzy sets and the ranges and interval ranges of their resulted fuzzy sets under certain conditions. Finally, we draw our conclusions in Section 7.

2. Fuzzy Sets, Gradual Sets and Gradual Numbers

In this section, we recall some basic concepts about fuzzy sets, gradual sets and gradual numbers. For details, we refer the readers to [2,5,6,18].

Let X be a topology space; that is, X is a non-empty set equipped with a topological structure. A fuzzy set A in X can be identified with its membership function $A : X \to [0,1]$. We use $F(X)$ to denote the set of all fuzzy sets in X.

For a subset S of X, we use $\overline{S}_X$ to denote the characteristic function of S on X, meaning that:

$$\overline{S}_X(x) = \begin{cases} 1, & \text{if } x \in S, \\ 0, & \text{if } x \in X \setminus S. \end{cases}$$

$\overline{S}_X$ can be regarded as a fuzzy set in X.

Let A be a fuzzy set in X. Then, the α-cut set of A is defined as

$$[A]_\alpha = \begin{cases} \{x : A(x) \leq \alpha\}, & \alpha > 0, \\ \mathbf{cl}_X\{x : A(x) > 0\}, & \alpha = 0, \end{cases}$$

where $\mathbf{cl}_X\{S\}$ denotes the closure of S in X.

Let $\mathbb{R}$ be the real number space. Let S be a set of real numbers. In this paper, we use $\mathbf{cl}\{S\}$ to denote $\mathbf{cl}_\mathbb{R}\{S\}$ and use $\overline{S}$ to denote the fuzzy set $\overline{S}_\mathbb{R}$ in $\mathbb{R}$.

Fuzzy set A is said to be *normal*if $[A]_1 \neq \emptyset$. The fuzzy number is a type of normal fuzzy set which has been exhaustively studied [5,19,20]. It should be mentioned that, in theoretical research and practical applications, discussions about fuzzy sets are often in frameworks containing non-normal fuzzy sets [10,15,21]. In this paper, the discussion is on general fuzzy sets including normal fuzzy sets and non-normal fuzzy sets.

Let A be a fuzzy set in X. The range of A is denoted by $\mathcal{R}(A)$. Wu [15] introduced the **interval range**I_A of A, which is defined by

$$I_A := \begin{cases} [0, \sup \mathcal{R}(A)], & \text{if } \sup \mathcal{R}(A) \text{ is attained;} \\ [0, \sup \mathcal{R}(A)), & \text{if } \sup \mathcal{R}(A) \text{ is not attained.} \end{cases}$$

That is to say $I_A = \{\alpha \in [0,1] : \alpha \leq \beta \text{ for some } \beta \in \mathcal{R}(A)\}$.

Dubois and Prade [6] first introduced the concept of gradual set and proposed the idea of using a gradual set to induce a fuzzy set (see Definitions 2 and 4 in [6]).

The definition of a gradual set according to Wu [15] is different from the definition of a gradual set according to Dubois and Prade [6]. Wu [15] defined the gradual set as follows (see Definition 3.1 in [15]):

- A gradual set μ of X is an assignment function $\mu : I \to P(X) \setminus \{\emptyset\}$, where I is a subset of $[0,1]$.

A gradual set of X is a special type of assignment function on $I \subseteq [0,1]$.

We introduce the fuzzy set F_μ induced by an assignment function $\mu : I \to P(X)$, $I \subseteq [0,1]$, which is defined by

$$F_\mu(x) = \sup\{\alpha \in (0,1] \cap I : x \in \mu(\alpha)\}. \tag{1}$$

In this paper, we set $\mathbf{sup}\,\emptyset = \mathbf{0}$ by default.

It can be seen that F_μ can also be written as

$$F_\mu(x) = \sup_{\alpha \in I} \alpha \cdot \overline{\mu(\alpha)}_X(x). \tag{2}$$

Wu [15] defined the fuzzy set induced from a gradual set by (2) (see Equation (10) in [15]).

Remark 1. *The way to induce a fuzzy set from an assignment function given by (1) is consistent with the ways to induce a fuzzy set from a gradual set given by Wu [15] and Dubois and Prade [6], respectively. These ways of inducing fuzzy sets essentially come from [6].*

Fortin, Dubois and Fargier [5] first introduced the concept of gradual number. They proposed how to represent a 1-dimensional fuzzy number as a crisp interval of gradual numbers.

Wu [15] slightly modified the concept of gradual number as follows:

- A gradual number a in $\mathbb{R}$ is an assignment function $a : I \to \mathbb{R}$, where I is a subset of $[0,1]$.

Wu's modified definition of a gradual number is more convenient to deal with non-normal fuzzy sets.

A gradual number a is said to be an element of a fuzzy set A in $F(\mathbb{R})$, denoted by $\boldsymbol{a \in A}$, if a is defined on I_A and $a(\alpha) \in [A]_\alpha$ for each $\alpha \in I_A$.

3. Arithmetic of Fuzzy Sets in $F(\mathbb{R})$ Using Gradual Number

In this section, we review the arithmetic of fuzzy sets in $F(\mathbb{R})$ introduced by Wu [15]. Furthermore, we give properties of the arithmetic and the composites of finite arithmetic, which are useful in this paper.

Wu [15] introduced the arithmetic operations of two gradual numbers which are elements of two fuzzy sets in $F(\mathbb{R})$, respectively.

Based on this, Wu [15] defined the arithmetic operations of two fuzzy sets in $F(\mathbb{R})$ as follows.

First, a gradual set or an assignment function, which is not a gradual set, is generated from the corresponding arithmetic of gradual numbers in these two fuzzy sets, respectively. The domain of this gradual set or this assignment function is the intersection of interval ranges of these two fuzzy sets. Then, the resulted fuzzy set of the arithmetic operation is induced from this gradual set or this assignment function.

In this paper, we still use $+, - \times, /$ to denote the arithmetic operations of fuzzy sets in $F(\mathbb{R})$ introduced by Wu [15].

Suppose that A and B are two fuzzy sets in $\mathbb{R}$, and that $\circ \in \{+, -, \times, /\}$ is an arithmetic operation. For two gradual numbers $a \in A$ and $b \in B$, we use $\boldsymbol{I^\cap}$ **to denote** $\boldsymbol{I_A \cap I_B}$, and the arithmetic operation $a \circ b$ is defined by

$$(a \circ b)(\alpha) = a(\alpha) \circ b(\alpha)$$

for all $\alpha \in I^\cap$.

$a \circ b$ is said to be meaningful if $\{\alpha \in I^\cap : a(\alpha) \circ b(\alpha) \text{ is meaningful}\} = I^\cap$, at this time, and $a \circ b$ is a gradual number on $I^\cap$, otherwise $a \circ b$ is meaningless.

Therefore, taking $\circ \in \{+, -, \times\}$, then $a \circ b$ is meaningful; taking $\circ = /$, then $a \circ b$ is meaningful if and only if the set $M_b := \{\alpha \in I^\cap : b(\alpha) \neq 0\}$ satisfies that $M_b = I^\cap$.

The family $\{a \circ b : a \in A,\ b \in B \text{ with } a \circ b \text{ is meaningful}\}$ induces an assignment function $\sigma : I^\cap \to P(\mathbb{R})$ given by

$$\sigma(\alpha) = \{(a \circ b)(\alpha) : a \in A, b \in B \text{ with } a \circ b \text{ is meaningful}\}. \quad (3)$$

Using this assignment function σ, $A \circ B$ is defined by

$$(A \circ B)(x) = \sup_{\alpha \in I^\cap} \alpha \cdot \overline{\sigma(\alpha)}(x). \quad (4)$$

From the above definition, A/B is given by

$$\begin{aligned}(A/B)(x) &= \sup_{\alpha \in I^\cap} \alpha \cdot \overline{\sigma(\alpha)}(x) \\ &= \sup_{\alpha \in I^\cap} \alpha \cdot \overline{\{a(\alpha)/b(\alpha) : a \in A, b \in B \text{ with } b(\alpha) \neq 0 \text{ for all } \alpha \in I^\cap\}}(x). \quad (5)\end{aligned}$$

Remark 2. *Here, we slightly adjust the presentation of the definition of $A \circ B$ to improve the accuracy and clarity of the presentation. Readers may compare it with the corresponding contents in Section 4 of [15].*

Remark 3. *From (4), we know that $A \circ B = F_\sigma$, where σ is given by (3).*

If the family $\{a \circ b : a \in A,\ b \in B \text{ with } a \circ b \text{ is meaningful}\} \neq \emptyset$, then σ is a gradual set from $I^\cap$ to $P(\mathbb{R}) \setminus \emptyset$; if the family $\{a \circ b : a \in A,\ b \in B \text{ with } a \circ b \text{ is meaningful}\} = \emptyset$, then $\sigma(\alpha) = \emptyset$ for all $\alpha \in I^\cap$.

Let $[A]_\alpha \circ [B]_\alpha := \{z : z = x \circ y \text{ with } x \in [A]_\alpha, y \in [B]_\alpha\}$. Then:

(i) If $\circ \in \{+, -, \times\}$, then $\sigma(\alpha) = [A]_\alpha \circ [B]_\alpha$ for all $\alpha \in I^\cap$.

(ii) *If* $\circ = /$ *and there exists a* $b \in B$ *satisfying that* $b(\alpha) \neq 0$ *for all* $\alpha \in I^{\cap}$, *then* $\sigma(\alpha) = [A]_\alpha / [B]_\alpha$ *for all* $\alpha \in I^{\cap}$.
(iii) *If* $\circ = /$ *and there is no* $b \in B$ *satisfying that* $b(\alpha) \neq 0$ *for all* $\alpha \in I^{\cap}$, *then* $\sigma(\alpha) = \emptyset$ *for all* $\alpha \in I^{\cap}$.

Proposition 1. *Let A and B be fuzzy sets in* $F(\mathbb{R})$. *Then,* $A \circ B = \overline{\emptyset}$ *if and only if* A, B *satisfy one of the following conditions.*
(i) $A = \overline{\emptyset}$ *or* $B = \overline{\emptyset}$; *that is* $I^{\cap} = \{0\}$.
(ii) $A \neq \overline{\emptyset}$, $B \neq \overline{\emptyset}$, $\circ = /$, *and there is no* $b \in B$ *satisfying that* $b(\alpha) \neq 0$ *for all* $\alpha \in I^{\cap}$.

Proof. The desired result follows from the definition of $A \circ B$. □

Suppose that f is a composite of the finite arithmetic operations of real numbers. The symbol $\widetilde{f}$ denotes the corresponding composite of finite arithmetic operations of fuzzy sets in $\mathbb{R}$.

For instance, if $f(x,y,z) = (x-y) \times z$ for $x, y, z \in \mathbb{R}$, then $\widetilde{f}(u,v,w) = (u-v) \times w$ for $u, v, w \in F(\mathbb{R})$.

For $\widetilde{f}(A_1, \ldots, A_n)$, we use $I^{\cap}$ **to denote** $\bigcap_{i=1}^{n} I_{A_i}$. If $\widetilde{f}(A_1, \ldots, A_n)$ is $A \circ B$, then $I^{\cap} = I_A \cap I_B$.

The symbol $f([A_1]_\alpha, \ldots, [A_n]_\alpha)$ is used to denote the set:

$$\{z : z = f(x_1, \ldots, x_n) \text{ with } x_i \in [A_i]_\alpha,\ i = 1, \ldots, n\}.$$

Theorem 1. *Let f be a composite of finite arithmetic operations of real numbers and for each* $i = 1, \ldots, n$, *let* A_i *be a fuzzy set in* $F(\mathbb{R})$. *Furthermore, let* μ *be an assignment function from* $I^{\cap}$ *to* $P(\mathbb{R})$ *defined as* $\mu(\alpha) = f([A_1]_\alpha, \ldots, [A_n]_\alpha)$.
(i) *If* $A_i = \overline{\emptyset}$ *for some* $i \in \{1, \ldots, n\}$, *i.e,* $I^{\cap} = \{0\}$, *then* $\widetilde{f}(A_1, \ldots, A_n) = F_\mu = \overline{\emptyset}$.
(ii) *If* $\widetilde{f}(A_1, \ldots, A_n) \neq \overline{\emptyset}$, *then* $\widetilde{f}(A_1, \ldots, A_n) = F_\mu$.
(iii) *If f is a composition of* $\circ \in \{+, -, \times\}$, *then* $\widetilde{f}(A_1, \ldots, A_n) = F_\mu$.

Proof. From Proposition 1, (i) is true.

If $\widetilde{f}(A_1, \ldots, A_n) \neq \overline{\emptyset}$, then $I^{\cap} \supsetneqq \{0\}$. By using Remark 3, we can deduce that $\widetilde{f}(A_1, \ldots, A_n) = F_\mu$. So (ii) is true.

Suppose that f is a composition of $\circ \in \{+, -, \times\}$. Then $\widetilde{f}(A_1, \ldots, A_n) = \overline{\emptyset}$ if and only if $A_i = \overline{\emptyset}$ for some $i \in \{1, \ldots, n\}$. Thus, (iii) follows from (i) and (ii). □

4. Ranges and Interval Ranges of Induced Fuzzy Sets

It is claimed in [15] that under certain kinds of conditions, the domain I of a gradual set μ is equal to the range $\mathcal{R}(F_\mu)$ of its induced fuzzy set F_μ.

In this section, we point out by examples that these conclusions should be modified. $\mathcal{R}(F_\mu) \neq I$ can be divided into three cases: "$\mathcal{R}(F_\mu) \supsetneqq I$", "$\mathcal{R}(F_\mu) \subsetneqq I$", and "$\mathcal{R}(F_\mu) \nsubseteq I$ and $I \nsupseteq \mathcal{R}(F_\mu)$". In fact, we show that even under stronger conditions, all these three cases could happen.

Furthermore, we give the relationship between I and $\mathcal{R}(F_\mu)$, and the relationship between I and the interval range I_{F_μ} of the induced fuzzy set F_μ.

The conclusions in this section show that for a gradual set μ and its induced fuzzy set F_μ, the interval range I_{F_μ} and the range $\mathcal{R}(F_\mu)$ of F_μ vary in the scope determined by the domain I of μ, respectively. On the other hand, even under stronger conditions than in [15], both $\mathcal{R}(F_\mu) \neq I$ and $I_{F_\mu} \neq I$ could happen.

An assignment function $\mu : I \to P(X)$ is said to be *strictly nested* if $\mu(\alpha) \subsetneqq \mu(\beta)$ for $\alpha, \beta \in I$ with $\alpha > \beta$.

Affirmation 1 ([15]). *(See Remark 3.5 and Proposition 3.6 of [1]) Let* $\mu : I \to P(X) \setminus \{\emptyset\}$ *be a gradual set:*

(i) *If $I = [0, \alpha]$ or $I = [0, \alpha)$, $\alpha > 0$, then $\mathcal{R}(F_\mu) = I$.*
(ii) *If μ is strictly nested, then $\mathcal{R}(F_\mu) = I$.*

Remark 4. *Affirmation 1 is not valid. In fact, even if the gradual set $\mu : I \to P(\mathbb{R}) \setminus \{\emptyset\}$ satisfies:*
(i′) *$I = [0, \alpha]$ or $I = [0, \alpha)$, $\alpha > 0$, and*
(ii′) *μ is strictly nested,*
all the three cases "$\mathcal{R}(F_\mu) \supsetneqq I$", "$\mathcal{R}(F_\mu) \subsetneqq I$", and "$\mathcal{R}(F_\mu) \nsubseteq I$ and $I \nsupseteq \mathcal{R}(F_\mu)$" could happen. The following three examples correspond to these three cases, respectively.

Example 1. *Let the gradual set $\mu : [0, \frac{1}{2}) \to P(\mathbb{R}) \setminus \{\emptyset\}$ be defined as*

$$\mu(\alpha) = [\alpha - 1, 1 - \alpha] \text{ for } \alpha \in [0, \frac{1}{2}).$$

Then, μ is strictly nested and:

$$F_\mu(x) = \begin{cases} 1 + x, & x \in [-1, -1/2], \\ 1/2, & x \in [-1/2, 1/2], \\ 1 - x, & x \in [1/2, 1], \\ 0, & \text{otherwise.} \end{cases}$$

Thus, $[0, 1/2] = \mathcal{R}(F_\mu) \supsetneqq I = [0, 1/2)$.

Example 2. *Let the gradual set $\mu : [0, 1] \to P(\mathbb{R}) \setminus \{\emptyset\}$ be defined as*

$$[\mu]_\alpha = \begin{cases} [-1/2, 1 - \alpha), & \alpha \in [1/2, 1], \\ [\alpha - 1, 1/2), & \alpha \in [0, 1/2]. \end{cases}$$

Then, μ is strictly nested and:

$$F_\mu(x) = \begin{cases} 1 + x, & x \in (-1, -1/2), \\ 1, & x = [-1/2, 0], \\ 1 - x, & x \in (0, 1/2), \\ 0, & \text{otherwise.} \end{cases}$$

Thus, $[0, 1/2) \cup (1/2, 1] = \mathcal{R}(F_\mu) \subsetneqq I = [0, 1]$.

Example 3. *Let the gradual set $\mu : [0, 1) \to P(\mathbb{R}) \setminus \{\emptyset\}$ be defined as*

$$[\mu]_\alpha = \begin{cases} [-1/2, 1 - \alpha), & \alpha \in [1/2, 1), \\ [\alpha - 1, 1/2), & \alpha \in [0, 1/2]. \end{cases}$$

Then, $I = [0, 1)$, μ is strictly nested and:

$$F_\mu(x) = \begin{cases} 1 + x, & x \in (-1, -1/2), \\ 1, & x = [-1/2, 0], \\ 1 - x, & x \in (0, 1/2), \\ 0, & \text{otherwise.} \end{cases}$$

Thus, $\mathcal{R}(F_\mu) = [0, 1/2) \cup (1/2, 1]$ and $I_{F_\mu} = [0, 1]$. Therefore, $\mathcal{R}(F_\mu) \nsubseteq I$ and $I \nsubseteq \mathcal{R}(F_\mu)$.

We have the following conclusions on the relationship between $\mathcal{R}(F_\mu)$ and I, and the relationship between I_{F_μ} and I.

Proposition 2. *Given a gradual set $\mu : I \to P(X) \setminus \{\emptyset\}$, then:*

$$\mathcal{R}(F_\mu) \setminus \{0\} \subseteq \mathbf{cl}\{I \setminus \{0\}\},$$

$$\widehat{I} \subseteq I_{F_\mu} \subseteq [0, \sup I],$$

where $\widehat{I} = \{\alpha \in [0,1] : \alpha \leq \beta$ *for some* $\beta \in I\}$.

Proof. If $F_\mu(x) > 0$, then there exists $\{\alpha_n\} \subseteq I \setminus \{0\}$, such that $x \in \mu(\alpha_n)$ and $\sup \alpha_n = F_\mu(x)$. Thus, we know $\mathcal{R}(F_\mu) \setminus \{0\} \subseteq \mathbf{cl}\{I \setminus \{0\}\}$ and $I_{F_\mu} \subseteq [0, \sup I]$.

If $x \in \mu(\alpha)$ for some $\alpha \in I$, then $F_\mu(x) \geq \alpha$. Therefore, we have $\widehat{I} \subseteq I_{F_\mu}$. □

Remark 5. *Note that* $[0, \sup I] \setminus \widehat{I} \subseteq \{\sup I\}$. *Thus, by Proposition 2,* I_{F_μ} *has two possibilities:* $\widehat{I}$ *and* $[0, \sup I]$ *(in some cases, these two are the same).*

Clearly, if $\sup I \in I$, *then* $\sup I \in \mathcal{R}(F_\mu)$.

Even gradual set μ *satisfies the conditions (i′) and (ii′) in Remark 4, and each "*$\subseteq$*" in Proposition 2 could be "*$\subsetneqq$*".*

Example 3 shows that $\mathcal{R}(F_\mu) \setminus \{0\} \subsetneqq \mathbf{cl}\{I \setminus \{0\}\}$ *and* $\widehat{I} \subsetneqq I_{F_\mu}$ *could happen. The following Example 4 shows that* $I_{F_\mu} \subsetneqq [0, \sup I]$ *could happen.*

Example 4. *Let the gradual set* $\mu : [0,1) \to P(\mathbb{R}) \setminus \{\emptyset\}$ *be defined as* $\mu(\alpha) = [\alpha, 1)$. *Then,*

$$F_\mu(x) = \begin{cases} x, & x \in [0,1), \\ 0, & \textit{otherwise}. \end{cases}$$

So $\mathcal{R}(F_\mu) = I_{F_\mu} = [0,1) \subsetneqq [0,1] = [0, \sup I]$.

Remark 6. *Let* $\mu : I \to P(X) \setminus \{\emptyset\}$ *be a gradual set. Note that* $I \subseteq \widehat{I}$. *Thus, by Proposition 2,* $I \subseteq I_{F_\mu}$.

From Remark 4, we can see that even conditions (i′) and (ii′) are satisfied, then $I \subsetneqq I_{F_\mu}$ *could happen. Clearly, at this time,* $\mathcal{R}(F_\mu) \nsubseteq I$.

In fact, if μ *satisfies condition (i′), then* $I \subsetneqq I_{F_\mu} \Leftrightarrow \mathcal{R}(F_\mu) \nsubseteq I \Leftrightarrow \sup I \notin I$ *and* $\sup I \in I_{F_\mu}$ $\Leftrightarrow \sup I \notin I$ *and* $\sup I \in \mathcal{R}(F_\mu)$.

5. Interval Ranges of Compositions of Arithmetic

It was claimed in [15] that $I_{A \circ B}$ is equal to $I^{\cap}$. In this section, we point out by examples that this conclusion should be modified. Furthermore, based on the results in Sections 3 and 4, we discussed the relationship between $I_{\widetilde{f}(A_1,\ldots,A_n)}$ and $I^{\cap}$.

Note that either $I^{\cap} \subseteq I_{\widetilde{f}(A_1,\ldots,A_n)}$ or $I^{\cap} \supsetneqq I_{\widetilde{f}(A_1,\ldots,A_n)}$. We give the characterizations of all these two cases, respectively. Moreover, we give the relationship between $I_{\widetilde{f}(A_1,\ldots,A_n)}$ and $I^{\cap}$ in some particular cases. As a corollary of the above results, we illustrate the relationship between $I_{A \circ B}$ and $I^{\cap}$.

The conclusions in this section show that $I_{\widetilde{f}(A_1,\ldots,A_n)}$ vary in the scope determined by $I^{\cap}$.

Affirmation 2 ([15]). *(see Section 4 of [15]) Suppose that* $A, B \in F(\mathbb{R})$ *and* $\circ \in \{+, -, \times, /\}$. *Then,* $I_{A \circ B} = I^{\cap}$ *(here* $I^{\cap} = I_A \cap I_B$*).*

It can be checked that the Affirmation 2 is equivalent to the following Affirmation 3.

Affirmation 3. *Supposing that* f *is a composite of the finite arithmetic operations of real numbers and that* $A_i \in F(\mathbb{R})$, $i = 1, \ldots, n$. *Then,* $I_{\widetilde{f}(A_1,\ldots,A_n)} = I^{\cap}$ *(here* $I^{\cap} = \bigcap_{i=1}^{n} I_{A_i}$*).*

Affirmation 2 is not valid. The following example shows that, for each $\circ \in \{+, -, \times, /\}$, $I^{\cap} \subsetneqq I_{A \circ B}$ could happen.

Example 5. *Let A be a fuzzy set in $\mathbb{R}$ defined as*

$$A(x) = \begin{cases} -0.5x + 1.5, & x \in (1,3), \\ 0, & \textit{otherwise}, \end{cases}$$

and let B be a fuzzy set in $\mathbb{R}$ defined as

$$B(x) = \begin{cases} 0.5x + 1.5, & x \in (-3,-1), \\ -0.5x + 1.5, & x \in (1,3), \\ 1.5 - 50/x, & x \in (100/3, 100), \\ 0, & \textit{otherwise}. \end{cases}$$

Then, $I_A = I_B = [0,1)$ and hence $I^{\cap} = [0,1)$.
Since:

$$[A]_\alpha = \begin{cases} (1, 3-2\alpha], & \alpha \in (0,1), \\ \varnothing, & \alpha = 1, \end{cases}$$

and:

$$[B]_\alpha = \begin{cases} [2\alpha - 3, -1) \cup (1, 3-2\alpha] \cup [100/(3-2\alpha), 100), & \alpha \in (0,1), \\ \varnothing, & \alpha = 1, \end{cases}$$

we have for all $\alpha \in (0,1)$:

$$\begin{aligned} 0 &\in [A]_\alpha + [B]_\alpha, \\ 0 &\in [A]_\alpha - [B]_\alpha, \\ 100 &\in [A]_\alpha \times [B]_\alpha, \\ 1 &\in [A]_\alpha / [B]_\alpha. \end{aligned}$$

So by Remark 3:

$$\begin{aligned} (A+B)(0) &= 1, \\ (A-B)(0) &= 1, \\ (A \times B)(100) &= 1, \\ (A/B)(1) &= 1, \end{aligned}$$

and thus:

$$\begin{aligned} I^{\cap} &= [0,1) \subsetneqq [0,1] = I_{A+B}, \\ I^{\cap} &= [0,1) \subsetneqq [0,1] = I_{A-B}, \\ I^{\cap} &= [0,1) \subsetneqq [0,1] = I_{A\times B}, \\ I^{\cap} &= [0,1) \subsetneqq [0,1] = I_{A/B}. \end{aligned}$$

Now we discuss the relationship between $I^{\cap}$ and $I_{\tilde{f}(A_1,\ldots,A_n)}$. Their relationship can be divided into two cases: $I^{\cap} \subseteq I_{\tilde{f}(A_1,\ldots,A_n)}$ or $I^{\cap} \supsetneqq I_{\tilde{f}(A_1,\ldots,A_n)}$. The following Theorems 2 and 3 give characterizations of these two cases, respectively.

Theorem 2. *Let f be a composite of finite arithmetic operations of real numbers and for each $i = 1,\ldots,n$, let A_i be a fuzzy set in $F(\mathbb{R})$. Then, $I_{\tilde{f}(A_1,\ldots,A_n)} \subseteq \mathbf{cl}\{I^{\cap}\}$ and the following statements are equivalent:*

(i) $I^{\cap} \subseteq I_{\tilde{f}(A_1,\ldots,A_n)}$.

(ii) $I^{\cap} \subseteq I_{\tilde{f}(A_1,\ldots,A_n)} \subseteq \mathbf{cl}\{I^{\cap}\}$.

(iii) $A_i = \overline{\varnothing}$ *for some* $i \in \{1,\ldots,n\}$ *or* $\tilde{f}(A_1,\ldots,A_n) \neq \overline{\varnothing}$.

Proof. If $\tilde{f}(A_1,\ldots,A_n) = \overline{\varnothing}$, then $I_{\tilde{f}(A_1,\ldots,A_n)} = \{0\} \subseteq \mathbf{cl}\{I^{\cap}\}$.

Let μ be the assignment function from $I^{\cap}$ to $P(\mathbb{R})$ defined by $\mu(\alpha) = f([A_1]_\alpha, \ldots, [A_n]_\alpha)$.

If $\widetilde{f}(A_1, \ldots, A_n) \neq \overline{\emptyset}$, then by Theorem 1, $\widetilde{f}(A_1, \ldots, A_n) = F_\mu$. Thus, by Proposition 2, $I_{\widetilde{f}(A_1,\ldots,A_n)} \subseteq [0, \sup I^{\cap}] = \mathbf{cl}\{I^{\cap}\}$.

Therefore, $I_{\widetilde{f}(A_1,\ldots,A_n)} \subseteq \mathbf{cl}\{I^{\cap}\}$, and from this fact, we have (i) $\Leftrightarrow$ (ii).

If (iii) holds, then by Theorem 1, $\widetilde{f}(A_1, \ldots, A_n) = F_\mu$. Note that $\widehat{I^{\cap}} = I^{\cap}$ and $[0, \sup I^{\cap}] = \mathbf{cl}\{I^{\cap}\}$. Therefore, by Proposition 2, (ii) holds. Thus, (iii) $\Rightarrow$ (ii).

Suppose $I^{\cap} \subseteq I_{\widetilde{f}(A_1,\ldots,A_n)}$. If $A_i \neq \overline{\emptyset}$ for all $i \in \{1, \ldots, n\}$, then $\{0\} \subsetneqq I^{\cap}$, and therefore $\{0\} \subsetneqq I_{\widetilde{f}(A_1,\ldots,A_n)}$. Thus, $\widetilde{f}(A_1, \ldots, A_n) \neq \overline{\emptyset}$. Therefore, we have (i) $\Rightarrow$ (iii). □

Theorem 3. *Let f be a composite of finite arithmetic operations of real numbers and for each $i = 1, \ldots, n$, let A_i be a fuzzy set in $F(\mathbb{R})$. Then, the following statements are equivalent:*

(i) $I^{\cap} \supsetneqq I_{\widetilde{f}(A_1,\ldots,A_n)}$.

(ii) $I^{\cap} \supsetneqq \{0\}$ *and* $I_{\widetilde{f}(A_1,\ldots,A_n)} = \{0\}$.

(iii) $A_i \neq \overline{\emptyset}$ *for all* $i \in \{1, \ldots, n\}$ *and* $\widetilde{f}(A_1, \ldots, A_n) = \overline{\emptyset}$.

Proof. The desired results follow immediately from Theorem 2. □

The situation described in clause (iii) of Theorem 3 could happen. The following Example 6 gives an example of $A \neq \overline{\emptyset}$, $B \neq \overline{\emptyset}$ and $A/B = \overline{\emptyset}$.

Example 6. *Let A and B be fuzzy sets in $F(\mathbb{R})$ defined as $A = \overline{\{1\}}$ and:*

$$B(x) = \begin{cases} 1, & x = 0, \\ 0.6, & x = 1, \\ 0, & \text{otherwise}. \end{cases}$$

Then, $I^{\cap} = [0, 1]$. Note that there is no gradual number a, b with $a \in A$ and $b \in B$ with a/b being meaningful. Hence, the assignment function σ corresponding to A/B satisfies $\sigma(\alpha) = \emptyset$ for all $\alpha \in [0, 1]$. Therefore, from (4) or (5), $A/B = \overline{\emptyset}$.

Clearly, at this time, $I^{\cap} = [0, 1] \supsetneqq \{0\} = I_{A/B}$.

Theorem 4 illustrates the relationship between $I^{\cap}$ and $I_{\widetilde{f}(A_1,\ldots,A_n)}$ in some particular cases.

Theorem 4. *Let f be a composite of finite arithmetic operations of real numbers and for each $i = 1, \ldots, n$, let A_i be a fuzzy set in $F(\mathbb{R})$:*

(i) *If $A_i = \overline{\emptyset}$ for some $i \in \{1, \ldots, n\}$, then $\widetilde{f}(A_1, \ldots, A_n) = \overline{\emptyset}$. At this time, $I^{\cap} = I_{\widetilde{f}(A_1,\ldots,A_n)} = \mathbf{cl}\{I^{\cap}\} = \{0\}$.*

(ii) *If f is a composition of $\circ \in \{+, -, \times\}$, then $I^{\cap} \subseteq I_{\widetilde{f}(A_1,\ldots,A_n)} \subseteq \mathbf{cl}\{I^{\cap}\}$.*

Proof. (i) obviously holds.

Suppose that f is a composition of $\circ \in \{+, -, \times\}$. Let μ be the assignment function from $I^{\cap}$ to $P(\mathbb{R})$ defined by $\mu(\alpha) = f([A_1]_\alpha, \ldots, [A_n]_\alpha)$. Then, by Theorem 1, $\widetilde{f}(A_1, \ldots, A_n) = F_\mu$. Note that $\widehat{I^{\cap}} = I^{\cap}$ and $[0, \sup I^{\cap}] = \mathbf{cl}\{I^{\cap}\}$. Thus, by Proposition 2, (ii) holds. □

Remark 7. *(ii) of Theorem 4 can also be proved as follows:*

Suppose that f is a composition of $\circ \in \{+, -, \times\}$ and that for each $i = 1, \ldots, n$, A_i is a fuzzy set in $F(\mathbb{R})$. Note that either $A_i = \overline{\emptyset}$ for some $i \in \{1, \ldots, n\}$ or $\widetilde{f}(A_1, \ldots, A_n) \neq \overline{\emptyset}$. Thus, by Theorem 2, the (ii) of Theorem 4 holds.

We have the following conclusions on $I_{A \circ B}$ and $I^{\cap}$.

Corollary 1. *Let A and B be fuzzy sets in $F(\mathbb{R})$:*

(i) If $\circ \in \{+, -, \times\}$, then $I^{\cap} \subseteq I_{A \circ B} \subseteq \mathbf{cl}\{I^{\cap}\}$.
(ii) $I^{\cap} \subseteq I_{A/B} \subseteq \mathbf{cl}\{I^{\cap}\}$ if and only if at least one of the following three items holds: $A = \overline{\emptyset}$, $B = \overline{\emptyset}$, and $A/B \neq \overline{\emptyset}$.

Proof. The desired results follow from Theorems 2 and 4. □

For each $\circ \in \{+, -, \times, /\}$, $I_{A \circ B} = I^{\cap}$ and $I_{A \circ B} \neq \mathbf{cl}\{I^{\cap}\}$ could happen. The following Example 7 is a such example.

Example 7. *Consider A in $F(\mathbb{R})$ defined by*

$$A(x) = \begin{cases} x, & x \in [0,1), \\ 0, & \textit{otherwise}. \end{cases}$$

Then, $I_A = [0,1)$.
It can be checked that:

$$I_{A+\overline{\{1\}}} = I_{A-\overline{\{1\}}} = I_{A\times\overline{\{1\}}} = I_{A/\overline{\{1\}}} = I^{\cap} = [0,1) \subsetneqq [0,1] = \mathbf{cl}\{I^{\cap}\}.$$

Remark 8. *From Theorems 2 and 3, examples in this section and the fact $\mathbf{cl}\{I^{\cap}\} \setminus I^{\cap} \subseteq \{\sup I\}$, we have:*

(i) $I_{\widetilde{f}(A_1,\ldots,A_n)}$ *has three possibilities: $\{0\}$, $I^{\cap}$ and $\mathbf{cl}\{I^{\cap}\}$ (in some cases, all or part of these three are the same).*
(ii) $I^{\cap} \subsetneqq I_{\widetilde{f}(A_1,\ldots,A_n)} \subsetneqq \mathbf{cl}\{I^{\cap}\}$ *is impossible.*
(iii) *If $I^{\cap} = \mathbf{cl}\{I^{\cap}\}$ and $I_{\widetilde{f}(A_1,\ldots,A_n)} \neq \{0\}$, then $I^{\cap} = I_{\widetilde{f}(A_1,\ldots,A_n)} = \mathbf{cl}\{I^{\cap}\}$.*

(ii) and (iii) can also be seen as corollaries of (i).

6. Discussions on Ranges and Interval Ranges of Compositions of Arithmetic

In [15], it is claimed that $I_{\widetilde{f}(A_1,\ldots,A_n)}$, $\mathcal{R}(\widetilde{f}(A_1,\ldots,A_n))$ and $I^{\cap}$ are equal under certain conditions. In this section, we show by examples that this affirmation is not valid.

$\mathcal{R}(\widetilde{f}(A_1,\ldots,A_n)) \neq I^{\cap}$ can be divided into three cases: "$\mathcal{R}(\widetilde{f}(A_1,\ldots,A_n)) \supsetneqq I^{\cap}$", "$\mathcal{R}(\widetilde{f}(A_1,\ldots,A_n)) \subsetneqq I^{\cap}$" and "$\mathcal{R}(\widetilde{f}(A_1,\ldots,A_n)) \nsubseteq I^{\cap}$ and $I^{\cap} \nsupseteq \mathcal{R}(\widetilde{f}(A_1,\ldots,A_n))$".

In this section, it is shown that even under stronger conditions than in [15], all the above three cases of $\mathcal{R}(\widetilde{f}(A_1,\ldots,A_n)) \neq I^{\cap}$ could happen.

$I_{\widetilde{f}(A_1,\ldots,A_n)}$ has three possibilities: $\{0\}$, $I^{\cap}$ and $\mathbf{cl}\{I^{\cap}\}$ (see Remark 8). $I_{\widetilde{f}(A_1,\ldots,A_n)} \neq I^{\cap}$ can be divided into two cases: "$I_{\widetilde{f}(A_1,\ldots,A_n)} \neq I^{\cap}$ and $I_{\widetilde{f}(A_1,\ldots,A_n)} = \{0\}$" and "$I_{\widetilde{f}(A_1,\ldots,A_n)} \neq I^{\cap}$ and $I_{\widetilde{f}(A_1,\ldots,A_n)} = \mathbf{cl}\{I^{\cap}\}$".

In this section, it is shown that under the conditions in [15], both the above two cases of $I_{\widetilde{f}(A_1,\ldots,A_n)} \neq I^{\cap}$ could happen, and even under stronger conditions than in [15], "$I_{\widetilde{f}(A_1,\ldots,A_n)} \neq I^{\cap}$ and $I_{\widetilde{f}(A_1,\ldots,A_n)} = \mathbf{cl}\{I^{\cap}\}$" could happen.

The conclusions in this section show that even under stronger conditions than in [15], $I_{\widetilde{f}(A_1,\ldots,A_n)}$, $\mathcal{R}(\widetilde{f}(A_1,\ldots,A_n))$ and $I^{\cap}$ may still be unequal.

A fuzzy set A is said to be a *canonical fuzzy set* if $[A]_\alpha \neq [A]_\beta$ for all $\alpha, \beta \in [0,1]$ with $\alpha \neq \beta$.

Affirmation 4 ([15]). *(See Proposition 4.5 of [15].) Let f be a composite of finite arithmetic operations of real numbers and for each $i = 1,\ldots,n$, let A_i be a fuzzy set in $\mathbb{R}$. If some of $A_1,\ldots,A_n$ are canonical fuzzy sets in $\mathbb{R}$, then the assignment function $\mu : I^{\cap} \to P(\mathbb{R})$ defined by $\mu(\alpha) = f([A_1]_\alpha,\ldots,[A_n]_\alpha)$ is strictly nested, and therefore:*

$$\mathcal{R}(\widetilde{f}(A_1,\ldots,A_n)) = I_{\widetilde{f}(A_1,\ldots,A_n)} = I^{\cap}. \tag{6}$$

Affirmation 4 is not valid. This can be drawn from the facts listed below. Note that if A is a canonical fuzzy set, then $I_A = [0,1]$ or $[0,1)$.

- There exist canonical fuzzy sets $A_1, \ldots, A_n$ such that $\widetilde{f}(A_1, \ldots, A_n) = \overline{\emptyset}$. In this case, $I^{\cap} = [0,1]$ or $[0,1)$, and $\mathcal{R}(\widetilde{f}(A_1, \ldots, A_n)) = I_{\widetilde{f}(A_1,\ldots,A_n)} = \{0\}$. Therefore, $\mathcal{R}(\widetilde{f}(A_1, \ldots, A_n)) \neq I^{\cap}$ and $I_{\widetilde{f}(A_1,\ldots,A_n)} \neq I^{\cap}$.
- Let A be a canonical fuzzy set. Then, $A \circ \overline{\mathbb{R}} = \overline{\mathbb{R}}$, where $\circ \in \{+, -, \times, /\}$. In this case, $I^{\cap} = I_A \cap I_{\overline{\mathbb{R}}} = [0,1]$ or $[0,1)$, $\mathcal{R}(A \circ \overline{\mathbb{R}}) = \{1\}$ and $I_{A \circ \overline{\mathbb{R}}} = [0,1]$. Thus, $\mathcal{R}(A \circ \overline{\mathbb{R}}) \neq I^{\cap}$. Furthermore, $I_{A \circ \overline{\mathbb{R}}} \neq I^{\cap}$ when $I^{\cap} = [0,1)$.

Affirmation 4 is still not valid when the above cases are excluded.

Consider conditions:

(a) $\widetilde{f}(A_1, \ldots, A_n) \neq \overline{\emptyset}$, and
(b) for each $i = 1, \ldots, n$, A_i is a canonical fuzzy set in $\mathbb{R}$.

Then, conditions (a) and (b) are stronger than that in Affirmation 4. If $A_1, \ldots, A_n$ and f satisfy conditions (a) and (b), then the above cases are excluded.

We will show that even $A_1, \ldots, A_n$ and f satisfy conditions (a) and (b), the conclusions in Affirmation 4 do not necessarily hold.

The following Example 8 shows that μ is not necessarily strictly nested even if $A_1, \ldots, A_n$ and f satisfy conditions (a) and (b).

Example 8. *Let A be a fuzzy set defined as*

$$[A]_\alpha = [2, 6 - 3\alpha] \cup [6,7]$$

for $\alpha \in [0,1]$, and let B be a fuzzy set defined as

$$[B]_\alpha = [1, 3 - \alpha] \cup [3,9]$$

for $\alpha \in [0,1]$. Then, A, B are canonical fuzzy sets in $\mathbb{R}$.

Note that for all $\alpha \in [0,1]$:

$$[A]_\alpha \times [B]_\alpha = [2, 63].$$

Thus, the gradual set $\mu : [0,1] \to P(\mathbb{R}) \setminus \{\emptyset\}$ given by $\mu(\alpha) = [A]_\alpha \times [B]_\alpha$ is not strictly nested.

Remark 9. *For A, B in Example 8, by Remark 3, we have:*

$$A \times B = F_\mu = \overline{[2, 63]}.$$

Thus, $\{1\} = \mathcal{R}(A \times B) \subsetneqq I_{A \times B} = I_A \cap I_B = [0,1]$. Therefore, (6) does not necessarily hold that $A_1, \ldots, A_n$ and f satisfy conditions (a) and (b).

Remark 10. *In fact, even $A_1, \ldots, A_n$ and f satisfy:*

(a) $\widetilde{f}(A_1, \ldots, A_n) \neq \overline{\emptyset}$,
(b) for each $i = 1, \ldots, n$, A_i is a canonical fuzzy set in $\mathbb{R}$, and
(c) μ is strictly nested,

all the three cases "$\mathcal{R}(\widetilde{f}(A_1, \ldots, A_n)) \supsetneqq I^{\cap}$", "$\mathcal{R}(\widetilde{f}(A_1, \ldots, A_n)) \subsetneqq I^{\cap}$" and "$\mathcal{R}(\widetilde{f}(A_1, \ldots, A_n)) \nsubseteq I^{\cap}$ and $I^{\cap} \nsupseteq \mathcal{R}(\widetilde{f}(A_1, \ldots, A_n))$" of $\mathcal{R}(\widetilde{f}(A_1, \ldots, A_n)) \neq I^{\cap}$ could happen. The following three examples correspond to these three cases, respectively.

Example 9. *Let A be a fuzzy set in $\mathbb{R}$ defined as*

$$A(x) = \begin{cases} x, & x \in [0,1), \\ 0, & \textit{otherwise}. \end{cases}$$

Then, $I_A = [0,1)$ *and:*

$$[A]_\alpha = \begin{cases} \emptyset, & \alpha = 1, \\ [\alpha,1), & 0 < \alpha < 1, \\ [0,1], & \alpha = 0. \end{cases}$$

Hence, A is a canonical fuzzy set, and the gradual set $\mu : [0,1) \to P(\mathbb{R}) \setminus \{\emptyset\}$ *given by*

$$\mu(\alpha) = [A]_\alpha - [A]_\alpha = \begin{cases} (\alpha-1, 1-\alpha), & 0 < \alpha < 1, \\ [-1,1], & \alpha = 0, \end{cases}$$

is strictly nested.

So, by Remark 3,

$$(A-A)(x) = F_\mu(x) = \begin{cases} 1+x, & x \in [-1,0], \\ 1-x, & x \in [0,1], \\ 0, & otherwise, \end{cases}$$

and thus

$$[0,1] = \mathcal{R}(A-A) = I_{A-A} \supsetneqq I^\cap = [0,1).$$

Example 10. *Let A be the fuzzy set derived from the gradual set given in Example 3; that is:*

$$[A]_\alpha = \begin{cases} [-1/2, 1-\alpha], & \alpha \in (1/2,1], \\ [\alpha-1, 1/2), & \alpha \in [0,1/2]. \end{cases}$$

Hence, A is a canonical fuzzy set and

$$[A]_\alpha + [A]_\alpha = \begin{cases} [-1, 2-2\alpha], & \alpha \in (1/2,1], \\ [2\alpha-2, 1), & \alpha \in [0,1/2]. \end{cases}$$

Therefore, we have the gradual set $\mu : [0,1] \to P(\mathbb{R}) \setminus \{\emptyset\}$ *given by* $\mu(\alpha) = [A]_\alpha + [A]_\alpha$ *is strictly nested and by Remark 3:*

$$(A+A)(x) = F_\mu(x) = \begin{cases} 1+x/2, & x \in (-2,-1), \\ 1, & x = [-1,0], \\ 1-x/2, & x \in (0,1), \\ 0, & otherwise. \end{cases}$$

Thus, $[0,1/2) \cup (1/2,1] = \mathcal{R}(A+A) \subsetneqq I^\cap = I_{A+A} = [0,1]$.

Example 11. *Let A be the fuzzy set in* $\mathbb{R}$ *defined as*

$$A(x) = \begin{cases} x+1, & -1 < x < -0.5, \\ 0.6, & x = -0.5, \\ -x+0.5, & -0.5 < x < 0, \\ 0, & otherwise, \end{cases}$$

and let B be the fuzzy set in $\mathbb{R}$ *defined as*

$$B(x) = \begin{cases} -x+0.5, & -0.5 < x < 0, \\ 0.6, & x = 0, \\ -0.5x+0.5, & 0 < x < 1, \\ 0, & otherwise. \end{cases}$$

Then:

$$[A]_\alpha = \begin{cases} [-1,0], & \alpha = 0, \\ [\alpha-1,0), & 0 < \alpha \le 0.5, \\ [-0.5,-\alpha+0.5], & 0.5 < \alpha \le 0.6, \\ (-0.5,-\alpha+0.5], & 0.6 < \alpha < 1, \\ \emptyset, & \alpha = 1, \end{cases}$$

and:

$$[B]_\alpha = \begin{cases} [-0.5,1], & \alpha = 0, \\ (-0.5,1-2\alpha], & 0 < \alpha \le 0.5, \\ (-0.5,-\alpha+0.5] \cup \{0\}, & 0.5 < \alpha \le 0.6, \\ (-0.5,-\alpha+0.5], & 0.6 < \alpha < 1, \\ \emptyset, & \alpha = 1. \end{cases}$$

Hence, A and B are canonical fuzzy sets, and:

$$[A]_\alpha - [B]_\alpha = \begin{cases} [-2,0.5], & \alpha = 0, \\ [-2+3\alpha,0.5), & 0 < \alpha \le 0.5, \\ [-0.5,-\alpha+1), & 0.5 < \alpha \le 0.6, \\ (-1+\alpha,-\alpha+1), & 0.6 < \alpha < 1, \\ \emptyset, & \alpha = 1. \end{cases}$$

Therefore, we have the gradual set: $\mu : [0,1) \to P(\mathbb{R}) \setminus \{\emptyset\}$ *given by* $\mu(\alpha) = [A]_\alpha - [B]_\alpha$ *which is strictly nested and:*

$$(A-B)(x) = F_\mu(x) = \begin{cases} (x+2)/3, & -2 < x < -0.5, \\ 0.6, & -0.5 \le x \le -0.4, \\ x+1, & -0.4 < x \le 0, \\ 1-x, & 0 < x < 0.5, \\ 0, & \text{otherwise}. \end{cases}$$

Thus, $\mathcal{R}(A-B) = [0,0.5) \cup (0.5,1]$. *Note that* $I^\cap = [0,1)$. *Therefore, we have* $\mathcal{R}(A-B) \nsubseteq I^\cap$ *and* $I^\cap \nsupseteq \mathcal{R}(A-B)$.

Remark 11. *From the facts listed at the beginning of this section, we know that under the assumption of Affirmation 4, both the two cases "*$I_{\widetilde{f}(A_1,\ldots,A_n)} \neq I^\cap$ *and* $I_{\widetilde{f}(A_1,\ldots,A_n)} = \{0\}$*" and "*$I_{\widetilde{f}(A_1,\ldots,A_n)} \neq I^\cap$ *and* $I_{\widetilde{f}(A_1,\ldots,A_n)} = \mathbf{cl}\{I^\cap\}$*" of* $I_{\widetilde{f}(A_1,\ldots,A_n)} \neq I^\cap$ *could happen.*

We mention that, even $A_1, \ldots, A_n$ *and* f *satisfy:*

(a) $\widetilde{f}(A_1, \ldots, A_n) \neq \overline{\emptyset}$,
(b) for each $i = 1, \ldots, n$, A_i *is a canonical fuzzy set in* $\mathbb{R}$, *and*
(c) μ *is strictly nested,*

$I_{\widetilde{f}(A_1,\ldots,A_n)} \neq I^\cap$ *could happen, and clearly at this time,* $I_{\widetilde{f}(A_1,\ldots,A_n)} = \mathbf{cl}\{I^\cap\}$, *because condition (a) is equivalent to* $I_{\widetilde{f}(A_1,\ldots,A_n)} \neq \{0\}$. *See Example 9 for such an example.*

In fact, under conditions (a), (b) and (c), $I_{\widetilde{f}(A_1,\ldots,A_n)}$ has two possibilities: $I^\cap$ and $\mathbf{cl}\{I^\cap\}$.

It can be seen that $I_{\widetilde{f}(A_1,\ldots,A_n)} = I^\cap = \mathbf{cl}\{I^\cap\}$ could happen under conditions (a), (b) and (c).

The following Example 12 shows that $I_{\widetilde{f}(A_1,\ldots,A_n)} \neq \mathbf{cl}\{I^\cap\}$ could happen while conditions (a), (b) and (c) are satisfied. Clearly, at this time, $I_{\widetilde{f}(A_1,\ldots,A_n)} = I^\cap$.

Example 12. *Consider A given in Example 9. Then, A is a canonical fuzzy set, and the gradual set* $\mu : [0,1) \to P(\mathbb{R}) \setminus \{\emptyset\}$ *given by*

$$\mu(\alpha) = [A]_\alpha + [A]_\alpha = \begin{cases} [2\alpha,2), & 0 < \alpha < 1, \\ [0,2], & \alpha = 0, \end{cases}$$

is strictly nested.

Therefore, by Remark 3:

$$(A + A)(x) = F_\mu(x) = \begin{cases} x/2, & x \in [0,2), \\ 0, & \text{otherwise}, \end{cases}$$

and thus:

$$[0,1) = I^\cap = I_{A+A} = \mathcal{R}(A + A) \neq \mathbf{cl}\{I^\cap\} = [0,1].$$

Remark 12. *When $\mathcal{R}(\widetilde{f}(A_1, \ldots, A_n)) = I_{\widetilde{f}(A_1,\ldots,A_n)} = I^\cap$, $I^\cap \neq \mathbf{cl}\{I^\cap\}$ could happen, and clearly in this case, $I^\cap \neq \{0\}$. See Example 12 for a such example.*

We can see that when $\mathcal{R}(\widetilde{f}(A_1, \ldots, A_n)) = I_{\widetilde{f}(A_1,\ldots,A_n)} = I^\cap$, $I^\cap = \mathbf{cl}\{I^\cap\}$ also could happen, and in this case, both $I^\cap = \{0\}$ and $I^\cap \neq \{0\}$ are possible.

Remark 13. *The content of this paper is interrelated. For instance, the conclusion in Remark 10 implies the conclusion in Remark 4. Examples 9, 10 and 11 also show the conclusion in Remark 4. Example 5 can be used to illustrate some of the relationship between $I_{\widetilde{f}(A_1,\ldots,A_n)}$ and $I^\cap$ given in this paper, and some of the relationship between $\mathcal{R}(\widetilde{f}(A_1, \ldots, A_n))$ and $I^\cap$ given in this paper.*

7. Conclusions

In this paper, we give general conclusions on the problems related to the interval ranges of the resulted fuzzy sets of compositions of finite arithmetic.

In [15], it is claimed that under certain conditions, the domain I of a gradual set μ is equal to the range $\mathcal{R}(F_\mu)$ of its induced fuzzy set F_μ.

We show by examples that this is not valid. In fact, even under stronger conditions than in [15], there are still various possibilities in the relationship between I and $\mathcal{R}(F_\mu)$. Moreover, we give the relationship between I and $\mathcal{R}(F_\mu)$, and the relationship between I and I_{F_μ}.

In [15], it is claimed that $I^\cap = I_{A \circ B}$ when $\circ$ is an arithmetic operation in $\{+, -, \times, /\}$.

We show by examples that this is not valid. Furthermore, we give the relationship between $I^\cap$ and $I_{\widetilde{f}(A_1,\ldots,A_n)}$. As a corollary, we give the relationship between $I^\cap$ and $I_{A \circ B}$ for $\circ \in \{+, -, \times, /\}$. We point out that $I_{\widetilde{f}(A_1,\ldots,A_n)}$ has three possibilities: $\{0\}$, $I^\cap$ and $\mathbf{cl}\{I^\cap\}$.

In [15], it is claimed that under certain conditions, $\mathcal{R}(\widetilde{f}(A_1, \ldots, A_n)) = I_{\widetilde{f}(A_1,\ldots,A_n)} = I^\cap$.

We show by examples that this is not valid. In fact, even under stronger conditions than in [15], there are still various possibilities in the relationship between $\mathcal{R}(\widetilde{f}(A_1, \ldots, A_n))$ and $I^\cap$, and there are still various possibilities in the relationship between $I_{\widetilde{f}(A_1,\ldots,A_n)}$ and $I^\cap$.

The conclusions of this paper show that $I_{\widetilde{f}(A_1,\ldots,A_n)}$ and $\mathcal{R}(\widetilde{f}(A_1, \ldots, A_n))$ vary in the scopes determined by $I^\cap$, respectively. On the other hand, even under stronger conditions than in [15], these and $I^\cap$ are not necessarily equal, and there exist various possibilities in their relationship with $I^\cap$.

The results in this paper can be used in the theoretical research and practical applications of gradual numbers, gradual sets and arithmetic using the gradual numbers and gradual sets introduced by Wu [15]. We will discuss the properties of this kind of arithmetic in the future.

Author Contributions: Formal analysis, Q.M.; methodology, H.H.; writing—original draft preparation, Q.M. and H.H.; writing—review and editing, H.H. All authors have read and agreed to the published version of the manuscript.

Funding: This research was funded by the Natural Science Foundation of Fujian Province of China (grant number 2020J01706).

Institutional Review Board Statement: Not applicable.

Informed Consent Statement: Not applicable.

Data Availability Statement: Not applicable.

Acknowledgments: The authors would like to thank the anonymous reviewers for their valuable comments and suggestions which greatly improved the presentation of this paper.

Conflicts of Interest: The authors declare no conflict of interest.

References

1. Zadeh, L.A. Fuzzy Sets. *Inform. Control* **1965**, *8*, 338–353. [CrossRef]
2. Dubois, D.; Prade, H. *Fundamentals of Fuzzy Sets, The Handbooks of Fuzzy Sets Series*; Kluwer Academic Publishers: London, UK, 2000; Volume 7.
3. Popa, L.; Sida, L. Fuzzy Inner Product Space: Literature Review and a New Approach. *Mathematics* **2021**, *9*, 765. [CrossRef]
4. Xing, Y.; Qiu, D. Solving Triangular Intuitionistic Fuzzy Matrix Game by Applying the Accuracy Function Method. *Symmetry* **2019**, *11*, 1258. [CrossRef]
5. Fortin, J.; Dubois, D.; Fargier, H. Gradual numbers and their application to fuzzy interval analysis. *IEEE Trans. Fuzzy Syst.* **2008**, *16*, 388–401. [CrossRef]
6. Dubois, D.; Prade, H. Gradual elements in a fuzzy set. *Soft Comput.* **2008**, *12*, 165–175. [CrossRef]
7. Dubois, D.; Prade, H. Gradualness, uncertainty and bipolarity: Making sense of fuzzy sets. *Fuzzy Sets Syst.* **2012**, *192*, 3–24 [CrossRef]
8. Boukezzoula, R.; Galichet, S.; Foulloy, L.; Elmasry, M. Extended gradual interval arithmetic and its application to gradual weighted averages. *Fuzzy Sets Syst.* **2014**, *257*, 67–84. [CrossRef]
9. Boukezzoula, R.; Coquin, D. Interval-valued fuzzy regression: Philosophical and methodological issues. *Appl. Soft Comput.* **2021**, *103*, 107145. [CrossRef]
10. Pourabdollaha, A.; Mendel, J.M.; John, R.I. Alpha-cut representation used for defuzzification in rule-based systems. *Fuzzy Sets Syst.* **2020**, *399*, 110–132. [CrossRef]
11. Klir, G.J. Fuzzy arithmetic with requisite constraints. *Fuzzy Sets Syst.* **1997**, *91*, 165–175. [CrossRef]
12. Stefanini, L. A generalization of Hukuhara difference and division for interval and fuzzy arithmetic. *Fuzzy Sets Syst.* **2010**, *161*, 1564–1584. [CrossRef]
13. Chalco-Cano, Y.; Lodwick, W.A.; Bede, B. Single level constraint interval arithmetic. *Fuzzy Sets Syst.* **2014**, *257*, 146–168. [CrossRef]
14. Ngan, S.-C. A concrete reformulation of fuzzy arithmetic. *Expert Syst. Appl.* **2021**, *167*, 113818. [CrossRef]
15. Wu, H.-C. Arithmetic operations of non-normal fuzzy sets using gradual numbers. *Fuzzy Sets Syst.* **2020**, *399*, 1–19. [CrossRef]
16. Nguyen, H.T. A note on the extension principle for fuzzy sets. *J. Math. Anal. Appl.* **1978**, *64*, 369–380. [CrossRef]
17. Zadeh, L.A. The concept of linguistic variable and its application to approximate reasoning. *Inform. Sci.* **1975**, *8*, 199–249. [CrossRef]
18. Wu, C.; Ma, M. *The Basic of Fuzzy Analysis*; National Defence Industry Press: Beijing, China, 1991. (In Chinese)
19. Gong, Z.; Hao, Y. Fuzzy Laplace transform based on the Henstock integral and its applications in discontinuous fuzzy systems. *Fuzzy Sets Syst.* **2019**, *358*, 1–28. [CrossRef]
20. Wang, G.; Shi, P.; Wang, B.; Zhang, J. Fuzzy n-ellipsoid numbers and representations of uncertain multichannel digital information. *IEEE Trans. Fuzzy Syst.* **2014**, *22*, 1113–1126. [CrossRef]
21. Huang, H. Characterizations of endograph metric and Γ-convergence on fuzzy sets. *Fuzzy Sets Syst.* **2018**, *350*, 55–84. [CrossRef]

mathematics

MDPI

Article

Application of Induced Preorderings in Score Function-Based Method for Solving Decision-Making with Interval-Valued Fuzzy Soft Information

Mabruka Ali [1], Adem Kiliçman [1,2,*] and Azadeh Zahedi Khameneh [2]

1 Department of Mathematics, Universiti Putra Malaysia, Serdang UPM 43400, Malaysia; altwer2016@gmail.com
2 Institute for Mathematical Research, Universiti Putra Malaysia, Serdang UPM 43400, Malaysia; zk.azadeh@upm.edu.my
* Correspondence:akilic@upm.edu.my

Abstract: Ranking interval-valued fuzzy soft sets is an increasingly important research issue in decision making, and provides support for decision makers in order to select the optimal alternative under an uncertain environment. Currently, there are three interval-valued fuzzy soft set-based decision-making algorithms in the literature. However, these algorithms are not able to overcome the issue of comparable alternatives and, in fact, might be ignored due to the lack of a comprehensive priority approach. In order to provide a partial solution to this problem, we present a group decision-making solution which is based on a preference relationship of interval-valued fuzzy soft information. Further, corresponding to each parameter, two crisp topological spaces, namely, lower topology and upper topology, are introduced based on the interval-valued fuzzy soft topology. Then, using the preorder relation on a topological space, a score function-based ranking system is also defined to design an adjustable multi-steps algorithm. Finally, some illustrative examples are given to compare the effectiveness of the present approach with some existing methods.

Keywords: interval-valued fuzzy soft sets; interval-valued fuzzy soft topology; preference relationship; decision-making

Citation: Ali, M.; Kılıçman, A.; Zahedi Khameneh, A. Application of Induced Preorderings in Score Function-Based Method for Solving Decision-Making with Interval-Valued Fuzzy Soft Information. *Mathematics* **2021**, *9*, 1575. https://doi.org/10.3390/math9131575

Academic Editor: Sorin Nadaban

Received: 18 May 2021
Accepted: 29 June 2021
Published: 4 July 2021

1. Introduction

Dealing with vagueness and uncertainty, rather than exactness, in most real-world situations is the main problem in data-analysis sciences and decision-making. Many mathematical theories and tools such as probability theory, fuzzy set theory [1], interval-valued fuzzy set theory [2], intuitionist fuzzy set theory [3], rough set theory [4] and soft set theory [5] have been implemented to handle this problem, with the latter allowing researchers to deal with parametric data. Nowadays, soft sets theory contributes to a vast range of applications, particularly in decision-making. In this regard, many important results have been achieved, from parameter reduction to new ranking models.

Many soft set extensions and their applications have been discussed in previous studies, such as fuzzy soft sets [6–13] intuitionistic fuzzy sets [14–17], rough soft sets [18,19] and fuzzy soft topology [20–23]. The interval-valued fuzzy soft method was first used for decision-making problems by Son [24]. He applied this method by using the comparison table. Yang et al. [25] developed the method presented in [7] for an interval-valued fuzzy soft set and then, applied the concept of interval-valued fuzzy choice values to propose an approach for solving decision-making problems. The notion of level set in decision-making based on interval-valued fuzzy soft sets was introduced by Feng et al. [26] and then, the level soft set for interval-valued fuzzy soft sets was developed, further see [27]. Khameneh et al. [28–30] introduced the preference relationship for both fuzzy soft sets and intuitionistic fuzzy soft sets and then selected an optimal option for group decision-making problems by defining a new function value. In addition, interval-valued fuzzy soft sets have also been applied to various fields, for example information measure [31–34], decision making [35–38], matrix theory [39–41], and parameter reduction [37,38,42].

Recently, Ma et al. [43] introduced an average and an antitheses table for interval-valued fuzzy soft sets and then selected an optimal option for group decision-making problems through the score value. Ma et al. [44] developed two methods [26,45] to solve decision-making problems by providing a new efficient decision-making algorithm and also considering added objects. However, these methods did not address the problem of incomparable alternatives because they lack a comprehensive priority approach. In order to solve these issues, this paper proposes an application of the induced preorderings based method for solving decision-making with interval-valued fuzzy soft information.Our contributions are as follows:

1 Proposing application of induced preordering based method for solving decision-making with interval-valued fuzzy soft information.
2 Proposing a novel score function of interval-valued fuzzy soft sets that selects an optimal option for group decision-making problems.
3 A real-life example is given to compare the effectiveness of this approach with some existing methods.

2. Preliminaries

In this section, we recall some definitions and properties of interval-valued fuzzy sets (IVF) and interval-valued fuzzy soft sets $(IVFS)$. Note that, throughout this paper, X and E denote the sets of objects and parameters, respectively. $\mathbb{I}^X$ and $[\mathbb{I}]^X$, where $\mathbb{I} = [0,1]$ and $[\mathbb{I}] = \{[a,b], a \leq b, a, b \in \mathbb{I}\}$ denote, respectively, the set of all fuzzy subsets and the set of all interval-valued fuzzy subsets of X.

Definition 1. *Ref. [2] A pair (f, X), is called an IVF subset of X if f is a mapping given by $f : X \to [\mathbb{I}]$ such that for any $x \in X, f(x) = [f^-(x), f^+(x)]$ is a closed subinterval of $[0,1]$ where $f^-(x)$ and $f^+(x)$ are referred to as the lower and upper degrees of membership x to f and $0 \leq f^-(x) \leq f^+(x) \leq 1$.*

In 1999, Molodtsov [5] defined the concept of soft sets (SS) for the first time as a pair of (f, E) or f_E such that E is a parameter set and f is the mapping $f : E \to 2^X$ where for any $e \in E, f(e)$ is a subset of X. By combining the concepts of soft sets and interval-valued fuzzy sets, a new hybrid tool was defined as the following.

Definition 2. *Ref. [25] A pair (f, E) is called an $IVFS$ set over X if the mapping f is given by $f : E \to [\mathbb{I}]^X$ where for any $e \in E$ and $x \in X$, $f(e)(x) = [f^-(e)(x), f^+(e)(x)]$.*

Consider two $IVFSs$ f_E, g_E over the common universe X. The union of f_E and g_E, denoted by $f_E \tilde{\vee} g_E$, is the $IVFSs$ $(f \tilde{\vee} g)_E$, where $\forall e \in E$ and any $x \in X$, we have $(f \tilde{\vee} g)(e)(x) = [\max\{f_e^-(x), g_e^-(x)\}, \max\{f_e^+(x), g_e^+(x)\}]$. The intersection of f_E and g_E, denoted by $f_E \tilde{\wedge} g_E$, is the $IVFSs$ $(f \tilde{\wedge} g)_E$, where $\forall e \in E$ and $\forall x \in X$, we have $(f \tilde{\wedge} g)(e)(x) = [\min\{f_e^-(x), g_e^-(x)\}, \min\{f_e^+(x), g_e^+(x)\}]$. The complement of f_E is denoted by f_E^c and is defined by $f^c : E \to [\mathbb{I}]^X$ where $\forall e \in E$ and any $x \in X$, $f^c(e)(x) = [1 - f_e^+(x), 1 - f_e^-(x)]$. The null $IVFSs$, denoted by $\emptyset_E$, is defined as an $IVFSs$ over X such that $f_e^-(x) = f_e^+(x) = 0$ for all $e \in E$ and any $x \in X$. The absolute $IVFSs$, denoted by X_E, is defined as an $IVFSs$ over X where $f_e^-(x) = f_e^+(x) = 1, \forall e \in E$ and any $x \in X$.

Using the matrix form of interval-valued fuzzy relations, authors in [39] represented a finite IVFSs f_E as the following $n \times m$ matrix

$$f_E = \left[[f_{ij}^-, f_{ij}^+] \right]_{n \times m} = \begin{bmatrix} [f_{e_1}^-(x_1), f_{e_1}^+(x_1)] & \ldots & [f_{e_1}^-(x_m), f_{e_1}^+(x_m)] \\ \vdots & \ldots & \vdots \\ [f_{e_n}^-(x_1), f_{e_1}^+(x_1)] & \ldots & [f_{e_n}^-(x_m), f_{e_1}^+(x_m)] \end{bmatrix}_{n \times m}$$

where $\mid E \mid = n$, $\mid X \mid = m$ and $f_{ij}^- = f_{e_i}^-(x_j)$, $f_{ij}^+ = f_{e_i}^+(x_j)$ for $i = 1, \ldots, n$ and $j = 1, \ldots, m$.

Accordingly, the concepts of union, intersection, complement, etc., can be represented in a matrix format in the finite case.

Definition 3. *Ref. [46] A triplet (X, E, τ) is called an interval-valued fuzzy soft topological space (IVFST) if τ is a collection of interval-valued fuzzy soft subsets of X containing absolute and null IVFSs and closed under arbitrary union and finite intersection.*

Preorders and Topologies

In this subsection, we present some basic properties about the connection between preorders and topologies proposed by [47].

Topological structures and classical order structures are well recognised to have close relationships, which can be summarised as follows:

(1) A subset A of X is called an upper set of X if $A =\uparrow A$, where $\uparrow A$ defined by $\uparrow A = \{y \in X : \exists x \in A, x \leq y\}$, and X is a preordered set, and B is called a lower set $B =\downarrow B = \{y \in X : \exists x \in X, y \leq x\}$.
(2) The family of all upper subsets of x is a topology for a preorder set $(X, \leq)$, which is called the Alexandrov topology induced in $(X, \leq)$.
(3) A topological space (X, τ) is defined by $x \leq y$ if and only if $x \in U$, then $y \in U$ for each open set U of X, or equivalently $x \in c\{y\}$, where $c\{y\}$, is the closure of $\{y\}$. Then, $\leq$ is a preorder on X, called the specialization order (X, τ) on X.

3. Construction Tow Preorderings in Lower and Upper Spaces

By using the notion of $[\alpha_1, \alpha_2]$-level sets of interval-valued fuzzy soft open sets in (X, E, τ), this section, introduces two topological spaces, known as lower and upper spaces, by which two preordering relations over the universal set X are investigated.

Definition 4. *Let f_E be an IVFS set over X. Corresponding to each parameter $e \in E$,, we define two crisp sets, called α-upper-e crisp set and β-lower-e crisp set, where $\alpha = [\alpha_1, \alpha_2] \subset \mathbb{I}, \beta = [\beta_1, \beta_2] \subset \mathbb{I}$ as the following:*

$$\begin{aligned}
U.C.S^f_\alpha(e) &= \{x \in X : [f^-_e(x), f^+_e(x)] > \alpha, \alpha \subseteq [0,1)\} \\
&= \{x \in X : f^-_e(x) > \alpha_1, f^+_e(x) > \alpha_2, \alpha_1, \alpha_2 \in [0,1)\} \\
L.C.S^f_\beta(e) &= \{x \in X : [f^-_e(x), f^+_e(x)] < \beta, \beta \subseteq (0,1]\} \\
&= \{x \in X : f^-_e(x) < \beta_1, f^+_e(x) < \beta_2, \beta_1, \beta_2 \in (0,1]\}
\end{aligned}$$

Proposition 1. *Let X be the set of objects, E be the set of parameters and f_E, g_E be two IVFSs over X. Suppose that the threshold intervals $\alpha_1, \alpha_2, \subseteq [0,1)$, and $\beta_1, \beta_2, \subseteq (0,1]$ are given such that $\alpha_1 = [\alpha^\star_1, \alpha^{\star\star}_1], \alpha_2 = [\alpha^\star_2, \alpha^{\star\star}_2], \beta_1 = [\beta^\star_1, \beta^{\star\star}_1]$ and $\beta_2 = [\beta^\star_2, \beta^{\star\star}_2]$. Consider the parameter $e \in E$.*

1. *If $\alpha_1 \geq \alpha_2$, then $U.C.S^f_{\alpha_1}(e) \subseteq U.C.S^f_{\alpha_2}(e)$. If $\beta_1 \geq \beta_2$, then $L.C.S^f_{\beta_2}(e) \subseteq L.C.S^f_{\beta_1}(e)$.*
2. *If $f_E \tilde{\leq} g_E$, then $U.CS^f_{\alpha_1}(e) \subseteq U.C.S^g_{\alpha_1}(e)$ and $L.C.S^f_{\beta_1}(e) \subseteq L.C.S^g_{\beta_1}(e)$.*
3. *If $f_E = X_E$, then $U.C.S^f_{\alpha_1}(e) = X$ and $L.C.S^f_{\beta_1}(e) = \emptyset$. Moreover, if $f_E = \emptyset_E$ then, $U.CS^f_{\alpha_1}(e) = \emptyset$ and $L.CS^f_{\beta_1}(e) = X$.*
4. *$U.C.S^f_{\alpha_1}(\neg e) = L.CS^f_{[1-\alpha^{\star\star}_1, 1-\alpha^\star_1]}(e)$ and $L.CS^f_{\alpha_1}(\neg e) = U.CS^f_{[1-\alpha^{\star\star}_1, 1-\alpha^\star_1]}(e)$.*
5. *$U.CS^{\neg f}_{\alpha_1}(e) = L.CS^f_{[1-\alpha^{\star\star}_1, 1-\alpha^\star_1]}(e)$ and $L.CS^{\neg f}_{\alpha_1}(\neg e) = U.Des^f_{[1-\alpha^{\star\star}_1, 1-\alpha^\star_1]}(e)$.*

Proof. It is straightforward. □

Theorem 1. *Let (X, E, τ) be an IVFSTS. Suppose that the threshold intervals $\alpha_1, \alpha_2, \subseteq [0,1)$, and $\beta_1, \beta_2, \subseteq (0,1]$ are given such that $\alpha = [\alpha_1, \alpha_2]$ and $\beta = [\beta_1, \beta_2]$, then*

1. *The collection $\{U.C.S^f_\alpha(e) : f_E \in \tau, e \in E, \alpha \subseteq [0,1)\}$, denoted by $\tau^u_{e,\alpha}$, is a topology over X.*
2. *The collection $\mathfrak{B}^l_\beta(e) = \{L.C.S^f_\beta(e) : f_E \in \tau, e \in E, \beta \subseteq (0,1]\}$, is a base for a topology over X, denoted by $\tau^l_{e,\beta}$.*

Proof. 1. (a) By Proposition 1, $X, \emptyset \in \tau^{u}_{e,\alpha}$, since $X_E, \emptyset_E \in \tau$.

(b) Let $\{U.C.S^{f_i}_{\alpha}(e)\}_{i\in I}$ be a subfamily of $\tau^{u}_{e,\alpha}$. Then, we have $\bigcup_i U.C.S^{f_i}_{\alpha}(e) = U.C.S^{(\tilde{\vee}_{i\in I} f_i)}_{\alpha}(e) \in \tau^{u}_{e,\alpha}$, since $\tilde{\vee}_{i\in I} f_{iE} \in \tau$.

(c) Let $U.C.S^{f}_{\alpha}(e)$ and $U.C.S^{g}_{\alpha}(e)$ be two open sets in $\tau^{u}_{e,\alpha}$. Then, we have $U.CS^{f}_{\alpha}(e) \cap U.C.S^{g}_{\alpha}(e) = U.C.S^{(f\tilde{\wedge}g)}_{\alpha}(e) \in \tau^{u}_{e,\alpha}$, since $f_E \tilde{\wedge} g_E \in \tau$. This completes the proof.

2. (a) That $X \in \mathfrak{B}^{l}_{\beta}(e)$ is implied from $\emptyset_E$ is in τ.

(b) Let $L.C.S^{f}_{\beta}(e)$ and $L.C.S^{g}_{\beta}(e)$ in $\mathfrak{B}^{l}_{\beta}(e)$. Then, we have $L.C.S^{f}_{\beta}(e) \cap L.C.S^{g}_{\beta}(e) = L.C.S^{f\tilde{\vee}g}_{\beta}(e) \in \mathfrak{B}^{l}_{\beta}(e)$ that is implied form $f\tilde{\vee}g \in \tau$.

□

Theorem 2. *Let (X, E, τ) be an $IVFSTS$. Suppose that the threshold intervals $\alpha_1, \alpha_2, \subseteq [0, 1)$, and $\beta_1, \beta_2, \subseteq (0, 1]$ are given such that $\alpha = [\alpha_1, \alpha_2]$ and $\beta = [\beta_1, \beta_2]$.*

1. *The binary relation $\succsim^{\tau}_{e,\alpha}$ on X defined by*

$$y \succeq^{\tau}_{e,\alpha} x \Leftrightarrow [\forall V \in \tau^{u}_{e,\alpha} : x \in V \Rightarrow y \in V]$$

is a preorder relation called α-upper-e preorder relation on X.

2. *The binary relation $\preceq^{\tau,\beta}_{e}$ on X defined by*

$$y \preceq^{\tau,\beta}_{e} x \Leftrightarrow [\forall U \in \tau^{l}_{e,\beta} : x \in U \Rightarrow y \in U]$$

is a preorder relation called β-lower-e preorder relation on X.

Proof. 1. For all $x \in X$, obviously, $x \succsim^{u}_{e,\alpha} x$, that is, "$\succsim^{u}_{e,\alpha}$" is reflexive. Now, for all $x, y, z \in X$, if $y \succsim^{u}_{e,\alpha} x$, and $z \succsim^{u}_{e,\alpha} y$, then, if for all $V \in \tau^{u}_{e,\alpha}$-open set, $x \in V$, then $y \in V$ and $z \in V$, so, $z \succsim^{u}_{e,\alpha} x$, that is, "$\succsim^{u}_{e,\alpha}$" is transitive. Thererfore, $(X, \succsim^{u}_{e,\alpha})$ is a preordered set.

2. A similar technique is used to prove the second part.

□

Theorem 3. *Let (X, E, τ) be an $IVFSTS$. Suppose that the threshold intervals $\alpha_1, \alpha_2, \subseteq [0, 1)$, and $\beta_1, \beta_2, \subseteq (0, 1]$ are given such that $\alpha = [\alpha_1, \alpha_2]$ and $\beta = [\beta_1, \beta_2]$.*

1. *The binary relation $\backsimeq^{\tau}_{e,\alpha}$,defined by*

$$y \backsimeq^{\tau}_{e,\alpha} x \Leftrightarrow [y \succsim^{\tau}_{e,\alpha} x, x \succsim^{\tau}_{e,\alpha} y]$$

is an equivalence relation over X. If $y \backsimeq^{\tau}_{e,\alpha} x$, then we say x and y are α-upper equivalent with to respect to the parameter e.
The equivalence relation $\backsimeq^{\tau}_{e,\alpha}$, generates the partition $P^{\tau}_{e,\alpha}$ of X where the equivalence classes are defined as $[x]^{\tau}_{e,\alpha} = \{z \in X : z \backsimeq^{\tau}_{e,\alpha} x$ and are called α-upper-e equivalence classes.

2. *The binary relation $\backsimeq^{\tau,\beta}_{e}$, where $\beta = [\beta_1, \beta_2]$,*

$$y \backsimeq^{\tau,\beta}_{e} x \Leftrightarrow [y \precsim^{\tau,\beta}_{e} x, x \precsim^{\tau,\beta}_{e} y]$$

is an equivalence relation over X. If $y \backsimeq^{\tau}_{e,\beta} x$, then we say x and y are $[\beta_1, \beta_2]$-lower equivalent with to respect to the parameter e. The equivalence relation $\backsimeq^{\tau,\beta}_{e}$, generates the partition $P^{\tau,\beta}_{e}$ of X where the equivalence classes are defined as $[x]^{\tau,\beta}_{e} = \{z \in X : z \backsimeq^{\tau,\beta}_{e} x$ and are called β-lower-e equivalence classes.

Proof. It is straightforward. □

Preorder and Equivalence Matrices

Now, let the finite sets $X = \{x_1, \cdots, x_m\}$ and $E = \{e_1, \cdots, e_n\}$ be given as the sets of objects and parameters. Then, the previous properties can be represented by using the matrix form of $IVFS$ sets as the following.

Take an $IVFS$ set f_E over X. First, for any $1 \le i \le m$ and $1 \le t \le n$, the concepts of α-upper-e_t and β-lower-e_t matrices of f_E, where $\alpha, \beta \subseteq \mathbb{I}$, can be formulated as the following two matrices (or row vectors)

$$U.C.S_\alpha e_t^f = [u_i^f(e_t, \alpha)]_{1\times m} = \begin{cases} 1 & \text{if } f_{e_t}^-(x_i) > \alpha_1, f_{e_t}^+(x_i) > \alpha_2 \\ 0 & \text{if } f_{e_t}^-(x_i) \le \alpha_1, f_{e_t}^+(x_i) \le \alpha_2 \end{cases} \tag{1}$$

and

$$L.C.S_\beta e_t^f = [l_i^f(e_t, \beta)]_{1\times m} = \begin{cases} 0 & \text{if } f_{e_t}^-(x_i) \ge \beta_1, f_{e_t}^+(x_i) \ge \beta_2 \\ 1 & \text{if } f_{e_t}^-(x_i) < \beta_1, f_{e_t}^+(x_i) < \beta_2 \end{cases} \tag{2}$$

where $\alpha = [\alpha_1, \alpha_1]$ and $\beta = [\beta_1, \beta_2]$ are the given threshold vectors.

Then, obviously, for any $e_t \in E$, the topologies $\tau^u_{e_t,\alpha}$ and $\tau^l_{e_t,\beta}$ can be represented by the collections

$$\tau^u_{e_t,\alpha} = \{[u_i^f(e_t,\alpha)]_{1\times m} : \alpha \subseteq [0,1), f_E \in \tau, 1 \le i \le m\}$$

and

$$\tau^l_{e_t,\beta} = \{[l_i^f(e_t,\beta)]_{1\times m} : \beta \subseteq (0,1], f_E \in \tau, 1 \le i \le m\}$$

where τ is the $IVFST$ on X.

Accordingly, the preorderings $\succeq^\tau_{e_t,\alpha}$ and $\preceq^{\tau,\beta}_{e_t}$ can be represented by

$$x_i \succeq^\tau_{e_t,\alpha} x_j \Leftrightarrow [\forall f_E \in \tau : u_j^f(e_t,\alpha) = 1 \Rightarrow u_i^f(e_t,\alpha) = 1]$$

and

$$x_i \preceq^{\tau,\beta}_{e_t} x_j \Leftrightarrow [\forall f_E \in \tau : l_j^f(e_t,\beta) = 1 \Rightarrow l_i^f(e_t,\beta) = 1]$$

where $x_i, x_j \in X$.

The matrix forms of the preorderings $\succsim^\tau_{e_t,\alpha}$ and $\precsim^{\tau,\beta}_{e_t}$ are used to define two comparison matrices $G_\alpha(e_t) = [g_\alpha(e_t)_{ij}]_{m\times m}$ and $S_\beta(e_t) = [s_\beta(e_t)_{ij}]_{m\times m}$, which are two square matrices whose rows and columns are labeled by the objects of X, as below.

Definition 5. *Consider the binary relations $\succsim^\tau_{e_t,\alpha}$ and $\precsim^{\tau,\beta}_{e_t}$ and threshold intervals $\alpha = [\alpha_1, \alpha_2]$, $\beta = [\beta_1, \beta_2 \subseteq \mathbb{I}$. Then, we define*

$$G_\alpha(e_t) = [g_\alpha(e_t)_{ij}]_{m\times m} : g_{[\alpha_1,\alpha_2]}(e_t)_{ij} = \begin{cases} 1 & \text{if } x_i \succsim^\tau_{e_t,\alpha} x_j \\ 0 & \text{otherwis} \end{cases} \tag{3}$$

and

$$S_\beta(e_t) = [s_\beta(e_t)_{ij}]_{m\times m} : s_{[\beta_1,\beta_2]}(e_t)_{ij} = \begin{cases} 1 & \text{if } x_i \precsim^{\tau,\beta}_{e_t} x_j \\ 0 & \text{otherwis} \end{cases} \tag{4}$$

Proposition 2. *Let (X, E, τ) be an IVFST and $G_\alpha(e)$ and $S_\beta(e)$ be two matrices defined in Equations (3) and (4). Then,*

1. *For $1 \le i \le m, g_\alpha(e_t)_{ii} = 1$ and $s_\beta(e_t)_{ii} = 1$,*
2. *If $g_\alpha(e_t)_{ij} = g_\alpha(e_t)_{jk} = 1$, then $g_\alpha(e_t)_{ik} = 1$. If $s_\beta(e_t)_{ij} = s_\beta(e_t)_{jk} = 1$, then $s_\beta(e_t)_{ik} = 1$.*
3. *$G_\alpha(e_t)$ and $S_\beta(e_t)$ are symmetric matrices.*

where $i, j, k \in \{1, \dots, m\}$

Proof. It is straightforward. □

Proposition 3. *Let (X, E, τ) be an IVFSTS and $\alpha, \beta \subseteq \mathbb{I}$, where $\alpha = [\alpha_1, \alpha_1]$ and $\beta = [\beta_1, \beta_2]$ are the threshold intervals, then*

1. *$G_\alpha(e_t)$ is an identity matrix if and only if $\neg(x_i \succeq^\tau_{e_t\alpha} x_j), \forall i, j = 1, \dots, m$ and $i \ne j$.*
2. *$S_\beta(e_t)$ is an identity matrix if and only if $\neg(x_i \preceq^{\tau,\beta}_{e_t} x_j), \forall i, j = 1, \dots, m$ and $i \ne j$.*
3. *$G_\alpha(e_t)$ is a unit matrix if and only if $x_i \succeq^\tau_{e_t,\alpha} x_j, \forall i, j = 1, \dots, m$ and $i \ne j$.*
4. *$S_\beta(e_t)$ is a unit matrix if and only if $x_i \preceq^{\tau,\beta}_{e_t} x_j, \forall i, j = 1, \dots, m$ and $i \ne j$.*

Proof. It is straightforward. □

Proposition 4. *Let (X, E, τ) be an IVFSTS and $\alpha, \beta \subseteq \mathbb{I}$, where $\alpha = [\alpha_1, \alpha_1], \beta = [\beta_1, \beta_2]$ are the threshold intervals, then*

1. $G_\alpha(e_t) = I_m^U$ *if and only if we have* $x_1 \succeq_{e_t,\alpha}^\tau \cdots \succeq_{e_t,\alpha}^\tau x_m$.
2. $S_\beta(e_t) = I_m^U$ *if and only if* $x_1 \preceq_{e_t}^{\tau,\beta} \cdots \preceq_{e_t}^{\tau,\beta} x_m$.
3. $G_\alpha(e_t) = I_m^L$ *if and only if* $x_m \succeq_{e_t,\alpha}^\tau \cdots \succeq_{e_t,\alpha}^\tau x_1$.
4. $S_\beta(e_t) = I_m^L$ *if and only if* $x_m \preceq_{e_t}^{\tau,\beta} \cdots \preceq_{e_t}^{\tau,\beta} x_1$.

where I_m^U, I_m^L are the upper and lower triangular matrix, respectively.

Proof. It is straightforward. □

Analogously, the equivalence relations $\simeq_{e_t,\alpha}^\tau$ and $\simeq_{e_t}^{\tau,\beta}$ can be applied to compute the following two square matrices $E_\alpha^U(e_t) = [e_\alpha^u(e_t)_{ij}]_{m\times m}$ and $E_\beta^L(e_t) = [e_\beta^l(e_t)_{ij}]_{m\times m}$, respectively, where $\alpha = [\alpha_1, \alpha_2]$, $\beta = [\beta_1, \beta_2] \subseteq \mathbb{I}$.

Definition 6. *Consider the binary relations $\simeq_{e_t,\alpha}^\tau$ and $\simeq_{e_t}^{\tau,\beta}$ and threshold intervals $\alpha = [\alpha_1, \alpha_2]$, $\beta = [\beta_1, \beta_2 \subseteq \mathbb{I}$. We define*

$$E_\alpha^U(e_t) = [e_\alpha^u(e_t)_{ij}]_{m\times m} : e_\alpha^u(e_t)_{ij} = \begin{cases} 1 & \text{if } x_i \simeq_{e_t,\alpha}^\tau x_j \\ 0 & \text{otherwis} \end{cases} \tag{5}$$

and

$$E_\beta^L(e_t) = [e_\beta^l(e_t)_{ij}]_{m\times m} : e_\beta^l(e_t)_{ij} = \begin{cases} 1 & \text{if } x_i \simeq_{e_t}^{\tau,\beta} x_j \\ 0 & \text{otherwis} \end{cases} \tag{6}$$

Proposition 5. *Let (X, E, τ) be an IVFST and $E_\alpha^U(e_t)$ and $E_\beta^L(e_t)$ be the comparison matrices defined in Equations (5) and (6). Then,*

1. *For any $1 \le i \le m$: $e_\alpha^u(e_t)_{ii} = 1$ and $e_\beta^l(e_t)_{ii} = 1$.*
2. *$E^U(e_t)$ and $E^L(e_t)$ are symmetric matrices.*
3. *If $e_\alpha^u(e_t)_{ik} = e_\alpha^u(e_t)_{jk} = 1$, then $e_\alpha^u(e_t)_{ij} = e_\alpha^u(e_t)_{ji} = 1$. If $e_\beta^l(e_t)_{ik} = e_\beta^l(e_t)_{jk} = 1$, then $e_\beta^l(e_t)_{ij} = e_\beta^l(e_t)_{ji} = 1$.*
4. *If $e_\alpha^u(e_t)_{ki} = e_\alpha^u(e_t)_{kj} = 1$, then $e_\alpha^u(e_t)_{ij} = e_\alpha^u(e_t)_{ji} = 1$. If $e_\beta^l(e_t)_{ki} = e_\beta^l(e_t)_{kj} = 1$, then $e_\beta^l(e_t)_{ij} = e_\beta^l(e_t)_{ji} = 1$.*

where $i, j, k \in \{1, \cdots, m\}$

Proof. It is straightforward. □

4. An Application in Decision-Making Problems

The main task in decision making methods is to rank the given candidates to find the optimum choice. Since the proposed preorderings, given in Section 3, are not total or linear, we define a score function S based on the entries of defined comparison matrices to obtain a new ranking system of objects according to preorderings $\succsim_{e_t,\alpha}^\tau$ and $\precsim_{e_t}^{\tau,\beta}$.

Definition 7. *Let X and E be the universal sets of objects and parameters, respectively, and $\alpha, \beta \subseteq \mathbb{I}$, where $\alpha = [\alpha_1, \alpha_1]$ and $\beta = [\beta_1, \beta_2]$, are the threshold intervals. The mapping $S = X \to \mathbb{R}$ defined by*

$$S(x_i) = S_i = \sum_{t=1}^{n}\left(\left[\sum_{j=1}^{m} g_\alpha(e_t)_{ij} - \sum_{j=1}^{m} e_\alpha^u(e_t)_{ij}\right] - \left[\sum_{j=1}^{m} s_\beta(e_t)_{ij} - \sum_{j=1}^{m} e_\beta^l(e_t)_{ij}\right]\right)$$

where $x_i \in X$ and S_i is score value of object x_i.

Example 1. *Suppose that* $X = \{o_1, o_2, o_3, o_5\}$ *be a set of 5 hotels in Langkawi and* $E = \{e_1, \ldots, e_4\}$ *be a set of parameters where for any* $t = 1, \ldots, 4$ *the parameter* e_t *stands for "location", "cleanliness", "facilities", and " food", respectively. Reviewers are classified into three groups: couples, solo travelers, and a group of friends. We consider these groups of reviewers as three different decision-makers,* f_1, f_2, f_3, *characterized based on the criteria* $e_t \in E$. *These three groups provide the following three IVFS matrices* f_{1E}, f_{2E}, f_{3E}.

Step 1. The following three interval-valued fuzzy soft set $f_{sE}(s = 1,2,3)$ that are given in Tables 1–3.

Table 1. f_{1E}.

f_{1E}	o_1	o_2	o_3	o_4	o_5
e_1	[0.1, 0.4]	[0.4, 0.4]	[0.4, 0.5]	[0, 0.5]	[0.0, 0.0]
e_2	[0.5, 0.6]	[0.3, 0.6]	[0.3, 1.0]	[0.7, 1.0]	[0.0, 0.7]
e_3	[0.0, 0.5]	[0.5, 0.8]	[0.1, 0.8]	[0.1, 0.9]	[0.3, 0.9]
e_4	[0.0, 0.8]	[0.7, 0.8]	[0.1, 0.7]	[0.1, 1.0]	[0.6, 1.0]

Table 2. f_{2E}.

f_{2E}	o_1	o_2	o_3	o_4	o_5
e_1	[0.2, 0.6]	[0.2, 0.6]	[0.6, 0.6]	[0.5, 0.6]	[0.5, 0.6]
e_2	[0.4, 0.8]	[0.0, 0.8]	[0.0, 0.6]	[0.6, 0.9]	[0.6, 0.9]
e_3	[0.1, 0.5]	[0.1, 0.8]	[0.6, 0.8]	[0.5, 0.6]	[0.5, 0.9]
e_4	[0.3, 0.6]	[0.3, 0.3]	[0.2, 0.3]	[0.2, 0.7]	[0.7, 0.7]

Table 3. f_{3E}.

f_{3E}	o_1	o_2	o_3	o_4	o_5
e_1	[0.5, 0.6]	[0.6, 0.5]	[0.1, 0.8]	[0.1, 0.5]	[0.5, 0.6]
e_2	[0.2, 0.8]	[0.2, 0.2]	[0.2, 0.2]	[0.2, 0.6]	[0.1, 0.6]
e_3	[0.1, 0.3]	[0.1, 0.2]	[0.2, 0.3]	[0.2, 0.3]	[0.2, 0.8]
e_4	[0.7, 0.8]	[0.0, 0.8]	[0.0, 0.1]	[0.1, 1.0]	[0.1, 0.3]

Step 2. Assume that $[\alpha_1, \alpha_2] = [0.3, 0.6]$ and $[\beta, \beta_2 = [0.2, 0.4]$.
Step 3. The upper crisp matrices and lower crisp matrices, as below:

$$f_1 = \begin{bmatrix} 0 & 0 & 0 & 0 & 0 \\ 0 & 0 & 0 & 1 & 0 \\ 0 & 1 & 0 & 0 & 0 \\ 0 & 1 & 0 & 0 & 1 \end{bmatrix}, f_2 = \begin{bmatrix} 0 & 0 & 0 & 0 & 0 \\ 1 & 0 & 0 & 1 & 1 \\ 0 & 0 & 1 & 0 & 1 \\ 0 & 0 & 0 & 0 & 1 \end{bmatrix}, f_3 = \begin{bmatrix} 0 & 1 & 0 & 0 & 0 \\ 0 & 0 & 0 & 0 & 0 \\ 0 & 0 & 0 & 0 & 0 \\ 1 & 0 & 0 & 0 & 0 \end{bmatrix}$$

$$f_1 = \begin{bmatrix} 0 & 0 & 0 & 0 & 1 \\ 0 & 0 & 0 & 0 & 0 \\ 0 & 0 & 0 & 0 & 0 \\ 0 & 0 & 0 & 0 & 0 \end{bmatrix}, f_2 = \begin{bmatrix} 0 & 0 & 0 & 0 & 0 \\ 0 & 0 & 0 & 0 & 0 \\ 0 & 0 & 0 & 0 & 0 \\ 0 & 0 & 0 & 0 & 0 \end{bmatrix}, f_3 = \begin{bmatrix} 0 & 0 & 0 & 0 & 0 \\ 0 & 0 & 0 & 0 & 0 \\ 1 & 1 & 0 & 0 & 0 \\ 0 & 0 & 1 & 0 & 1 \end{bmatrix}$$

Step 4. The upper topology and lower topology are shown in Tables 4 and 5.

Table 4. α-Upper-e_t topology; $\alpha = [\alpha_1, \alpha_2], t = 1, \ldots, 4$.

		$\tau^{u}_{e_t,\alpha}$										
e_1	$\{[0]_{1\times5}$	$[1]_{1\times5}$	[0	1	0	0	0]}					
e_2	$\{[0]_{1\times5}$	$[1]_{1\times5}$	[0	0	0	1	0]	[1	0	0	1	1]}
e_3	$\{[0]_{1\times5}$	$[1]_{1\times5}$	[0	1	0	0	0]	[0	0	1	0	1]
	[0	1	1	0	1] }							
e_4	$\{[0]_{1\times5}$	$[1]_{1\times5}$	[0	1	0	0	1]	[0	0	0	0	1]
	[1	0	0	0	0]	[1	0	0	0	1]		
	[1	1	0	0	1] }							

Table 5. β-Lower-e_t topology; $\beta = [\beta_1, \beta_2], t = 1, \ldots, 4$.

		$\tau^{L}_{e_t,\beta}$					
e_1	$\{[0]_{1\times5}$	$[1]_{1\times5}$	[0	0	0	0	1]}
e_2	$\{[0]_{1\times5}$	$[1]_{1\times5}\}$					
e_3	$\{[0]_{1\times5}$	$[1]_{1\times5}$	[1	1	0	0	0] }
e_4	$\{[0]_{1\times5}$	$[1]_{1\times5}$	[0	0	1	0	1]}

Step 5. The comparison matrices $G(e_t, \alpha), S(e_t, \beta), E^{U}(e_t, \alpha)$ and $E^{U}(e_t, \alpha)$, over X where $\alpha = [\alpha_1, \alpha_2], \beta = [\beta_1, \beta_2], t = 1, \ldots, 4$ as below:

$$G(e_1, [0.3, 0.6]) = \begin{bmatrix} 1 & 0 & 1 & 1 & 1 \\ 1 & 1 & 1 & 1 & 1 \\ 1 & 0 & 1 & 1 & 1 \\ 1 & 0 & 1 & 1 & 1 \\ 1 & 0 & 1 & 1 & 1 \end{bmatrix} S(e_1[0.2, 0.4) = \begin{bmatrix} 1 & 1 & 1 & 1 & 1 \\ 1 & 1 & 1 & 1 & 1 \\ 1 & 1 & 1 & 1 & 1 \\ 1 & 1 & 1 & 1 & 1 \\ 0 & 0 & 0 & 0 & 1 \end{bmatrix}$$

$$G(e_2, [0.3, 0.6]) = \begin{bmatrix} 1 & 1 & 1 & 0 & 1 \\ 0 & 1 & 1 & 0 & 0 \\ 0 & 1 & 1 & 0 & 0 \\ 1 & 1 & 1 & 1 & 1 \\ 1 & 1 & 1 & 1 & 1 \end{bmatrix} S(e_2[0.2, 0.4) = \begin{bmatrix} 1 & 1 & 1 & 1 & 1 \\ 1 & 1 & 1 & 1 & 1 \\ 1 & 1 & 1 & 1 & 1 \\ 1 & 1 & 1 & 1 & 1 \\ 1 & 1 & 1 & 1 & 1 \end{bmatrix}$$

$$G(e_3, [0.3, 0.6]) = \begin{bmatrix} 1 & 0 & 0 & 1 & 0 \\ 1 & 1 & 0 & 1 & 0 \\ 1 & 0 & 1 & 1 & 1 \\ 1 & 0 & 0 & 1 & 0 \\ 1 & 0 & 1 & 1 & 1 \end{bmatrix} S(e_3, [0.2, 0.4) = \begin{bmatrix} 1 & 1 & 0 & 0 & 0 \\ 1 & 1 & 0 & 0 & 0 \\ 1 & 1 & 1 & 1 & 1 \\ 1 & 1 & 1 & 1 & 1 \\ 1 & 1 & 1 & 1 & 1 \end{bmatrix}$$

$$G(e_4, [0.3, 0.6]) = \begin{bmatrix} 1 & 0 & 1 & 1 & 0 \\ 0 & 1 & 1 & 1 & 0 \\ 0 & 0 & 1 & 1 & 0 \\ 0 & 0 & 1 & 1 & 0 \\ 0 & 1 & 1 & 1 & 1 \end{bmatrix} S(e_4, [0.2, 0.4) = \begin{bmatrix} 1 & 1 & 1 & 1 & 1 \\ 1 & 1 & 1 & 1 & 1 \\ 0 & 0 & 1 & 0 & 1 \\ 1 & 1 & 1 & 1 & 1 \\ 0 & 0 & 1 & 0 & 1 \end{bmatrix}$$

$$E^{U}(e_1, [0.3, 0.6]) = \begin{bmatrix} 1 & 0 & 1 & 1 & 1 \\ 0 & 1 & 0 & 0 & 0 \\ 1 & 0 & 1 & 1 & 1 \\ 1 & 0 & 1 & 1 & 1 \\ 1 & 0 & 1 & 1 & 1 \end{bmatrix} E^{L}(e_1, [0.2, 0.4) = \begin{bmatrix} 1 & 1 & 1 & 1 & 0 \\ 1 & 1 & 1 & 1 & 0 \\ 1 & 1 & 1 & 1 & 0 \\ 1 & 1 & 1 & 1 & 0 \\ 0 & 0 & 0 & 0 & 1 \end{bmatrix}$$

$$E^U(e_2,[0.3,0.6]) = \begin{bmatrix} 1 & 0 & 0 & 0 & 1 \\ 0 & 1 & 1 & 0 & 0 \\ 0 & 1 & 1 & 0 & 0 \\ 0 & 0 & 0 & 0 & 1 \\ 1 & 0 & 0 & 1 & 1 \end{bmatrix} E^L(e_2,[0.2,0.4) = \begin{bmatrix} 1 & 1 & 1 & 1 & 1 \\ 1 & 1 & 1 & 1 & 1 \\ 1 & 1 & 1 & 1 & 1 \\ 1 & 1 & 1 & 1 & 1 \\ 1 & 1 & 1 & 1 & 1 \end{bmatrix}$$

$$E^U(e_3,[0.3,0.6]) = \begin{bmatrix} 1 & 0 & 0 & 1 & 0 \\ 0 & 1 & 0 & 0 & 0 \\ 0 & 0 & 1 & 0 & 1 \\ 1 & 0 & 0 & 1 & 0 \\ 0 & 0 & 1 & 0 & 1 \end{bmatrix} E^L(e_3,[0.2,0.4) = \begin{bmatrix} 1 & 1 & 0 & 0 & 0 \\ 1 & 1 & 0 & 0 & 0 \\ 0 & 0 & 1 & 1 & 1 \\ 0 & 0 & 1 & 1 & 1 \\ 0 & 0 & 1 & 1 & 1 \end{bmatrix}$$

$$E^U(e_4,[0.3,0.6]) = \begin{bmatrix} 1 & 0 & 0 & 0 & 0 \\ 0 & 1 & 0 & 0 & 0 \\ 0 & 0 & 1 & 1 & 0 \\ 0 & 0 & 1 & 1 & 0 \\ 0 & 0 & 0 & 0 & 1 \end{bmatrix} E^L(e_4,[0.2,0.4) = \begin{bmatrix} 1 & 1 & 0 & 1 & 0 \\ 1 & 1 & 0 & 1 & 0 \\ 0 & 0 & 1 & 0 & 1 \\ 1 & 1 & 0 & 1 & 0 \\ 0 & 0 & 1 & 0 & 1 \end{bmatrix}$$

Step 6. By using Definition (2), we have,

$$S_1 = r_1(e_1;[0.3,0.6],[0.2,0.4]) + r_1(e_2,[0.3,0.6],[0.2,0.4]) + r_1(e_3,[0.3,0.6],[0.2,0.4]) \\ +r_1(e_4,[0.3,0.6],[0.2,0.4] = 1.$$

Similarly, $S_2 = 5$, $S_3 = 0$, $S_4 = -1$, $S_6 = 5$.

Step 7. Then, the ordering is obtained as below

$$o_2 \backsimeq o_5 \succeq o_1 \succeq o_3 \succeq o_4$$

Steps 8 and 9. Accordingly, o_2 and o_5 can be the best objects (Acceptance region), while o_4 not be selected(Rejection region), and o_1, o_3 cannot be judged(Boundary region).

4.1. Comparison with Existing Methods

In this section, we will apply and compare present method and other methods [25,43,44] using real-life example via datasets given in [47] Table 8 from the www.weather.com.cn website. (accessed on 15 May 2021).

Example 2. *Let an IFVSs f_E describes a family who wants to go to a city in China. Suppose that the weather provides a forecast for fifteen cities in China during the holiday, $X = \{o_1, \ldots, o_{15}\}$, which is shown in Table 6. Suppose that the data of weather forecast describes five parameters $E = \{e_1, e_2, e_3, e_4, e_5\}$. Parameters $e_t, t = 1, \ldots, 5$, stand for "temperature", "air quality index", "levels of ultraviolet radiation", "wind speed", "precipitation", respectively.*

Step 1. The *IVFSs* f_E is given in Table 6.
Step 2. Suppose that
$\alpha = [0.67, 0.92], [0.75, 0.94], [0.66, 0.92], [0.49, 0.75], [0.96, 0.99]$
$\beta = [0.14, 0.8], [0.37, 0.77], [0.25, 0.76], [0.26, 0.76], [0.67, 1],$
where $\alpha = [\alpha_1, \alpha_2], \beta = [\beta_1, \beta_2]$

Table 6. Table for f_E.

f_E	e_1	e_2	e_3	e_4	e_5
o_1	[0.14, 0.86]	[0.21, 0.97]	[0.0, 0.47]	[0.25, 1.0]	[1.0, 1.0]
o_2	[0.43 ,0.82]	[0.45, 0.78]	[0.0, 0.33]	[0.25, 1.0]	[0.83, 1]
o_3	[0.64, 1.0]	[0.26, 0.63]	[0.0, 0.73]	[0.5, 1.0]	[1.0, 1.0]
o_4	[0.5, 0.82]	[0.45, 0.82]	[0.6, 0.93]	[0.0, 1.0]	[0.97, 1]
o_5	[0.39, 0.68]	[0.79, 0.88]	[0.67, 1.0]	[0.25, 1.0]	[0.83, 1]
o_6	[0.68, 0.93]	[0.6, 0.77]	[0.6, 0.93]	[0.5, 1.0]	[0.58, 1]
o_7	[0.36, 0.71]	[0.37, 0.96]	[0.67, 0.93]	[0.0, 0.75]	[0.96, 1]
o_8	[0.5, 0.89]	[0.76, 0.95]	[0.67, 1.0]	[0.5, 1.0]	[0.89, 1]
o_9	[0.25, 0.71]	[0.02, 1.0]	[0.67, 1.0]	[0.0, 0.75]	[0.58, 1]
o_{10}	[0.0, 0.71]	[0.53, 0.92]	[0.6, 0.93]	[0.5, 0.75]	[1.0, 1.0]
o_{11}	[0.0, 0.54]	[0.58, 1.0]	[0.73, 1.0]	[0.0, 0.75]	[0.67, 1]
o_{12}	[0.34, 0.89]	[0.0, 1.0]	[0.67, 1.0]	[0.25, 0.75]	[1.0, 1.0]
o_{13}	[0.25, 0.71]	[0.58, 1.0]	[0.73, 1.0]	[0.0, 0.75]	[0.67, 1]
o_{14}	[0.34, 0.89]	[0.53, 0.95]	[0.67, 0.93]	[0.25, 0.75]	[1.0, 1.0]
o_{15}	[0.25, 0.71]	[0.66, 0.97]	[0.6, 0.93]	[0.25, 1.0]	[0.0, 1.0]

Steps 3 and 4. The α-Upper-e_t Crisp and β-Lower-e_t Crisp; the α-Upper-e_t Topology and β-Lower-e_t Topology (where $(t = 1, \ldots, 5)$) as shown in Tables 7–10.

Table 7. α-Upper-e_t; $t = 1, \ldots, 5$.

	Upper-e_t Crisp														
x	x_1	x_2	x_3	x_4	x_5	x_6	x_7	x_8	x_9	x_{10}	x_{11}	x_{12}	x_{13}	x_{14}	x_{15}
e_1	[0	0	0	0	0	1	0	0	0	0	0	0	0	0	0]
e_2	[0	0	0	0	0	0	0	1	0	0	0	0	0	0	0]
e_3	[0	0	0	0	1	1	1	1	1	0	1	1	1	1	0]
e_4	[0	0	1	0	0	0	0	1	0	1	0	0	0	0	0]
e_5	[1	0	1	1	0	0	0	0	0	1	0	1	0	1	0]

Table 8. β-Lower-e_t; $t = 1, \ldots, 5$.

	Lower-e_t Crisp														
x	x_1	x_2	x_3	x_4	x_5	x_6	x_7	x_8	x_9	x_{10}	x_{11}	x_{12}	x_{13}	x_{14}	x_{15}
e_1	[0	0	0	0	0	0	0	0	0	1	1	0	0	0	0]
e_2	[0	0	1	0	0	0	0	0	0	0	0	0	0	0	0]
e_3	[1	1	1	0	0	0	0	0	0	0	0	0	0	0	0]
e_4	[0	0	0	0	0	0	1	0	1	0	1	1	1	1	0]
e_5	[0	0	0	0	0	1	0	0	1	0	0	0	0	0	0]

Table 9. α-Upper-e_t topology; $t = 1, \ldots, 5$.

		$\tau^u_{e_t,\alpha}$															
e_1	$\{[0]_{1\times15}$,	$[1]_{1\times15}$,	[0	0	0	0	0	1	0	0	0	0	0	0	0	0	0]}
e_2	$\{[0]_{1\times15}$,	$[1]_{1\times15}$,	[0	0	0	0	0	0	0	1	0	0	0	0	0	0	0]}
e_3	$\{[0]_{1\times15}$,	$[1]_{1\times15}$,	[0	0	0	0	1	1	1	1	1	0	1	1	1	1	0]}
e_4	$\{[0]_{1\times15}$,	$[1]_{1\times15}$,	[0	0	1	0	0	0	0	1	0	1	0	0	0	0	0]}
e_5	$\{[0]_{1\times15}$,	$[1]_{1\times15}$,	[1	0	1	1	0	0	0	0	0	1	0	1	0	1	0]}

Table 10. β-Lower-e_t topology; $t = 1, \ldots, 5$.

	$\tau^{l}_{e_t,\beta}$
e_1	$\{[0]_{1\times15}, [1]_{1\times15}, [0\ 0\ 0\ 0\ 0\ 0\ 0\ 0\ 0\ 1\ 1\ 0\ 0\ 0\ 0]\}$
e_2	$\{[0]_{1\times15}, [1]_{1\times15}, [0\ 0\ 1\ 0\ 0\ 0\ 0\ 0\ 0\ 0\ 0\ 0\ 0\ 0\ 0]\}$
e_3	$\{[0]_{1\times15}, [1]_{1\times15}, [1\ 1\ 1\ 0\ 0\ 0\ 0\ 0\ 0\ 0\ 0\ 0\ 0\ 0\ 0]\}$
e_4	$\{[0]_{1\times15}, [1]_{1\times15}, [0\ 0\ 0\ 0\ 0\ 0\ 1\ 0\ 1\ 0\ 1\ 1\ 1\ 1\ 0]\}$
e_5	$\{[0]_{1\times15}, [1]_{1\times15}, [0\ 0\ 0\ 0\ 0\ 1\ 0\ 0\ 1\ 0\ 0\ 0\ 0\ 0\ 0]\}$

Step 5. The comparison matrices $G(e_t, \alpha), S(e_t, \beta), E^U(e_t, \alpha)$ and $E^L(e_t, \beta)$, where $\alpha = [\alpha_1, \alpha_2], \beta = [\beta_1, \beta_2], t = 1, \ldots, 5$ are below:

$G(e_1, [0.67, 0.92])$ and $L(e_1, [0.13, 0.8])$

$$\begin{bmatrix}
1&1&1&1&1&0&1&1&1&1&1&1&1&1&1\\
1&1&1&1&1&0&1&1&1&1&1&1&1&1&1\\
1&1&1&1&1&0&1&1&1&1&1&1&1&1&1\\
1&1&1&1&1&0&1&1&1&1&1&1&1&1&1\\
1&1&1&1&1&0&1&1&1&1&1&1&1&1&1\\
1&1&1&1&1&1&1&1&1&1&1&1&1&1&1\\
1&1&1&1&1&0&1&1&1&1&1&1&1&1&1\\
1&1&1&1&1&0&1&1&1&1&1&1&1&1&1\\
1&1&1&1&1&0&1&1&1&1&1&1&1&1&1\\
1&1&1&1&1&0&1&1&1&1&1&1&1&1&1\\
1&1&1&1&1&0&1&1&1&1&1&1&1&1&1\\
1&1&1&1&1&0&1&1&1&1&1&1&1&1&1\\
1&1&1&1&1&0&1&1&1&1&1&1&1&1&1\\
1&1&1&1&1&0&1&1&1&1&1&1&1&1&1\\
1&1&1&1&1&0&1&1&1&1&1&1&1&1&1
\end{bmatrix} \& \begin{bmatrix}
1&1&1&1&1&1&1&1&1&0&0&1&1&1&1\\
1&1&1&1&1&1&1&1&1&0&0&1&1&1&1\\
1&1&1&1&1&1&1&1&1&0&0&1&1&1&1\\
1&1&1&1&1&1&1&1&1&0&0&1&1&1&1\\
1&1&1&1&1&1&1&1&1&0&0&1&1&1&1\\
1&1&1&1&1&1&1&1&1&0&0&1&1&1&1\\
1&1&1&1&1&1&1&1&1&0&0&1&1&1&1\\
1&1&1&1&1&1&1&1&1&0&0&1&1&1&1\\
1&1&1&1&1&1&1&1&1&0&0&1&1&1&1\\
1&1&1&1&1&1&1&1&1&1&1&1&1&1&1\\
1&1&1&1&1&1&1&1&1&1&1&1&1&1&1\\
1&1&1&1&1&1&1&1&1&0&0&1&1&1&1\\
1&1&1&1&1&1&1&1&1&0&0&1&1&1&1\\
1&1&1&1&1&1&1&1&1&0&0&1&1&1&1\\
1&1&1&1&1&1&1&1&1&0&0&1&1&1&1
\end{bmatrix}$$

$G(e_2, [0.75, 0.94])$ and $L(e_2, [0.37, 0.77])$

$$\begin{bmatrix}
1&1&1&1&1&1&1&0&1&1&1&1&1&1&1\\
1&1&1&1&1&1&1&0&1&1&1&1&1&1&1\\
1&1&1&1&1&1&1&0&1&1&1&1&1&1&1\\
1&1&1&1&1&1&1&0&1&1&1&1&1&1&11\\
1&1&1&1&1&1&1&0&1&1&1&1&1&1&1\\
1&1&1&1&1&1&1&0&1&1&1&1&1&1&1\\
1&1&1&1&1&1&1&0&1&1&1&1&1&1&1\\
1&1&1&1&1&1&1&1&1&1&1&1&1&1&1\\
1&1&1&1&1&1&1&0&1&1&1&1&1&1&1\\
1&1&1&1&1&1&1&0&1&1&1&1&1&1&1\\
1&1&1&1&1&1&1&0&1&1&1&1&1&1&1\\
1&1&1&1&1&1&1&0&1&1&1&1&1&1&1\\
1&1&1&1&1&1&1&0&1&1&1&1&1&1&1\\
1&1&1&1&1&1&1&0&1&1&1&1&1&1&1\\
1&1&1&1&1&1&1&0&1&1&1&1&1&1&1
\end{bmatrix} \& \begin{bmatrix}
1&1&0&1&1&1&1&1&1&1&1&1&1&1&1\\
1&1&0&1&1&1&1&1&1&1&1&1&1&1&1\\
1&1&1&1&1&1&1&1&1&1&1&1&1&1&1\\
1&1&0&1&1&1&1&1&1&1&1&1&1&1&1\\
1&1&0&1&1&1&1&1&1&1&1&1&1&1&1\\
1&1&0&1&1&1&1&1&1&1&1&1&1&1&1\\
1&1&0&1&1&1&1&1&1&1&1&1&1&1&1\\
1&1&0&1&1&1&1&1&1&1&1&1&1&1&1\\
1&1&0&1&1&1&1&1&1&1&1&1&1&1&1\\
1&1&0&1&1&1&1&1&1&1&1&1&1&1&1\\
1&1&0&1&1&1&1&1&1&1&1&1&1&1&1\\
1&1&0&1&1&1&1&1&1&1&1&1&1&1&1\\
1&1&0&1&1&1&1&1&1&1&1&1&1&1&1\\
1&1&0&1&1&1&1&1&1&1&1&1&1&1&1\\
1&1&0&1&1&1&1&1&1&1&1&1&1&1&1
\end{bmatrix}$$

$G(e_3, [0.66, 0.92])$ and $L(e_3, [0.25, 0.76])$

$$\begin{bmatrix} 1 & 1 & 1 & 1 & 0 & 0 & 0 & 0 & 0 & 1 & 0 & 0 & 0 & 0 & 1 \\ 1 & 1 & 1 & 1 & 0 & 0 & 0 & 0 & 0 & 1 & 0 & 0 & 0 & 0 & 1 \\ 1 & 1 & 1 & 1 & 0 & 0 & 0 & 0 & 0 & 1 & 0 & 0 & 0 & 0 & 1 \\ 1 & 1 & 1 & 1 & 0 & 0 & 0 & 0 & 0 & 1 & 0 & 0 & 0 & 0 & 1 \\ 1 & 1 & 1 & 1 & 1 & 1 & 1 & 1 & 1 & 1 & 1 & 1 & 1 & 1 & 1 \\ 1 & 1 & 1 & 1 & 1 & 1 & 1 & 1 & 1 & 1 & 1 & 1 & 1 & 1 & 1 \\ 1 & 1 & 1 & 1 & 1 & 1 & 1 & 1 & 1 & 1 & 1 & 1 & 1 & 1 & 1 \\ 1 & 1 & 1 & 1 & 1 & 1 & 1 & 1 & 1 & 1 & 1 & 1 & 1 & 1 & 1 \\ 1 & 1 & 1 & 1 & 1 & 1 & 1 & 1 & 1 & 1 & 1 & 1 & 1 & 1 & 1 \\ 1 & 1 & 1 & 1 & 0 & 0 & 0 & 0 & 0 & 1 & 0 & 0 & 0 & 0 & 1 \\ 1 & 1 & 1 & 1 & 1 & 1 & 1 & 1 & 1 & 1 & 1 & 1 & 1 & 1 & 1 \\ 1 & 1 & 1 & 1 & 1 & 1 & 1 & 1 & 1 & 1 & 1 & 1 & 1 & 1 & 1 \\ 1 & 1 & 1 & 1 & 1 & 1 & 1 & 1 & 1 & 1 & 1 & 1 & 1 & 1 & 1 \\ 1 & 1 & 1 & 1 & 1 & 1 & 1 & 1 & 1 & 1 & 1 & 1 & 1 & 1 & 1 \\ 1 & 1 & 1 & 1 & 0 & 0 & 0 & 0 & 0 & 1 & 0 & 0 & 0 & 0 & 1 \end{bmatrix} \& \begin{bmatrix} 1 & 1 & 1 & 1 & 1 & 1 & 1 & 1 & 1 & 1 & 1 & 1 & 1 & 1 & 1 \\ 1 & 1 & 1 & 1 & 1 & 1 & 1 & 1 & 1 & 1 & 1 & 1 & 1 & 1 & 1 \\ 1 & 1 & 1 & 1 & 1 & 1 & 1 & 1 & 1 & 1 & 1 & 1 & 1 & 1 & 1 \\ 0 & 0 & 0 & 1 & 1 & 1 & 1 & 1 & 1 & 1 & 1 & 1 & 1 & 1 & 1 \\ 0 & 0 & 0 & 1 & 1 & 1 & 1 & 1 & 1 & 1 & 1 & 1 & 1 & 1 & 1 \\ 0 & 0 & 0 & 1 & 1 & 1 & 1 & 1 & 1 & 1 & 1 & 1 & 1 & 1 & 1 \\ 0 & 0 & 0 & 1 & 1 & 1 & 1 & 1 & 1 & 1 & 1 & 1 & 1 & 1 & 1 \\ 0 & 0 & 0 & 1 & 1 & 1 & 1 & 1 & 1 & 1 & 1 & 1 & 1 & 1 & 1 \\ 0 & 0 & 0 & 1 & 1 & 1 & 1 & 1 & 1 & 1 & 1 & 1 & 1 & 1 & 1 \\ 0 & 0 & 0 & 1 & 1 & 1 & 1 & 1 & 1 & 1 & 1 & 1 & 1 & 1 & 1 \\ 0 & 0 & 0 & 1 & 1 & 1 & 1 & 1 & 1 & 1 & 1 & 1 & 1 & 1 & 1 \\ 0 & 0 & 0 & 1 & 1 & 1 & 1 & 1 & 1 & 1 & 1 & 1 & 1 & 1 & 1 \\ 0 & 0 & 0 & 1 & 1 & 1 & 1 & 1 & 1 & 1 & 1 & 1 & 1 & 1 & 1 \\ 0 & 0 & 0 & 1 & 1 & 1 & 1 & 1 & 1 & 1 & 1 & 1 & 1 & 1 & 1 \\ 0 & 0 & 0 & 1 & 1 & 1 & 1 & 1 & 1 & 1 & 1 & 1 & 1 & 1 & 1 \end{bmatrix}$$

$G(e_4, [0.48, 0.74])$ and $L(e_4, [0.26, 0.76])$

$$\begin{bmatrix} 1 & 1 & 0 & 1 & 1 & 1 & 1 & 0 & 1 & 0 & 1 & 1 & 1 & 1 & 1 \\ 1 & 1 & 0 & 1 & 1 & 1 & 1 & 0 & 1 & 0 & 1 & 1 & 1 & 1 & 1 \\ 1 & 1 & 1 & 1 & 1 & 1 & 1 & 1 & 1 & 1 & 1 & 1 & 1 & 1 & 1 \\ 1 & 1 & 0 & 1 & 1 & 1 & 1 & 0 & 1 & 0 & 1 & 1 & 1 & 1 & 1 \\ 1 & 1 & 0 & 1 & 1 & 1 & 1 & 0 & 1 & 0 & 1 & 1 & 1 & 1 & 1 \\ 1 & 1 & 0 & 1 & 1 & 1 & 1 & 0 & 1 & 0 & 1 & 1 & 1 & 1 & 1 \\ 1 & 1 & 0 & 1 & 1 & 1 & 1 & 0 & 1 & 0 & 1 & 1 & 1 & 1 & 1 \\ 1 & 1 & 1 & 1 & 1 & 1 & 1 & 1 & 1 & 1 & 1 & 1 & 1 & 1 & 1 \\ 1 & 1 & 0 & 1 & 1 & 1 & 1 & 0 & 1 & 0 & 1 & 1 & 1 & 1 & 1 \\ 1 & 1 & 1 & 1 & 1 & 1 & 1 & 1 & 1 & 1 & 1 & 1 & 1 & 1 & 1 \\ 1 & 1 & 0 & 1 & 1 & 1 & 1 & 0 & 1 & 0 & 1 & 1 & 1 & 1 & 1 \\ 1 & 1 & 0 & 1 & 1 & 1 & 1 & 0 & 1 & 0 & 1 & 1 & 1 & 1 & 1 \\ 1 & 1 & 0 & 1 & 1 & 1 & 1 & 0 & 1 & 0 & 1 & 1 & 1 & 1 & 1 \\ 1 & 1 & 0 & 1 & 1 & 1 & 1 & 0 & 1 & 0 & 1 & 1 & 1 & 1 & 1 \\ 1 & 1 & 0 & 1 & 1 & 1 & 1 & 0 & 1 & 0 & 1 & 1 & 1 & 1 & 1 \end{bmatrix} \& \begin{bmatrix} 1 & 1 & 1 & 1 & 1 & 1 & 0 & 1 & 0 & 1 & 0 & 0 & 0 & 0 & 1 \\ 1 & 1 & 1 & 1 & 1 & 1 & 0 & 1 & 0 & 1 & 0 & 0 & 0 & 0 & 1 \\ 1 & 1 & 1 & 1 & 1 & 1 & 0 & 1 & 0 & 1 & 0 & 0 & 0 & 0 & 1 \\ 1 & 1 & 1 & 1 & 1 & 1 & 0 & 1 & 0 & 1 & 0 & 0 & 0 & 0 & 1 \\ 1 & 1 & 1 & 1 & 1 & 1 & 0 & 1 & 0 & 1 & 0 & 0 & 0 & 0 & 1 \\ 1 & 1 & 1 & 1 & 1 & 1 & 0 & 1 & 0 & 1 & 0 & 0 & 0 & 0 & 1 \\ 1 & 1 & 1 & 1 & 1 & 1 & 1 & 1 & 1 & 1 & 1 & 1 & 1 & 1 & 1 \\ 1 & 1 & 1 & 1 & 1 & 1 & 0 & 1 & 0 & 1 & 0 & 0 & 0 & 0 & 1 \\ 1 & 1 & 1 & 1 & 1 & 1 & 1 & 1 & 1 & 1 & 1 & 1 & 1 & 1 & 1 \\ 1 & 1 & 1 & 1 & 1 & 1 & 0 & 1 & 0 & 1 & 0 & 0 & 0 & 0 & 1 \\ 1 & 1 & 1 & 1 & 1 & 1 & 1 & 1 & 1 & 1 & 1 & 1 & 1 & 1 & 1 \\ 1 & 1 & 1 & 1 & 1 & 1 & 1 & 1 & 1 & 1 & 1 & 1 & 1 & 1 & 1 \\ 1 & 1 & 1 & 1 & 1 & 1 & 1 & 1 & 1 & 1 & 1 & 1 & 1 & 1 & 1 \\ 1 & 1 & 1 & 1 & 1 & 1 & 1 & 1 & 1 & 1 & 1 & 1 & 1 & 1 & 1 \\ 1 & 1 & 1 & 1 & 1 & 1 & 0 & 1 & 0 & 1 & 0 & 0 & 0 & 0 & 1 \end{bmatrix}$$

$G(e_5, [0.96, 0.99])$ and $L(e_5, [0.67, 1])$

$$\begin{bmatrix} 1 & 1 & 1 & 1 & 1 & 1 & 1 & 1 & 1 & 1 & 1 & 1 & 1 & 1 & 1 \\ 0 & 1 & 0 & 0 & 1 & 1 & 1 & 1 & 1 & 0 & 1 & 0 & 1 & 0 & 1 \\ 1 & 1 & 1 & 1 & 1 & 1 & 1 & 1 & 1 & 1 & 1 & 1 & 1 & 1 & 1 \\ 1 & 1 & 1 & 1 & 1 & 1 & 1 & 1 & 1 & 1 & 1 & 1 & 1 & 1 & 1 \\ 0 & 1 & 0 & 0 & 1 & 1 & 1 & 1 & 1 & 0 & 1 & 0 & 1 & 0 & 1 \\ 0 & 1 & 0 & 0 & 1 & 1 & 1 & 1 & 1 & 0 & 1 & 0 & 1 & 0 & 1 \\ 0 & 1 & 0 & 0 & 1 & 1 & 1 & 1 & 1 & 0 & 1 & 0 & 1 & 0 & 1 \\ 0 & 1 & 0 & 0 & 1 & 1 & 1 & 1 & 1 & 0 & 1 & 0 & 1 & 0 & 1 \\ 0 & 1 & 0 & 0 & 1 & 1 & 1 & 1 & 1 & 0 & 1 & 0 & 1 & 0 & 1 \\ 1 & 1 & 1 & 1 & 1 & 1 & 1 & 1 & 1 & 1 & 1 & 1 & 1 & 1 & 1 \\ 0 & 1 & 0 & 0 & 1 & 1 & 1 & 1 & 1 & 0 & 1 & 0 & 1 & 0 & 1 \\ 1 & 1 & 1 & 1 & 1 & 1 & 1 & 1 & 1 & 1 & 1 & 1 & 1 & 1 & 1 \\ 0 & 1 & 0 & 0 & 1 & 1 & 1 & 1 & 1 & 0 & 1 & 0 & 1 & 0 & 1 \\ 1 & 1 & 1 & 1 & 1 & 1 & 1 & 1 & 1 & 1 & 1 & 1 & 1 & 1 & 1 \\ 0 & 1 & 0 & 0 & 1 & 1 & 1 & 1 & 1 & 0 & 1 & 0 & 1 & 0 & 1 \end{bmatrix} \& \begin{bmatrix} 1 & 1 & 1 & 1 & 1 & 0 & 1 & 1 & 0 & 1 & 1 & 1 & 1 & 1 & 1 \\ 1 & 1 & 1 & 1 & 1 & 0 & 1 & 1 & 0 & 1 & 1 & 1 & 1 & 1 & 1 \\ 1 & 1 & 1 & 1 & 1 & 0 & 1 & 1 & 0 & 1 & 1 & 1 & 1 & 1 & 1 \\ 1 & 1 & 1 & 1 & 1 & 0 & 1 & 1 & 0 & 1 & 1 & 1 & 1 & 1 & 1 \\ 1 & 1 & 1 & 1 & 1 & 0 & 1 & 1 & 0 & 1 & 1 & 1 & 1 & 1 & 1 \\ 1 & 1 & 1 & 1 & 1 & 1 & 1 & 1 & 1 & 1 & 1 & 1 & 1 & 1 & 1 \\ 1 & 1 & 1 & 1 & 1 & 0 & 1 & 1 & 0 & 1 & 1 & 1 & 1 & 1 & 1 \\ 1 & 1 & 1 & 1 & 1 & 0 & 1 & 1 & 0 & 1 & 1 & 1 & 1 & 1 & 1 \\ 1 & 1 & 1 & 1 & 1 & 1 & 1 & 1 & 1 & 1 & 1 & 1 & 1 & 1 & 1 \\ 1 & 1 & 1 & 1 & 1 & 0 & 1 & 1 & 0 & 1 & 1 & 1 & 1 & 1 & 1 \\ 1 & 1 & 1 & 1 & 1 & 0 & 1 & 1 & 0 & 1 & 1 & 1 & 1 & 1 & 1 \\ 1 & 1 & 1 & 1 & 1 & 0 & 1 & 1 & 0 & 1 & 1 & 1 & 1 & 1 & 1 \\ 1 & 1 & 1 & 1 & 1 & 0 & 1 & 1 & 0 & 1 & 1 & 1 & 1 & 1 & 1 \\ 1 & 1 & 1 & 1 & 1 & 0 & 1 & 1 & 0 & 1 & 1 & 1 & 1 & 1 & 1 \\ 1 & 1 & 1 & 1 & 1 & 0 & 1 & 1 & 0 & 1 & 1 & 1 & 1 & 1 & 1 \end{bmatrix}$$

Now, we compute matrices $E^U(e_t, \alpha), E^L(e_t, \beta) \alpha = [\alpha_1, \alpha_2], \beta = [\beta_1, \beta_2], t = 1, \ldots, 5$
$E^U(e_1, [0.67, 0.92])$ and $E^L(e_1, [0.13, 0.8])$

$$\begin{bmatrix} 1&1&1&1&1&0&1&1&1&1&1&1&1&1&1\\ 1&1&1&1&1&0&1&1&1&1&1&1&1&1&1\\ 1&1&1&1&1&0&1&1&1&1&1&1&1&1&1\\ 1&1&1&1&1&0&1&1&1&1&1&1&1&1&1\\ 1&1&1&1&1&0&1&1&1&1&1&1&1&1&1\\ 0&0&0&0&0&1&0&0&0&0&0&0&0&0&0\\ 1&1&1&1&1&0&1&1&1&1&1&1&1&1&1\\ 1&1&1&1&1&0&1&1&1&1&1&1&1&1&1\\ 1&1&1&1&1&0&1&1&1&1&1&1&1&1&1\\ 1&1&1&1&1&0&1&1&1&1&1&1&1&1&1\\ 1&1&1&1&1&0&1&1&1&1&1&1&1&1&1\\ 1&1&1&1&1&0&1&1&1&1&1&1&1&1&1\\ 1&1&1&1&1&0&1&1&1&1&1&1&1&1&1\\ 1&1&1&1&1&0&1&1&1&1&1&1&1&1&1\\ 1&1&1&1&1&0&1&1&1&1&1&1&1&1&1 \end{bmatrix} \& \begin{bmatrix} 1&1&1&1&1&1&1&1&1&0&0&1&1&1&1\\ 1&1&1&1&1&1&1&1&1&0&0&1&1&1&1\\ 1&1&1&1&1&1&1&1&1&0&0&1&1&1&1\\ 1&1&1&1&1&1&1&1&1&0&0&1&1&1&1\\ 1&1&1&1&1&1&1&1&1&0&0&1&1&1&1\\ 1&1&1&1&1&1&1&1&1&0&0&1&1&1&1\\ 1&1&1&1&1&1&1&1&1&0&0&1&1&1&1\\ 1&1&1&1&1&1&1&1&1&0&0&1&1&1&1\\ 1&1&1&1&1&1&1&1&1&0&0&1&1&1&1\\ 0&0&0&0&0&0&0&0&0&1&1&0&0&0&0\\ 0&0&0&0&0&0&0&0&0&1&1&0&0&0&0\\ 1&1&1&1&1&1&1&1&1&0&0&1&1&1&1\\ 1&1&1&1&1&1&1&1&1&0&0&1&1&1&1\\ 1&1&1&1&1&1&1&1&1&0&0&1&1&1&1\\ 1&1&1&1&1&1&1&1&1&0&0&1&1&1&1 \end{bmatrix}$$

$E^U(e_2,[0.75,0.94])$ and $E^L(e_2,[0.37,0.77])$

$$\begin{bmatrix} 1&1&1&1&1&1&1&0&1&1&1&1&1&1&1\\ 1&1&1&1&1&1&1&0&1&1&1&1&1&1&1\\ 1&1&1&1&1&1&1&0&1&1&1&1&1&1&1\\ 1&1&1&1&1&1&1&0&1&1&1&1&1&1&11\\ 1&1&1&1&1&1&1&0&1&1&1&1&1&1&1\\ 1&1&1&1&1&1&1&0&1&1&1&1&1&1&1\\ 1&1&1&1&1&1&1&0&1&1&1&1&1&1&1\\ 0&0&0&0&0&0&0&1&0&0&0&0&0&0&0\\ 1&1&1&1&1&1&1&0&1&1&1&1&1&1&1\\ 1&1&1&1&1&1&1&0&1&1&1&1&1&1&1\\ 1&1&1&1&1&1&1&0&1&1&1&1&1&1&1\\ 1&1&1&1&1&1&1&0&1&1&1&1&1&1&1\\ 1&1&1&1&1&1&1&0&1&1&1&1&1&1&1\\ 1&1&1&1&1&1&1&0&1&1&1&1&1&1&1\\ 1&1&1&1&1&1&1&0&1&1&1&1&1&1&1 \end{bmatrix} \& \begin{bmatrix} 1&1&0&1&1&1&1&1&1&1&1&1&1&1&1\\ 1&1&0&1&1&1&1&1&1&1&1&1&1&1&1\\ 0&0&1&0&0&0&0&0&0&0&0&0&0&0&0\\ 1&1&0&1&1&1&1&1&1&1&1&1&1&1&1\\ 1&1&0&1&1&1&1&1&1&1&1&1&1&1&1\\ 1&1&0&1&1&1&1&1&1&1&1&1&1&1&1\\ 1&1&0&1&1&1&1&1&1&1&1&1&1&1&1\\ 1&1&0&1&1&1&1&1&1&1&1&1&1&1&1\\ 1&1&0&1&1&1&1&1&1&1&1&1&1&1&1\\ 1&1&0&1&1&1&1&1&1&1&1&1&1&1&1\\ 1&1&0&1&1&1&1&1&1&1&1&1&1&1&1\\ 1&1&0&1&1&1&1&1&1&1&1&1&1&1&1\\ 1&1&0&1&1&1&1&1&1&1&1&1&1&1&1\\ 1&1&0&1&1&1&1&1&1&1&1&1&1&1&1\\ 1&1&0&1&1&1&1&1&1&1&1&1&1&1&1 \end{bmatrix}$$

$E^U(e_3,[0.66,0.92])$ and $E^L(e_3,[0.25,0.76])$

$$\begin{bmatrix} 1&1&1&1&0&0&0&0&0&1&0&0&0&0&1\\ 1&1&1&1&0&0&0&0&0&1&0&0&0&0&1\\ 1&1&1&1&0&0&0&0&0&1&0&0&0&0&1\\ 1&1&1&1&0&0&0&0&0&1&0&0&0&0&1\\ 0&0&0&0&1&1&1&1&1&0&1&1&1&1&0\\ 0&0&0&0&1&1&1&1&1&0&1&1&1&1&0\\ 0&0&0&0&1&1&1&1&1&0&1&1&1&1&0\\ 0&0&0&0&1&1&1&1&1&0&1&1&1&1&0\\ 0&0&0&0&1&1&1&1&1&0&1&1&1&1&0\\ 1&1&1&1&0&0&0&0&0&1&0&0&0&0&1\\ 0&0&0&0&1&1&1&1&1&0&1&1&1&1&0\\ 0&0&0&0&1&1&1&1&1&0&1&1&1&1&0\\ 0&0&0&0&1&1&1&1&1&0&1&1&1&1&0\\ 0&0&0&0&1&1&1&1&1&0&1&1&1&1&0\\ 1&1&1&1&0&0&0&0&0&1&0&0&0&0&1 \end{bmatrix} \& \begin{bmatrix} 1&1&1&0&0&0&0&0&0&0&0&0&0&0&0\\ 1&1&1&0&0&0&0&0&0&0&0&0&0&0&0\\ 1&1&1&0&0&0&0&0&0&0&0&0&0&0&0\\ 0&0&0&1&1&1&1&1&1&1&1&1&1&1&1\\ 0&0&0&1&1&1&1&1&1&1&1&1&1&1&1\\ 0&0&0&1&1&1&1&1&1&1&1&1&1&1&1\\ 0&0&0&1&1&1&1&1&1&1&1&1&1&1&1\\ 0&0&0&1&1&1&1&1&1&1&1&1&1&1&1\\ 0&0&0&1&1&1&1&1&1&1&1&1&1&1&1\\ 0&0&0&1&1&1&1&1&1&1&1&1&1&1&1\\ 0&0&0&1&1&1&1&1&1&1&1&1&1&1&1\\ 0&0&0&1&1&1&1&1&1&1&1&1&1&1&1\\ 0&0&0&1&1&1&1&1&1&1&1&1&1&1&1\\ 0&0&0&1&1&1&1&1&1&1&1&1&1&1&1\\ 0&0&0&1&1&1&1&1&1&1&1&1&1&1&1 \end{bmatrix}$$

$E^U(e_4,[0.48,0.74])$ and $E^L(e_4,[0.26,0.76])$

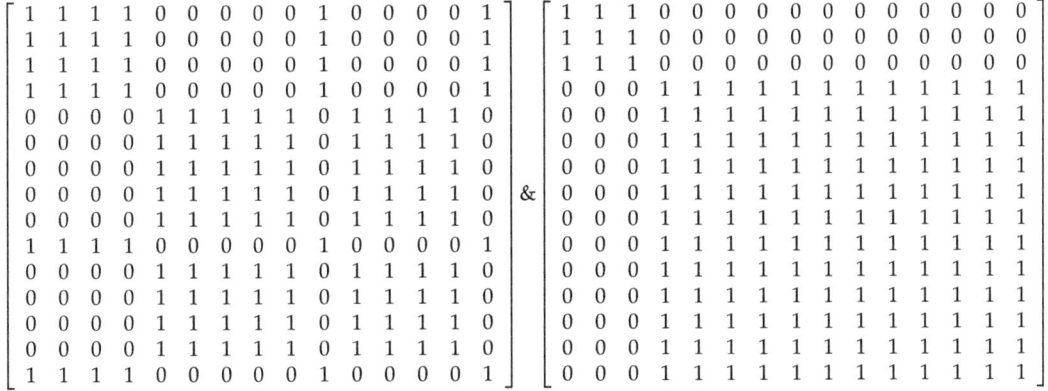

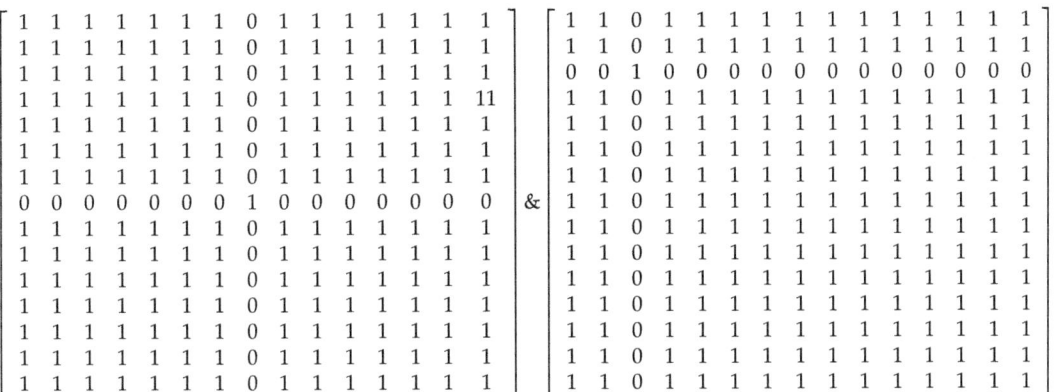

$$
\begin{bmatrix}
1 & 1 & 0 & 1 & 1 & 1 & 1 & 0 & 1 & 0 & 1 & 1 & 1 & 1 & 1 \\
1 & 1 & 0 & 1 & 1 & 1 & 1 & 0 & 1 & 0 & 1 & 1 & 1 & 1 & 1 \\
0 & 0 & 1 & 0 & 0 & 0 & 0 & 1 & 0 & 1 & 0 & 0 & 0 & 0 & 0 \\
1 & 1 & 0 & 1 & 1 & 1 & 1 & 0 & 1 & 0 & 1 & 1 & 1 & 1 & 1 \\
1 & 1 & 0 & 1 & 1 & 1 & 1 & 0 & 1 & 0 & 1 & 1 & 1 & 1 & 1 \\
1 & 1 & 0 & 1 & 1 & 1 & 1 & 0 & 1 & 0 & 1 & 1 & 1 & 1 & 1 \\
1 & 1 & 0 & 1 & 1 & 1 & 1 & 0 & 1 & 0 & 1 & 1 & 1 & 1 & 1 \\
0 & 0 & 1 & 0 & 0 & 0 & 0 & 1 & 0 & 1 & 0 & 0 & 0 & 0 & 0 \\
1 & 1 & 0 & 1 & 1 & 1 & 1 & 0 & 1 & 0 & 1 & 1 & 1 & 1 & 1 \\
0 & 0 & 1 & 0 & 0 & 0 & 0 & 1 & 0 & 1 & 0 & 0 & 0 & 0 & 0 \\
1 & 1 & 0 & 1 & 1 & 1 & 1 & 0 & 1 & 0 & 1 & 1 & 1 & 1 & 1 \\
1 & 1 & 0 & 1 & 1 & 1 & 1 & 0 & 1 & 0 & 1 & 1 & 1 & 1 & 1 \\
1 & 1 & 0 & 1 & 1 & 1 & 1 & 0 & 1 & 0 & 1 & 1 & 1 & 1 & 1 \\
1 & 1 & 0 & 1 & 1 & 1 & 1 & 0 & 1 & 0 & 1 & 1 & 1 & 1 & 1 \\
1 & 1 & 0 & 1 & 1 & 1 & 1 & 0 & 1 & 0 & 1 & 1 & 1 & 1 & 1
\end{bmatrix}
\&
\begin{bmatrix}
1 & 1 & 1 & 1 & 1 & 1 & 0 & 1 & 0 & 1 & 0 & 0 & 0 & 0 & 1 \\
1 & 1 & 1 & 1 & 1 & 1 & 0 & 1 & 0 & 1 & 0 & 0 & 0 & 0 & 1 \\
1 & 1 & 1 & 1 & 1 & 1 & 0 & 1 & 0 & 1 & 0 & 0 & 0 & 0 & 1 \\
1 & 1 & 1 & 1 & 1 & 1 & 0 & 1 & 0 & 1 & 0 & 0 & 0 & 0 & 1 \\
1 & 1 & 1 & 1 & 1 & 1 & 0 & 1 & 0 & 1 & 0 & 0 & 0 & 0 & 1 \\
1 & 1 & 1 & 1 & 1 & 1 & 0 & 1 & 0 & 1 & 0 & 0 & 0 & 0 & 1 \\
0 & 0 & 0 & 0 & 0 & 0 & 1 & 0 & 1 & 0 & 1 & 1 & 1 & 1 & 0 \\
1 & 1 & 1 & 1 & 1 & 1 & 0 & 1 & 0 & 1 & 0 & 0 & 0 & 0 & 1 \\
0 & 0 & 0 & 0 & 0 & 0 & 1 & 0 & 1 & 0 & 1 & 1 & 1 & 1 & 0 \\
1 & 1 & 1 & 1 & 1 & 1 & 0 & 1 & 0 & 1 & 0 & 0 & 0 & 0 & 1 \\
0 & 0 & 0 & 0 & 0 & 0 & 1 & 0 & 1 & 0 & 1 & 1 & 1 & 1 & 0 \\
0 & 0 & 0 & 0 & 0 & 0 & 1 & 0 & 1 & 0 & 1 & 1 & 1 & 1 & 0 \\
0 & 0 & 0 & 0 & 0 & 0 & 1 & 0 & 1 & 0 & 1 & 1 & 1 & 1 & 0 \\
0 & 0 & 0 & 0 & 0 & 0 & 1 & 0 & 1 & 0 & 1 & 1 & 1 & 1 & 0 \\
1 & 1 & 1 & 1 & 1 & 1 & 0 & 1 & 0 & 1 & 0 & 0 & 0 & 0 & 1
\end{bmatrix}
$$

$E^{U}(e_5, [0.96, 0.99])$ and $E^{L}(e_5, [0.67, 1])$

$$
\begin{bmatrix}
1 & 0 & 1 & 1 & 0 & 0 & 0 & 0 & 0 & 1 & 0 & 1 & 0 & 1 & 0 \\
0 & 1 & 0 & 0 & 1 & 1 & 1 & 1 & 1 & 0 & 1 & 0 & 1 & 0 & 1 \\
1 & 0 & 1 & 1 & 0 & 0 & 0 & 0 & 0 & 1 & 0 & 1 & 0 & 1 & 0 \\
1 & 0 & 1 & 1 & 0 & 0 & - & 0 & 0 & 1 & 0 & 1 & 0 & 1 & 0 \\
0 & 1 & 0 & 0 & 1 & 1 & 1 & 1 & 1 & 0 & 1 & 0 & 1 & 0 & 1 \\
0 & 1 & 0 & 0 & 1 & 1 & 1 & 1 & 1 & 0 & 1 & 0 & 1 & 0 & 1 \\
0 & 1 & 0 & 0 & 1 & 1 & 1 & 1 & 1 & 0 & 1 & 0 & 1 & 0 & 1 \\
0 & 1 & 0 & 0 & 1 & 1 & 1 & 1 & 1 & 0 & 1 & 0 & 1 & 0 & 1 \\
0 & 1 & 0 & 0 & 1 & 1 & 1 & 1 & 1 & 0 & 1 & 0 & 1 & 0 & 1 \\
1 & 0 & 1 & 1 & 0 & 0 & 0 & 0 & 0 & 1 & 0 & 1 & 0 & 1 & 0 \\
0 & 1 & 0 & 0 & 1 & 1 & 1 & 1 & 1 & 0 & 1 & 0 & 1 & 0 & 1 \\
1 & 0 & 1 & 1 & 0 & 0 & 0 & 0 & 0 & 1 & 0 & 1 & 0 & 1 & 0 \\
0 & 1 & 0 & 0 & 1 & 1 & 1 & 1 & 1 & 0 & 1 & 0 & 1 & 0 & 1 \\
1 & 0 & 1 & 1 & 0 & 0 & 0 & 0 & 0 & 1 & 0 & 1 & 0 & 1 & 0 \\
0 & 1 & 0 & 0 & 1 & 1 & 1 & 1 & 1 & 0 & 1 & 0 & 1 & 0 & 1
\end{bmatrix}
\&
\begin{bmatrix}
1 & 1 & 1 & 1 & 1 & 0 & 1 & 1 & 0 & 1 & 1 & 1 & 1 & 1 & 1 \\
1 & 1 & 1 & 1 & 1 & 0 & 1 & 1 & 0 & 1 & 1 & 1 & 1 & 1 & 1 \\
1 & 1 & 1 & 1 & 1 & 0 & 1 & 1 & 0 & 1 & 1 & 1 & 1 & 1 & 1 \\
1 & 1 & 1 & 1 & 1 & 0 & 1 & 1 & 0 & 1 & 1 & 1 & 1 & 1 & 1 \\
1 & 1 & 1 & 1 & 1 & 0 & 1 & 1 & 0 & 1 & 1 & 1 & 1 & 1 & 1 \\
0 & 0 & 0 & 0 & 0 & 1 & 0 & 0 & 1 & 0 & 0 & 0 & 0 & 0 & 0 \\
1 & 1 & 1 & 1 & 1 & 0 & 1 & 1 & 0 & 1 & 1 & 1 & 1 & 1 & 1 \\
1 & 1 & 1 & 1 & 1 & 0 & 1 & 1 & 0 & 1 & 1 & 1 & 1 & 1 & 1 \\
0 & 0 & 0 & 0 & 0 & 1 & 0 & 0 & 1 & 0 & 0 & 0 & 0 & 0 & 0 \\
1 & 1 & 1 & 1 & 1 & 0 & 1 & 1 & 0 & 1 & 1 & 1 & 1 & 1 & 1 \\
1 & 1 & 1 & 1 & 1 & 0 & 1 & 1 & 0 & 1 & 1 & 1 & 1 & 1 & 1 \\
1 & 1 & 1 & 1 & 1 & 0 & 1 & 1 & 0 & 1 & 1 & 1 & 1 & 1 & 1 \\
1 & 1 & 1 & 1 & 1 & 0 & 1 & 1 & 0 & 1 & 1 & 1 & 1 & 1 & 1 \\
1 & 1 & 1 & 1 & 1 & 0 & 1 & 1 & 0 & 1 & 1 & 1 & 1 & 1 & 1 \\
1 & 1 & 1 & 1 & 1 & 0 & 1 & 1 & 0 & 1 & 1 & 1 & 1 & 1 & 1
\end{bmatrix}
$$

Step 6. By using Definition (7), we have:

$$
\begin{aligned}
S_1 = {} & r_1(e_1; [0.67, 0.92], [0.13, 0.8]) + r_1(e_2, [0.75, 0.94], [0.37, 0.77]) + r_1 \\
& (e_3, [0.66, 0.92], [0.25, 0.76]) + r_1(e_4, [0.48, 0.74], [0.26, 0.76] + r_1(e_5, [0.96, 0.99], [0.67, 1] \\
& = 0 + 0 - 3 + 0 + 0 = -3
\end{aligned}
$$

Similarly, we have:

$$
\begin{aligned}
& S_2 = -12, S_3 = -5, S_4 = -9, S_5 = 6, S_6 = 7, S_7 = 15, S_8 = 42, S_9 = 2, \\
& S_8 = 42, S_9 = 2, S_{10} = 8, S_{11} = -16, S_{12} = 6, S_{13} = -3, S_{14} = -9, S_{15} = 0.
\end{aligned}
$$

Step 7. We have the following ordering system on X :

$$
o_8 \succeq o_7 \succeq o_{10} \succeq o_6 \succeq o_{12} \simeq o_5 \succeq o_9 \succeq o_{15} \succeq o_{13} \simeq o_1 \succeq o_{14} \simeq o_4 \succeq o_2 \succeq o_{11}.
$$

Steps 8 and 9. Then, from the corresponding object, we obtain, o_8 to be the best object (Acceptance region), while o_{11} is not selected (Rejection region) and others options $(o_7, o_{10}, o_6, o_{12}, o_5, o_9, o_{15},$
$o_{13}, o_1, o_{14}, o_4, o_2)$ cannot be judged(Boundary region).

Example 3. *(Example 2) Let us discuss Example 2 compared to existing methods proposed in [25,43,44] according to the ranking of objects.*

Yang et al. [25] defined the function score value as simply the total of lower and upper membership degrees of objects concerning each parameter. Ma et al. [44] applied Yang's Algorithm 1, which is given in [25] to solve Example 2 and showed the score value as follows: $o_8 \succeq o_6 \succeq o_{14} \succeq o_5 \succeq o_4 \succeq o_{10} \succeq o_{12} \succeq o_3 \succeq o_7 \succeq o_{13} \succeq o_{15} \succeq o_{11} \succeq o_9 \succeq o_2 \succeq o_1$.

Ma et al. [44] proposed a new efficient decision-making algorithm by using added objects. By using Algorithm 3 Section 4 in [44], Example 2 was solved and the score value for all objects was obtained as follows $o_8 \succeq o_6 \succeq o_{14} \succeq o_5 \succeq o_4 \succeq o_{10} \succeq o_{12} \succeq o_3 \succeq o_7 \succeq o_{13} \succeq o_{15} \succeq o_{11} \succeq o_9 \succeq o_2 \succeq o_1$.

Ma et al. [43] applied a new decision-making algorithm, based on the average table and the antithesis table—the antithesis the table has symmetry between the objects. Applying Algorithm in [43], Section 3, to solve the Example 2, the following ranking of objects is obtained $o_8 \succeq o_6 \succeq o_5 \succeq o_{14} \succeq o_4 \succeq o_{12} \succeq o_{10} \succeq o_3 \succeq o_{13} \succeq o_7 \succeq o_{15} \succeq o_{11} \succeq o_2 \succeq o_9 \succeq o_1$.

The comparison results among the present method and methods in [25,43,44] are given in Figure 1.

Algorithm 1: Rangking Objects by Interval-Valued Fuzzy Soft Topology

---

Input:
$|E| = n, |X| = m, |D| = k, f_{sE}, 1 \leq s \leq k$,
threshold intervals $\alpha, \beta \subseteq \mathbb{I}$,
where n, m the number of parameters, object, respectively, and k shows the number of decision makers, and f_{sE} shows the matrix of $IVFSs$.
Output: Optimal objects and worst objects.
begin
while $t = 1, 2, \ldots, n, i = 1, 2, \ldots m,$ *and* $s = 1, 2, \ldots k$ **do**
- Step 1. Compute crisp sets $U.C.S_\alpha f_s(e_t), L.C.S_\beta f_s(e_t)$ (see Matrices (1) and (2)).
- Step 2. Compute topological $(X, \tau^u_{e_t,\alpha}), (X, \tau^l_{e_t,\beta})$ ( ).
- Step 3. Compute $G_\alpha(e_t), S_\beta(e_t), E^U_\alpha(e_t)$, and $E^L_\beta(e_t)$(see (3)–(6)), for all $t = 1, 2, \ldots n$;
- **if** $G_\alpha(e_t) = I^U_m$ *and* $S_\beta(e_t) = I^L_m$ **then**
  - x_1 is the optimum decision and x_m is the worst one ;
- **else**
  - **if** $G_\alpha(e_t) = I^L_m$ *and* $S_\beta(e_t) = I^U_m$ **then**
    - x_m is the optimum decision and x_1 is the worst one;
  - **else**
    - **if** $E^U_\alpha(e_t) = E^L_\beta(e_t) = I_m$, *where* I_m *is the identiy matrix* **then**
      - there is no optimal over X;
    - **else**
      - **if** $E^U_\alpha(e_t) = E^L_\beta(e_t) = J_m$, *where* J_m *is the unit matrix* **then**
        - all objects of X can be selected as an optimal choice;
      - **else**
        - Go to the step 6.
      - Step 4. Calculate the score function $S_i \forall i$(Definition 7).
      - Step 5.Rank all objects according to the values S_i.
    - Step 6.The optimal alternative is to choose any one of the alternatives x_o such that $S_o = \max_i S_i$.
    - The alternative x_l such that $S_l = \min_i S_i$ should not be selected.
    - Step 7.**if** *the number of elements that* S_o *is maximum is more one* **then**
      - any one of x_o may be chosen.
    - **else**

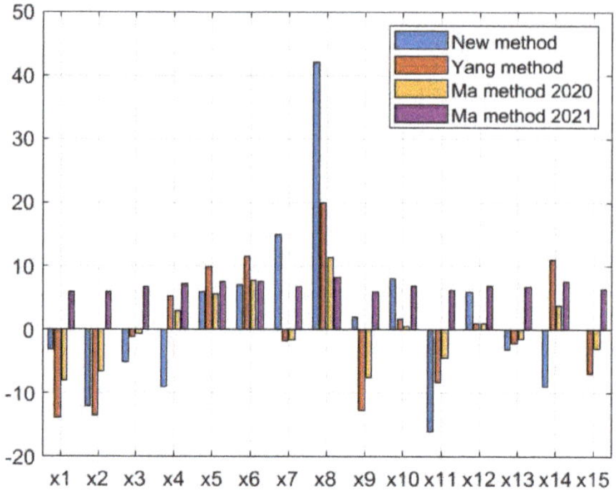

Figure 1. Comparison methods.

5. Discussion

According to the present method and the methods proposed in [25,43,44], to reach the process consensus, Yang et al. [25] use the "AND" operator, while methods in [43,44] did not discuss the aggregation problem. In addition, Example 2 shows that all the methods have the same option o_8, which is the best object. Consequently, algorithms in methods [25,43,44] select just one option, which is the optimum, and do not select the worst option, while the proposed algorithm selects two options—the optimum and as well as the worst option. However, the methods in [25,43,44], rank the objects based on a linear ordering system (see Example 3), while the present method ranks the objects based on preorder relation and a preference relationship, which allows one to have some incomparable objects (nonlinear ordering system). For example, in the Example 2, the objects o_{12} and o_5 have the same overall score values, which means that these objects cannot be compared with all of the others.

This is the same for the objects o_{13}, o_1 and o_{14}, o_4 (see Figure 2). The comparison results between the new proposed method and methods in [25,43,44] are also given in Table 11.

Table 11. Comparison of Existing Methods

Methods	Output Comparision	Aggregation Methodology	Ranking Methodology	Rank the Objects
[25]	optimal option	AND operator	fuzzy choice values	a linear ordering system
[44]	optimal option	Not discussed	choice values	a linear ordering system
[43]	optimal option	Not discussed	score function computed from an average and an antithesis tables	a linear ordering system
present method	optimal option and worst	$IVFST$	A collective preference relationship in topological space	a nonlinear ordering system

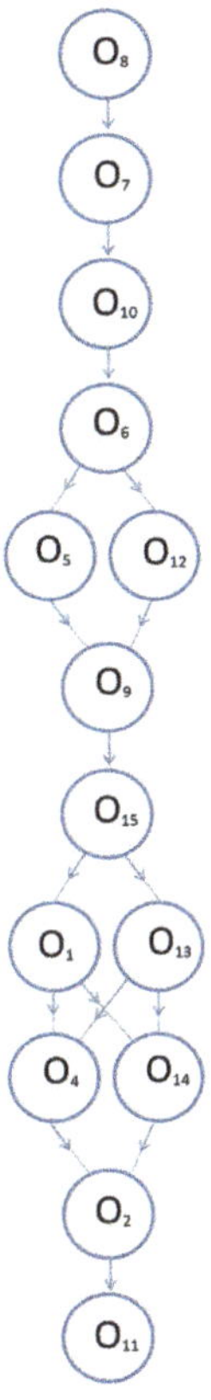

Figure 2. Nonlinear ordering system.

6. Conclusions

The interval-valued soft set is a useful tool to deal with fuzziness and uncertainties in decision-making problems. In this paper, we constructed two crisp topological spaces over the set of objects, and then presented two different preorder relations in these topological spaces. By using a new method for ranking data, we proposed an approach for solving multi-attribute group decision-making problems by using a new method for ranking data. Finally, a real-life example has been presented to verify the proposed method approach and to demonstrate the effectiveness by comparing the results with those of some of the existing approaches.

For future research, it would be of merit to apply the decision-making methods into practical applications such as evaluation systems, recommender systems, and conflict handling.

Author Contributions: Formal analysis, A.Z.K.; methodology, A.K. and A.Z.K.; supervision, A.K. and A.Z.K.; validation, A.K.; visualization, A.K.; writing—original draft, M.A.; writing—review & editing, A.K. All authors contributed equally to the writing of this paper. All authors read and approved the final manuscript.

Funding: The second and third authors gratefully acknowledge the Fundamental Research Grant Schemes, Reference No: FRGS/1/2018/STG06/UPM/01/3, awarded by the Ministry of Higher Education Malaysia.

Institutional Review Board Statement: Not applicable.

Informed Consent Statement: Not applicable.

Data Availability Statement: Not applicable.

Acknowledgments: The authors would like to thank the referees and editors for the useful comments and valuable remarks, which improved the current manuscript substantially.

Conflicts of Interest: The authors declare that they have no competing interests.

References

1. Zadeh, L.A. Fuzzy sets. *Inf. Control* **1965**, *8*, 338–353. [CrossRef]
2. Gorzałczany, M.B. A method of inference in approximate reasoning based on interval-valued fuzzy sets. *Fuzzy Sets Syst.* **1987**, *21*, 1–17. [CrossRef]
3. Atanassov, K.T. Intuitionistic fuzzy sets. In *Intuitionistic Fuzzy Sets*; Physica: Heidelberg, Germany, 1999; pp. 1–137.
4. Pawlak, Z. Rough sets. *Int. J. Comput. Inf. Sci.* **1982**, *11*, 341–356. [CrossRef]
5. Molodtsov, D. Soft set theory first results. *Comput. Math. Appl.* **1999,** *37*, 19–31 [CrossRef]
6. Maji, P.K.; Biswas, P.; Roy, A.R. A fuzzy soft sets. *J. Fuzzy Math.* **2001**, *9*, 89–602.
7. Roy, A.R.; Maji, P.K. A fuzzy soft set theoretic approach to decision making problems. *J. Comput. Appl. Math.* **2007**, *203*, 412–418. [CrossRef]
8. Kong, Z.; Gao, L.; Wang, L. Comment on "A fuzzy soft set theoretic approach to decision making problems". *J. Comput. Appl. Math.* **2009**, *223*, 540–542. [CrossRef]
9. Feng, F.; Jun, Y.B.; Liu, X.; Li, L. An adjustable approach to fuzzy soft set based decision making. *J. Comput. Appl. Math.* **2010**, *234*, 10–20. [CrossRef]
10. Alcantud, J.C.R. A novel algorithm for fuzzy soft set based decision making from multiobserver input parameter data set. *Inf. Fusion* **2016**, *29*, 142–148. [CrossRef]
11. Alcantud, J.C.R. Some formal relationships among soft sets, fuzzy sets, and their extensions. *Int. J. Approx. Reason.* **2016**, *68*, 45–53. [CrossRef]
12. Alcantud, J.C.R.; Mathew, T.J. Separable fuzzy soft sets and decision making with positive and negative attributes. *Appl. Soft Comput.* **2017**, *59*, 586–595. [CrossRef]
13. Aktaş, H.; Çağman, N. Soft decision making methods based on fuzzy sets and soft sets. *J. Intell. Fuzzy Syst.* **2016**, *30*, 2797–2803. [CrossRef]
14. Maji, P.K. More on intuitionistic fuzzy soft sets. In *International Workshop on Rough Sets, Fuzzy Sets, Data Mining, and Granular-Soft Computing*; Springer: Berlin/Heidelberg, Germany, 2009; pp. 231–240.
15. Maji, P.K.; Biswas, R.; Roy, A.R. Intuitionistic fuzzy soft sets. *J. Fuzzy Math.* **2001**, *9*, 677–692.
16. Jiang, Y.; Tang, Y.; Chen, Q.; Liu, H.; Tang, J. Interval-valued intuitionistic fuzzy soft sets and their properties. *Comput. Math. Appl.* **2010**, *60*, 906–918. [CrossRef]
17. Maji, P K.; Roy, A.R.; Biswas, R. On intuitionistic fuzzy soft sets. *J. Fuzzy Math.* **2004**, *12*, 669–684.
18. Liu, Y.; Qin, K.; Martínez, L. Improving decision making approaches based on fuzzy soft sets and rough soft sets. *Appl. Soft Comput.* **2018**, *65*, 320–332. [CrossRef]
19. Feng, F.; Li, C.; Davvaz, B.; Ali, M. I. Soft sets combined with fuzzy sets and rough sets: A tentative approach. *Soft Comput.* **2010**, *14*, 899–911. [CrossRef]
20. Roy, S.; Samanta, T.K. A note on fuzzy soft topological spaces. *Ann. Fuzzy Math. Inform.* **2012**, *3*, 305–311.
21. Tanay, B.; Kandemir, M.B. Topological structure of fuzzy soft sets. *Comput. Math. Appl.* **2011**, *61*, 2952–2957. [CrossRef]
22. Khameneh, A.Z.; Kılıçman, A.; Salleh, A.R. Fuzzy soft boundary. *Ann. Fuzzy Math. Inform.* **2014**, *8*, 687–703.
23. Khameneh, A.Z.; Kılıçman, A.; Salleh, A.R. Fuzzy soft product topology. *Ann. Fuzzy Math. Inform.* **2014**, *7*, 935–947.
24. Son, M.-J. Interval-valued Fuzzy Soft Sets. *J. Korean Inst. Intell. Syst.* **2007**, *17*, 557–562. [CrossRef]
25. Yang, X.; Lin, T.Y.; Yang, J.; Li, Y.; Yu, D. Combination of interval-valued fuzzy set and soft set. *Comput. Math. Appl.* **2009**, *58*, 521–527. [CrossRef]
26. Feng, F.; Li, Y.; Leoreanu-Fotea, V. Application of level soft sets in decision making based on interval-valued fuzzy soft sets. *Comput. Math. Appl.* **2010**, *60*, 1756–1767. [CrossRef]
27. Xiao, Z.; Chen, W.; Li, L. A method based on interval-valued fuzzy soft set for multi-attribute group decision-making problems under uncertain environment. *Knowl. Inf. Syst.* **2013**, *34*, 653–669. [CrossRef]
28. Khameneh, A.Z.; Kılıçman, A.; Salleh, A.R. An adjustable approach to multi-criteria group decision-making based on a preference relationship under fuzzy soft information. *Int. J. Fuzzy Syst.* **2017**, *19*, 1840–1865. [CrossRef]
29. Khameneh, A.Z.; Kılıçman, A.; Salleh, A.R. Application of a preference relationship in decision-making based on intuitionistic fuzzy soft sets. *J. Intell. Fuzzy Syst.* **2018**, *34*, 123–139. [CrossRef]
30. Khameneh, A.Z.; Kılıçman, A.; Salleh, A.R. An adjustable method for data ranking based on fuzzy soft sets. *Indian J. Sci. Technol.* **2015**, *8*, 1–9.
31. Jiang, Y.; Tang, Y.; Liu, H.; Chen, Z. Entropy on intuitionistic fuzzy soft sets and on interval-valued fuzzy soft sets. *Inf. Sci.* **2013**, *240*, 95–114. [CrossRef]
32. Turksen, I.B.; Zhong, Z. An approximate analogical reasoning schema based on similarity measures and interval-valued fuzzy sets. *Fuzzy Sets Syst.* **1990**, *34*, 323–346. [CrossRef]
33. Peng, X.; Yang, Y. Algorithms for interval-valued fuzzy soft sets in stochastic multi-criteria decision making based on regret theory and prospect theory with combined weight. *Appl. Soft Comput.* **2017**, *54*, 415–430. [CrossRef]

34. Peng, X.D.; Yang, Y. Information measures for interval-valued fuzzy soft sets and their clustering algorithm. *J. Comput. Appl.* **2015**, *35*, 2350–2354.
35. Chen, W.J.; Zou, Y. Rational decision making models with incomplete information based on interval-valued fuzzy soft sets. *J. Comput.* **2017**, *28*, 193–207.
36. Yuan, F.; Hu, M.J. Application of interval-valued fuzzy soft sets in evaluation of teaching quality. *J. Hunan Inst. Sci. Technol.* **2012**, *25*, 28–30.
37. Esposito, C.; Moscato, V.; Sperlí, G. Trustworthiness Assessment of Users in Social Reviewing Systems. *IEEE Trans. Syst. Man Cybern. Syst.* **2021**. [CrossRef]
38. Han, Q.; Molinaro, C.; Picariello, A.; Sperli, G.; Subrahmanian, V. S.; Xiong, Y. Generating Fake Documents using Probabilistic Logic Graphs. *IEEE Trans. Dependable Secur. Comput.* **2021**. [CrossRef]
39. Rajarajeswari, P.; Dhanalakshmi, P. Interval-valued fuzzy soft matrix theory. *Ann. Pure Appl. Math.* **2014**, *7*, 61–72.
40. Basu, T.M.; Mahapatra, N.K.; Mondal, S.K. Matrices in interval-valued fuzzy soft set theory and their application. *S. Asian J. Math.* **2014**, *4*, 1–22.
41. Zhang, Q.; Sun, D. An Improved Decision-Making Approach Based on Interval-valued Fuzzy Soft Set. *J. Phys. Conf. Ser.* **2021**, *1828*, 012041. [CrossRef]
42. Ma, X.; Qin, H.; Sulaiman, N.; Herawan, T.; Abawajy, J.H. The parameter reduction of the interval-valued fuzzy soft sets and its related algorithms. *IEEE Trans. Fuzzy Syst.* **2013**, *22*, 57–71. [CrossRef]
43. Ma, X.; Wang, Y.; Qin, H.; Wang, J. A Decision-Making Algorithm Based on the Average Table and Antitheses Table for Interval-Valued Fuzzy Soft Set. *Symmetry* **2020**, *12*, 1131. [CrossRef]
44. Ma, X.; Fei, Q.; Qin, H.; Li, H.; Chen, W. A new efficient decision making algorithm based on interval-valued fuzzy soft set. *Appl. Intell.* **2021**, *51*, 3226–3240. [CrossRef]
45. Peng, X.; Garg, H. Algorithms for interval-valued fuzzy soft sets in emergency decision making based on WDBA and CODAS with new information measure. *Comput. Ind. Eng.* **2018**, *119*, 439–452. [CrossRef] [PubMed]
46. Ali, M.; Kılıçman, A.; Khameneh, A.Z. Separation Axioms of Interval-Valued Fuzzy Soft Topology via Quasi-Neighborhood Structure. *Mathematics* **2020**, *8*, 178. [CrossRef]
47. Lai, H.; Zhang, D. Fuzzy preorder and fuzzy topology. *Fuzzy Sets Syst.* **2006**, *157*, 1865–1885. [CrossRef]

mathematics

Article

L-Fuzzy Sub-Effect Algebras

Yan-Yan Dong * and Fu-Gui Shi

Beijing Key Laboratory on MCAACI, School of Mathematics and Statistics, Beijing Institute of Technology, Beijing 102488, China; fuguishi@bit.edu.cn
* Correspondence: 3120195725@bit.edu.cn

Abstract: In this paper, the notions of L-fuzzy subalgebra degree and L-subalgebras on an effect algebra are introduced and some characterizations are given. We use four kinds of cut sets of L-subsets to characterize the L-fuzzy subalgebra degree. We induce an L-fuzzy convexity by the L-fuzzy subalgebra degree, and we prove that a morphism between two effect algebras is an L-fuzzy convexity preserving mapping and a monomorphism is an L-fuzzy convex-to-convex mapping. Finally, it is proved that the set of all L-subalgebras on an effect algebra can form an L-convexity, and its L-convex hull formula is given.

Keywords: effect algebra; L-fuzzy subalgebra degree; L-subalgebra; L-fuzzy convexity; L-convex hull formula

Citation: Dong, Y.-Y.; Shi, F.-G. L-Fuzzy Sub-Effect Algebras. *Mathematics* **2021**, *9*, 1596. https://doi.org/10.3390/math9141596

Academic Editor: Sorin Nadaban

Received: 13 May 2021
Accepted: 29 June 2021
Published: 7 July 2021

1. Introduction

Effect algebras were introduced by Foulis and Bennett to axiomatize quantum logic effects on a Hilbert space [1]. The elements of effect algebras represent events that may be unsharp or imprecise. Effect algebras are partial algebras with one partial binary operation that can be converted into bounded posets in general and into lattices in some cases. Since Zadeh introduced the concept of fuzzy sets [2], the theory of fuzzy sets has become a vigorous area of research in different disciplines. In recent years, many scholars have studied (fuzzy) ideals [3,4] and (fuzzy) filters [5,6] (the lattice background is $[0,1]$) on effect algebras, but there is no research on the fuzzy subalgebras (L-subalgebras). In order to fill this gap, we first want to extend the unit interval $[0,1]$ to a completely distributive lattice L and introduce the notion of L-subalgebras on effect algebras. As we all know, for a given L-subset A, A is either an L-subalgebra or not, so let us consider the question: what is the degree to which A is an L-subalgebra? To solve this question, we propose the concept of L-subalgebra degree on effect algebras and investigate their basic properties in this paper.

The notion of convexity [7] is inspired by the shape of some figures, such as circles and polyhedrons in Euclidean spaces. As we all know, a convex structure satisfying the Exchange Law [8] is a matroid, matroids are precisely the structures for which the very simple and efficient greedy algorithm works. Many real-world problems can be defined and solved by making use of matroid theory. So convexity theory has been regarded as an increasingly important role in solving problems. In fact, there exist convexities in many mathematical structures such as semigroup, ring, posets, graphs, convergence spaces and so on [9–13]. It is also natural to consider if there exist convex structures on effect algebras. By these motivations, we will try to prove the existence of convexity on effect algebras.

With the development of fuzzy sets, the notion of convexities has already been extended to fuzzy case. In 1994, Rosa [14] first proposed the notion of fuzzy convex structures with the unit interval $[0,1]$ as the lattice background. In 2009, Y. Maruyama [15] defined another more generalized fuzzy convex structure based on a completely distributive lattice L, which is called L-convex structure, some of the latest research related to L-convex structure can be found in [16–21]. In 2014, a new approach to the fuzzification of convex structures was introduced in [22]. It is called an M-fuzzifying convex structure, in which each subset can be regarded as a convex set to some degrees. Further, there are some studies about M-fuzzifying convex structures showed in [23–26]. In 2017, abstract convexity was extended to a more general case, which is called an (L, M)-fuzzy convex structure in [27]. Particularly, an (L, L)-fuzzy structure is briefly called an L-fuzzy convex structure. Many researchers investigated (L, M)-fuzzy convex structures from different aspects [28–32]. In our paper, we mainly discuss L-fuzzy convex structure and L-convex structure on an effect algebra.

The paper is organized in the following way. In Section 2, we will give some necessary notations and definitions. In Section 3, we propose the notion of L-fuzzy subalgebra degree on an effect algebra by means of the implication operator of L, and we provide their characterizations by cut sets of L-subsets. For instance, we give the notion of L-subalgebras on effect algebras. In Section 4, we obtain an L-fuzzy convexity induced by L-fuzzy subalgebra degree, and we analyze the corresponding L-fuzzy convexity preserving mappings and L-fuzzy convex-to-convex mappings. Finally, we prove that the set of all L-subalgebras on an effect algebra can form an L-convexity, and we provide its L-convex hull formula.

2. Preliminaries

In this section, we provide some notions and results that will be used in this paper.

Definition 1 ([1])**.** *An effect algebra is a system $(E, \oplus, 0, 1)$ consisting of a set E with two special elements $0, 1 \in E$, called the zero and the unit, and with a partially defined binary operation $\oplus$ satisfying the following conditions: for any $x, y, z \in E$,*

(E1) *(Commutative law) if $x \oplus y$ is defined, then $y \oplus x$ is defined and $x \oplus y = y \oplus x$,*

(E2) *(Associative law) if $x \oplus y$ is defined and $(x \oplus y) \oplus z$ is defined, then $y \oplus z$ and $x \oplus (y \oplus z)$ are defined and $(x \oplus y) \oplus z = x \oplus (y \oplus z)$,*

(E3) *(Orthosupplement law) for every $x \in E$ there exists a unique $x' \in E$ such that $x \oplus x'$ is defined and $x \oplus x' = 1$. The unique element $x' \in E$ called the orthosupplement of x,*

(E4) *(Zero-one law) if $x \oplus 1$ is defined, then $x = 0$.*

In the following, an effect algebra $(E, \oplus, 0, 1)$ is denoted by E unless otherwise specified.

Definition 2 ([33])**.** *A nonempty subset S of an effect algebra E is called a subalgebra if it satisfies the following conditions:*

(i) $0, 1 \in S$,

(ii) *$x \in S$ implies $x' \in S$,*

(iii) *if $x, y \in S$ and $x \oplus y$ is defined, then $x \oplus y \in S$.*

Let E be an effect algebra. If $x \oplus y$ is defined, then we say $x \perp y$ for all $x, y \in E$. Define a binary relation on E by $x \leq y$ if for some $z \in E$, $x \oplus z = y$, which turns out to be a partial ordering on E such that 0 and 1 is the smallest element and the greatest element of E, respectively. If the poset $(E, \leq)$ is a lattice, then E is called a lattice-ordered effect algebra.

Let L be a complete lattice. We denote the minimal element and the maximal element of L by $\perp_L$ and $\top_L$, respectively. An element $\lambda \in L$ is called co-prime if $\lambda \leq \delta \vee \xi$ implies $\lambda \leq \delta$ or $\lambda \leq \xi$. The set of nonzero co-prime elements in L is denoted by $J(L)$. An element $\lambda \in L$ is called prime if $\lambda \geq \delta \wedge \xi$ implies $\lambda \geq \delta$ or $\lambda \geq \xi$. The set of nonunit prime elements in L is denoted by $P(L)$. From [34], we know that each element of L is the sup of co-prime elements and the inf of prime elements.

Let $\delta, \xi \in L$, the symbol $\delta \prec \xi$ (δ is wedge below ξ) means that for every $H \subseteq L$, $\xi \leq \bigvee H$ implies the existence of $\eta \in H$ such that $\delta \leq \eta$. A complete lattice L is completely distributive [34] if and only if $\xi = \bigvee\{\delta \mid \delta \prec \xi\}$ for each $\xi \in L$. The set $\{\delta \mid \delta \prec \xi\}$, denoted by $\beta(\xi)$, is called the greatest minimal family of ξ in the sense of [34]. Let $\beta^*(\xi) = \beta(\xi) \cap J(L)$. Moreover, define a binary relation $\prec^{op}$ as follows: for $\xi, \delta \in L$, $\xi \prec^{op} \delta$ if and only if for every subset $H \subseteq L$, $\bigwedge H \leq \xi$ implies $\lambda \leq \delta$ for some $\lambda \in H$.

The set $\{\delta \in L \mid \xi \prec^{op} \delta\}$, denoted by $\alpha(\xi)$, is the greatest maximal family of ξ in the sense of [34]. Let $\alpha^*(\xi) = \alpha(\xi) \cap P(L)$. We know that α is an $\bigwedge - \bigcup$ mapping and β is a union-preserving mapping, it holds that $\xi = \bigvee \beta(\xi) = \bigvee \beta^*(\xi) = \bigwedge \alpha(\xi) = \bigwedge \alpha^*(\xi)$ for each $\xi \in L$ (see [34]). From [35], we have $\alpha(\top_L) = \emptyset$ and $\beta(\bot_L) = \emptyset$.

In the following, a completely distributive lattice $(L, \wedge, \vee)$ is denoted by L unless otherwise specified.

For an effect algebra E, each mapping $A : E \longrightarrow L$ is called an L-subset of E, and we denote the collection of all L-subsets of E by L^E. L^E is also a complete lattice by defining "$\leq$" point-wisely. Furthermore, the smallest element and the largest element in L^E are denoted by $\underline{\bot}$ and $\overline{\top}$, respectively. The mapping $g_L^{\rightarrow} : L^{E_1} \longrightarrow L^{E_2}$ is induced by $g : E_1 \longrightarrow E_2$ as follows:

$$\forall A \in L^{E_1}, \forall y \in E_2, \ g_L^{\rightarrow}(A)(y) = \bigvee_{g(x)=y} A(x).$$

$g_L^{\leftarrow} : L^{E_2} \longrightarrow L^{E_1}$ is induced by the mapping g as follows:

$$\forall B \in L^{E_2}, \forall x \in E_1, \ g_L^{\leftarrow}(B)(x) = B(g(x)).$$

Definition 3 ([36]). *Let E be an effect algebra and $A \in L^E$. For any $\lambda \in L$, define*

$$A_{[\lambda]} = \{x \in E \mid A(x) \geq \lambda\}, \qquad A^{(\lambda)} = \{x \in E \mid A(x) \not\leq \lambda\},$$
$$A_{(\lambda)} = \{x \in E \mid \lambda \in \beta(A(x))\}, \qquad A^{[\lambda]} = \{x \in E \mid \lambda \notin \alpha(A(x))\}.$$

The right adjoint $\rightarrow$ of the meet operation $\wedge$ is a mapping from $L \times L$ to L defined as

$$\lambda \rightarrow \mu = \bigvee\{\delta \in L | \lambda \wedge \delta \leq \mu\}.$$

Some basic properties of the operation $\rightarrow$ are listed in the following [37,38].

(1) $\lambda \rightarrow \mu = \top_L \Leftrightarrow \lambda \leq \mu$;
(2) $\lambda \wedge \delta \leq \mu \Leftrightarrow \lambda \leq \delta \rightarrow \mu$;
(3) $\lambda \leq \delta \Rightarrow \mu \rightarrow \lambda \leq \mu \rightarrow \delta$ and $\lambda \rightarrow \mu \geq \delta \rightarrow \mu$;
(4) $\lambda \rightarrow \bigwedge_{i \in I} \mu_i = \bigwedge_{i \in I}(\lambda \rightarrow \mu_i)$.

Let $\underline{\bot}, \overline{\top} \in L^E$ represent $\underline{\bot}(x) = \bot_L$ and $\overline{\top}(x) = \top_L$ for all $x \in E$. Next, we recall (L, M)-fuzzy convexities in [27], which are more general fuzzy convexities. Let L and M be two completely distributive lattices. An (L, M)-fuzzy convexity on an effect algebra E is defined as follows:

Definition 4 ([27]). *A mapping $\mathcal{C} : L^E \longrightarrow M$ is called an (L, M)-fuzzy convexity if it satisfies the following conditions:*

(LMC1) $\mathcal{C}(\underline{\bot}) = \mathcal{C}(\overline{\top}) = \top_M$;
(LMC2) *if $\{A_i \mid i \in I\} \subseteq L^E$ is nonempty, then $\mathcal{C}\left(\bigwedge_{i \in I} A_i\right) \geq \bigwedge_{i \in I} \mathcal{C}(A_i)$;*
(LMC3) *if $\{A_i \mid i \in I\} \subseteq L^E$ is nonempty and directed, then $\mathcal{C}\left(\bigvee_{i \in I} A_i\right) \geq \bigwedge_{i \in I} \mathcal{C}(A_i)$.*

In this case, the pair $(E, \mathcal{C})$ is called an (L, M)-fuzzy convex space. An (L, L)-fuzzy convexity is briefly called an L-fuzzy convexity.

An $(L, \mathbf{2})$-fuzzy convexity is an L-convexity in [15]. An $(I, \mathbf{2})$-fuzzy convexity is a fuzzy convexity in [14], where $I = [0, 1]$. A $(\mathbf{2}, M)$-fuzzy convexity is an M-fuzzifying convexity in [22]. A $(\mathbf{2}, \mathbf{2})$-fuzzy convexity is a convexity in [7].

Definition 5 ([27]). *Let $(E, \mathcal{C})$ and $(F, \mathcal{D})$ be two (L, M)-fuzzy convex spaces. If $g : E \longrightarrow F$ is a mapping between E and F, then*

(i) *$g : (E, \mathcal{C}) \longrightarrow (F, \mathcal{D})$ is called an (L, M)-fuzzy convexity preserving mapping provided that $\mathcal{C}(g_L^{\leftarrow}(A)) \geq \mathcal{D}(A)$ for all $A \in L^F$.*

(ii) $g : (E, \mathcal{C}) \longrightarrow (F, \mathcal{D})$ *is called an* (L, M)*-fuzzy convex-to-convex mapping provided that* $\mathcal{D}(g_L^{\rightarrow}(B)) \geq \mathcal{C}(B)$ *for all* $B \in L^E$.

An (L, L)*-fuzzy convexity preserving mapping and an* (L, L)*-fuzzy convex-to-convex mapping are briefly called an L-fuzzy convexity preserving mapping and an L-fuzzy convex-to-convex mapping, respectively.*

An $(L, \mathbf{2})$-fuzzy convexity preserving mapping is an L-convexity preserving mapping in [15]. An $(L, \mathbf{2})$-fuzzy convex-to-convex mapping is an L-convex-to-convex mapping in [15].

Definition 6 ([1]). *Let* $\{E_i\}_{i\in\Lambda}$ *be a family of effect algebras, define* $\prod\limits_{i\in\Lambda} E_i$ *as:*

$$\prod_{i\in\Lambda} E_i = \{x \mid x : \Lambda \longrightarrow \bigcup_{i\in\Lambda} E_i \text{ s.t.} \forall i \in \Lambda, x_i = x(i) \in E_i\}.$$

Define the operation $\oplus$ *on* $\prod\limits_{i\in\Lambda} E_i$ *as: for* $x, y \in \prod\limits_{i\in\Lambda} E_i$, $x \perp y$ *iff* $x_i \perp y_i$ *for all* $i \in \Lambda$. *In this case,* $(x \oplus y)(i) = x_i \oplus y_i$ *and* $x'(i) = x_i'$. *Further,* $\mathbf{0}_i = 0_i, \mathbf{1}_i = 1_i$ *where* 0_i *and* 1_i *are the minimal element and the maximal element of* E_i. *Then* $(\prod\limits_{i\in\Lambda} E_i, \oplus, \mathbf{0}, \mathbf{1})$ *is called the direct product of effect algebras.*

It is easy to check that $(\prod\limits_{i\in\Lambda} E_i, \oplus, \mathbf{0}, \mathbf{1})$ is an effect algebra.

Definition 7 ([39]). *Let* $\{E_i\}_{i=1}^n$ *be a family of effect algebras and* A_i *be an L-subset of* E_i *for all* $i \in \Lambda$, *then the L-subset* $\prod\limits_{i\in\Lambda} A_i$ *of* $\prod\limits_{i\in\Lambda} E_i$ *is defined by* $(\prod\limits_{i\in\Lambda} A_i)(x) = \bigwedge\limits_{i\in\Lambda} A_i(x_i)$.

3. *L*-Fuzzy Subalgebra Degree on Effect Algebras

In this section, we introduce the concept of L-fuzzy subalgebra degree on effect algebras by means of the implication operator of L. We define an L-fuzzy subalgebra provided that its L-fuzzy subalgebra degree is equal to $\top_L$. Moreover, we give some characterizations of L-subalgebra degree in terms of four kinds of cut sets of L-subsets.

Definition 8. *Let E be an effect algebra and* $A \in L^E$. *Then the L-fuzzy subalgebra degree* $\mathcal{E}(A)$ *of A is defined as:*

$$\mathcal{E}(A) = \bigwedge_{x,y\in E, x\perp y} \{[A(x) \rightarrow A(0)] \wedge [A(x) \rightarrow A(x')] \wedge [(A(x) \wedge A(y)) \rightarrow A(x \oplus y)]\}.$$

Example 1. *Let* $E = \{0, x, y, 1\}$ *and* $\oplus$ *be given by:*

$\oplus$	0	x	y	1
0	0	x	y	1
x	x	1	*	*
y	y	*	1	*
1	1	*	*	*

Then $(E, \oplus, 0, 1)$ *is an effect algebra. Let* $L = [0, 1]$ *and define the L-subsets of E as follows:*

(i) $A_1(0) = 0.8, A_1(x) = 0.5, A_1(y) = 0.5, A_1(1) = 0.8$. *By Definition 8, we can obtain* $\mathcal{E}(A_1) = 1$.
Indeed, since for any $z \in E$, $A_1(z) \leq A_1(0)$, *we have* $A_1(z) \rightarrow A_1(0) = 1$ *for all* $z \in E$. *We can easily know* $x' = x, y' = y, 0' = 1$ *and* $1' = 0$, *it follows from* $A_1(0) = A_1(1)$ *that* $A_1(z) = A_1(z')$, *so* $A_1(z) \rightarrow A_1(z') = 1$ *for all* $z \in E$. *We can routinely prove* $(A(z) \wedge A(w)) \rightarrow A(z \oplus w) = 1$ *for all* $z, w \in E$ *with* $z \perp w$. *Therefore,* $\mathcal{E}(A_1) = 1$.

(ii) $A_2(0) = 0.3, A_2(x) = 1, A_2(y) = 1, A_2(1) = 1$. *By Definition 8, we can obtain* $\mathcal{E}(A_2) = 0.3$.
Indeed, since $A_2(x) \to A_2(0) = A_2(1) \to A_2(0) = A_2(y) \to A_2(0) = 0.3$*, we can easily obtain* $\mathcal{E}(A_2) = 0.3$.

(iii) $A_3(0) = 0, A_2(x) = 0.5, A_3(y) = 1, A_3(1) = 0.6$. *By Definition 8, we can obtain* $\mathcal{E}(A_3) = 0$.
Indeed, since $A_3(y) \to A_3(0) = 1 \to 0 = 0$*, we have* $\mathcal{E}(A_3) = 0$.

From the properties (3) of the implication operator, the following lemma is obvious.

Lemma 1. *Let E be an effect algebra and $A \in L^E$. Then $\mathcal{E}(A) \geq \lambda$ ($\lambda \in L$) if and only if for any $x, y \in E$ satisfying $x \perp y$,*

$$A(x) \wedge A(y) \wedge \lambda \leq A(x \oplus y), A(x) \wedge \lambda \leq A(x') \text{ and } A(x) \wedge \lambda \leq A(0).$$

By Lemma 1, we can obtain the following theorem.

Theorem 1. *Let E be an effect algebra and $A \in L^E$. Then*

$$\mathcal{E}(A) = \bigvee\{\lambda \in L \mid \forall x, y \in E \text{ s.t. } x \perp y, A(x) \wedge A(y) \wedge \lambda \leq A(x \oplus y), A(x) \wedge \lambda \leq A(x') \wedge A(0)\}.$$

We use four kinds of cut sets of L-subsets to characterize the L-fuzzy subalgebra degree in the following Theorem.

Theorem 2. *Let E be an effect algebra and $A \in L^E$. Then*

(i) $\mathcal{E}(A) = \bigvee\{\lambda \in L \mid \forall \mu \leq \lambda, A_{[\mu]} = \emptyset \text{ or it is a subalgebra of } E\}$.

(ii) $\mathcal{E}(A) = \bigvee\{\lambda \in L \mid \forall \mu \in P(L), \mu \not\geq \lambda, A^{(\mu)} = \emptyset \text{ or it is a subalgebra of } E\}$.

(iii) $\mathcal{E}(A) = \bigvee\{\lambda \in L \mid \forall \mu \notin \alpha(\lambda), A^{[\mu]} = \emptyset \text{ or it is a subalgebra of } E\}$.

(iv) $\mathcal{E}(A) = \bigvee\{\lambda \in L \mid \forall \mu \in P(L), \mu \notin \alpha(\lambda), A^{[\mu]} = \emptyset \text{ or it is a subalgebra of } E\}$.

(v) $\mathcal{E}(A) = \bigvee\{\lambda \in L \mid \forall \mu \in \beta(\lambda), A_{(\mu)} = \emptyset \text{ or it is a subalgebra of } E\}$ *if* $\beta(\lambda \wedge \mu) = \beta(\lambda) \cap \beta(\mu)$ *for all* $\lambda, \mu \in L$.

Proof. (i) For any $x, y \in E$ satisfying $x \perp y$, assume that $\lambda \in L$ with the property of $A(x) \wedge A(y) \wedge \lambda \leq A(x \oplus y)$ and $A(x) \wedge \lambda \leq A(x') \wedge A(0)$. For any $\mu \leq \lambda$, suppose $A_{[\mu]} \neq \emptyset$ and let $x, y \in A_{[\mu]}$ with $x \perp y$, then

$$\mu = \mu \wedge \lambda \leq A(x) \wedge A(y) \wedge \lambda \leq A(x \oplus y) \text{ and } \mu = \mu \wedge \lambda \leq A(x) \wedge \lambda \leq A(x') \wedge A(0),$$

which implies $x \oplus y \in A_{[\mu]}$ and $x', 0 \in A_{[\mu]}$. Hence $A_{[\mu]}$ is a subalgebra of E. This gives that

$$\mathcal{E}(A) \leq \bigvee\{\lambda \in L \mid \forall \mu \leq \lambda, A_{[\mu]} = \emptyset \text{ or it is a subalgebra of } E\}.$$

Conversely, assume $\lambda \in L$ and for each $\mu \leq \lambda$, $A_{[\mu]} = \emptyset$ or it is a subalgebra of E. For any $x, y \in E$ satisfying $x \perp y$, let $\mu = A(x) \wedge A(y) \wedge \lambda$ and $\gamma = A(x) \wedge \lambda$, then $\mu \leq \lambda$ and $\gamma \leq \lambda$. It follows that $x, y \in A_{[\mu]}$ and $x \in A_{[\gamma]}$. Since $A_{[\mu]}$ and $A_{[\gamma]}$ are subalgebras of E, we have $x \oplus y \in A_{[\mu]}$ and $x', 0 \in A_{[\gamma]}$, i.e., $A(x) \wedge A(y) \wedge \lambda \leq A(x \oplus y)$ and $A(x) \wedge \lambda \leq A(x') \wedge A(0)$. So we have

$$\mathcal{E}(A) \geq \bigvee\{\lambda \in L \mid \forall \mu \leq \lambda, A_{[\mu]} = \emptyset \text{ or it is a subalgebra of } E\}.$$

(ii) For any $x, y \in E$ satisfying $x \perp y$, assume that $\lambda \in L$ with the property of $A(x) \wedge A(y) \wedge \lambda \leq A(x \oplus y)$ and $A(x) \wedge \lambda \leq A(x') \wedge A(0)$. For any $\mu \in P(L)$ with $\mu \not\geq \lambda$, suppose $A^{(\mu)} \neq \emptyset$ and let $x, y \in A^{(\mu)}$ with $x \perp y$, i.e., $A(x) \not\leq \mu$ and $A(y) \not\leq \mu$, then $A(x) \wedge A(y) \wedge \lambda \not\leq \mu$ since μ is prime. By $A(x) \wedge A(y) \wedge \lambda \leq A(x \oplus y)$ and $A(x) \wedge \lambda \leq$

$A(x') \wedge A(0)$, it follows that $A(x \oplus y) \not\leq \mu$ and $A(x') \wedge A(0) \not\leq \mu$, i.e., $x \oplus y, x', 0 \in A^{(\mu)}$. Hence $A^{(\mu)}$ is a subalgebra of E. This shows that

$$\mathcal{E}(A) \leq \bigvee\{\lambda \in L \mid \forall \mu \in P(L), \mu \not\geq \lambda, A^{(\mu)} = \emptyset \text{ or it is a subalgebra of } E\}.$$

Conversely, assume that $\lambda \in L$ and $A^{(\mu)} = \emptyset$ or it is a subalgebra of E for $\mu \in P(L)$ with $\mu \not\geq \lambda$. For any $x, y \in E$ satisfying $x \perp y$, let $\mu \in P(L)$ and $A(x) \wedge A(y) \wedge \lambda \not\leq \mu$, then $x, y \in A^{(\mu)}$ and $\lambda \not\leq \mu$. Since $A^{(\mu)}$ is a subalgebra of E, it holds that $x \oplus y \in A^{(\mu)}$, i.e., $A(x \oplus y) \not\leq \mu$. Hence $A(x) \wedge A(y) \wedge \lambda \leq A(x \oplus y)$. Similarly, we can prove $A(x) \wedge \lambda \leq A(x') \wedge A(0)$. This shows

$$\mathcal{E}(A) \geq \bigvee\{\lambda \in L \mid \forall \mu \in P(L), \mu \not\geq \lambda, A^{(\mu)} = \emptyset \text{ or it is a subalgebra of } E\}.$$

(iii) For any $x, y \in E$ satisfying $x \perp y$, assume that $\lambda \in L$ with the property of $A(x) \wedge A(y) \wedge \lambda \leq A(x \oplus y)$ and $A(x) \wedge \lambda \leq A(x') \wedge A(0)$. For any $\mu \notin \alpha(\lambda)$, suppose $A^{[\mu]} \neq \emptyset$ and let $x, y \in A^{[\mu]}$ with $x \perp y$, i.e., $\mu \notin \alpha(A(x))$ and $\mu \notin \alpha(A(y))$, then by $\mu \notin \alpha(\lambda)$, we have $\mu \notin \alpha(A(x)) \cup \alpha(A(y)) \cup \alpha(\lambda)$. From

$$\alpha(A(x)) \cup \alpha(A(y)) \cup \alpha(\lambda) = \alpha(A(x) \wedge A(y) \wedge \lambda) \text{ and } \alpha(A(x)) \cup \alpha(\lambda) = \alpha(A(x) \wedge \lambda),$$

we know $\mu \notin \alpha(A(x) \wedge A(y) \wedge \lambda)$ and $\mu \notin \alpha(A(x) \wedge \lambda)$. By $A(x) \wedge A(y) \wedge \lambda \leq A(x \oplus y)$ and $A(x) \wedge \lambda \leq A(x') \wedge A(0)$, we have $\mu \notin \alpha(A(x \oplus y))$, $\mu \notin \alpha(A(x'))$ and $\mu \notin \alpha(A(0))$, i.e., $x \oplus y, x', 0 \in A^{[\mu]}$. Hence $A^{[\mu]}$ is a subalgebra of E. This shows

$$\mathcal{E}(A) \leq \bigvee\{\lambda \in L \mid \forall \mu \notin \alpha(\lambda), A^{[\mu]} = \emptyset \text{ or it is a subalgebra of } E\}.$$

Conversely, assume that $\lambda \in L$ and $A^{[\mu]} = \emptyset$ or it is a subalgebra of E for $\mu \notin \alpha(\lambda)$. Now we prove $A(x) \wedge A(y) \wedge \lambda \leq A(x \oplus y)$ and $A(x) \wedge \lambda \leq A(x') \wedge A(0)$ for all $x, y \in E$ satisfying $x \perp y$. Let $\mu \in L$ and $\mu \notin \alpha(A(x) \wedge A(y) \wedge \lambda)$, then $\mu \notin \alpha(A(x)) \cup \alpha(A(y)) \cup \alpha(\lambda)$. Thus $x, y \in A^{[\mu]}$ and $\mu \notin \alpha(\lambda)$. Since $A^{[\mu]}$ is a subalgebra of E, it holds that $x \oplus y \in A^{[\mu]}$, i.e., $\mu \notin \alpha(A(x \oplus y))$. This means $\alpha(A(x) \wedge A(y) \wedge \lambda) \supseteq \alpha(A(x \oplus y))$. Hence

$$A(x) \wedge A(y) \wedge \lambda = \bigwedge \alpha(A(x) \wedge A(y) \wedge \lambda) \leq \bigwedge \alpha(A(x \oplus y)) = A(x \oplus y).$$

Similarly, we can prove $A(x) \wedge \lambda \leq A(x') \wedge A(0)$. This shows

$$\mathcal{E}(A) \geq \bigvee\{\lambda \in L \mid \forall \mu \notin \alpha(\lambda), A^{[\mu]} = \emptyset \text{ or it is a subalgebra of } E\}.$$

(iv) By (iii) we first obtain

$$\mathcal{E}(A) \leq \bigvee\{\lambda \in L \mid \forall \mu \in P(L), \mu \notin \alpha(\lambda), A^{[\mu]} = \emptyset \text{ or it is a subalgebra of } E\}.$$

Conversely, assume that $\lambda \in L$ and $A^{[\mu]} = \emptyset$ or it is a subalgebra of E for all $\mu \in P(L)$ satisfying $\mu \notin \alpha(\lambda)$. Now we prove $A(x) \wedge A(y) \wedge \lambda \leq A(x \oplus y)$ and $A(x) \wedge \lambda \leq A(x') \wedge A(0)$ for all $x, y \in E$ satisfying $x \perp y$. Let $\mu \in P(L)$ and $\mu \notin \alpha^*(A(x) \wedge A(y) \wedge \lambda)$. It follows that

$$\mu \notin \alpha(A(x) \wedge A(y) \wedge \lambda) = \alpha(A(x)) \cup \alpha(A(y)) \cup \alpha(\lambda).$$

Thus $x, y \in A^{[\mu]}$ and $\mu \notin \alpha(\lambda)$. Since $A^{[\mu]}$ is a subalgebra of E, it holds that $x \oplus y \in A^{[\mu]}$, i.e., $\mu \notin \alpha(A(x \oplus y))$. So $\mu \notin \alpha^*(A(x \oplus y))$, this means $\alpha^*(A(x) \wedge A(y) \wedge \lambda) \supseteq \alpha^*(A(x \oplus y))$. Hence

$$A(x) \wedge A(y) \wedge \lambda = \bigwedge \alpha^*(A(x) \wedge A(y) \wedge \lambda) \leq \bigwedge \alpha^*(A(x \oplus y)) = A(x \oplus y).$$

Similarly, we can prove $A(x) \wedge \lambda \leq A(x') \wedge A(0)$. This shows

$$\mathcal{E}(A) \geq \bigvee\{\lambda \in L \mid \forall \mu \in P(L), \mu \notin \alpha(\lambda), A^{[\mu]} = \emptyset \text{ or it is a subalgebra of } E\}.$$

(v) For any $x, y \in E$ satisfying $x \perp y$, assume that $\perp_L \neq \lambda \in L$ with the property of $A(x) \wedge A(y) \wedge \lambda \leq A(x \oplus y)$ and $A(x) \wedge \lambda \leq A(x') \wedge A(0)$. For any $\mu \in \beta(\lambda)$, suppose $A_{(\mu)} \neq \emptyset$ and let $x, y \in A_{(\mu)}$ with $x \perp y$, i.e., $\mu \in \beta(A(x))$ and $\mu \in \beta(A(y))$. Then by $\mu \in \beta(\lambda)$, we have $\mu \in \beta(A(x)) \cap \beta(A(y)) \cap \beta(\lambda)$. From

$$\beta(A(x)) \cap \beta(A(y)) \cap \beta(\lambda) = \beta(A(x) \wedge A(y) \wedge \lambda) \text{ and } \beta(A(x)) \cap \beta(\lambda) = \beta(A(x) \wedge \lambda),$$

we know $\mu \in \beta(A(x) \wedge A(y) \wedge \lambda)$ and $\mu \in \beta(A(x) \wedge \lambda)$. By $A(x) \wedge A(y) \wedge \lambda \leq A(x \oplus y)$ and $A(x) \wedge \lambda \leq A(x') \wedge A(0)$, we have $\mu \in \beta(A(x \oplus y))$, $\mu \in \beta(A(x'))$ and $\mu \in \beta(A(0))$, i.e., $x \oplus y, x', 0 \in A_{(\mu)}$. Hence $A_{(\mu)}$ is a subalgebra of E. This shows

$$\mathcal{E}(A) \leq \bigvee\{\lambda \in L \mid \forall \mu \in \beta(\lambda), A_{(\mu)} = \emptyset \text{ or it is a subalgebra of } E\}.$$

Conversely, assume that $\lambda \in L$ and $A_{(\mu)} = \emptyset$ or it is a subalgebra of E for $\mu \in \beta(\lambda)$. Now we prove $A(x) \wedge A(y) \wedge \lambda \leq A(x \oplus y)$ and $A(x) \wedge \lambda \leq A(x') \wedge A(0)$ for all $x, y \in E$ satisfying $x \perp y$. The statement holds obviously when $A(x) \wedge A(y) \wedge \lambda = \perp_L$. Assume $A(x) \wedge A(y) \wedge \lambda \neq \perp_L$, it implies $A(x) \neq \perp_L, A(y) \neq \perp_L$ and $\lambda \neq \perp_L$. Take $\mu \in L$ with $\mu \in \beta(A(x) \wedge A(y) \wedge \lambda)$, then $\mu \in \beta(A(x)) \cap \beta(A(y)) \cap \beta(\lambda)$. Thus $x, y \in A_{(\mu)}$ and $\mu \in \beta(\lambda)$. Since $A_{(\mu)}$ is a subalgebra of E, it holds that $x \oplus y \in A_{(\mu)}$, i.e., $\mu \in \beta(A(x \oplus y))$. This means $\beta(A(x) \wedge A(y) \wedge \lambda) \subseteq \beta(A(x \oplus y))$. Hence

$$A(x) \wedge A(y) \wedge \lambda = \bigvee \beta(A(x) \wedge A(y) \wedge \lambda) \leq \bigvee \beta(A(x \oplus y)) = A(x \oplus y).$$

Similarly, we can prove $A(x) \wedge \lambda \leq A(x') \wedge A(0)$. This shows

$$\mathcal{E}(A) \geq \bigvee\{\lambda \in L \mid \forall \mu \in \beta(\lambda), A_{(\mu)} = \emptyset \text{ or it is a subalgebra of } E\}.$$

This completes the proof. □

Definition 9. *Let E be an effect algebra and $A \in L^E$. Then A is called an L-subalgebra provided that $\mathcal{E}(A) = \top_L$. In particular, we say an L-subalgebra is a fuzzy subalgebra when $L = [0,1]$.*

Example 2. *In Example 1, it follows from Definition 9 that A_1 is a fuzzy subalgebra, but A_2 and A_3 are not fuzzy subalgebras; A_2 is a fuzzy subalgebra in the degree of 0.3.*

From Definition 9 and Lemma 1, the following proposition is obvious.

Proposition 1. *Let E be an effect algebra and $A \in L^E$. Then A is an L-subalgebra if and only if for any $x, y \in E$ with $x \perp y$,*

$$A(0) \geq A(x), A(x') \geq A(x) \text{ and } A(x \oplus y) \geq A(x) \wedge A(y).$$

Corollary 1. *Let A be an L-subalgebra of an effect algebra E. Then*

(i) *for each $x \in E$, $A(1) \geq A(x)$;*
(ii) *for each $x \in E$, $A(x) = A(x')$;*
(iii) *$A(1) = A(0)$.*

Proof. (i) Since $x \perp x'$ and A is an L-subalgebra of E, we have

$$A(1) = A(x \oplus x') \geq A(x) \wedge A(x') = A(x).$$

(ii) By Proposition 1 we first obtain $A(x') \geq A(x)$. On the other hand, by Definition 1 (E3), we know $x'' = x$ and thus $A(x) = A(x'') \geq A(x')$.

(iii) By (i) we know $A(1) \geq A(0)$. By Proposition 1, we have $A(1) \leq A(0)$, so $A(1) = A(0)$. □

By Definition 9 and Theorem 2, we can obtain the following two results

Corollary 2. *Let A be an L-subset of an effect algebra E. Then the following conditions are equivalent:*

(i) *A is an L-subalgebra;*
(ii) *for each* $\lambda \in J(L)$, $A_{[\lambda]} = \emptyset$ *or it is a subalgebra of E;*
(iii) *for each* $\lambda \in L$, $A_{[\lambda]} = \emptyset$ *or it is a subalgebra of E;*
(iv) *for each* $\lambda \in L$, $A^{[\lambda]} = \emptyset$ *or it is a subalgebra of E;*
(v) *for each* $\lambda \in P(L)$, $A^{[\lambda]} = \emptyset$ *or it is a subalgebra of E;*
(vi) *for each* $\lambda \in P(L)$, $A^{(\lambda)} = \emptyset$ *or it is a subalgebra of E.*

Corollary 3. *Let A be an L-subset of an effect algebra E. If* $\beta(\lambda \wedge \mu) = \beta(\lambda) \cap \beta(\mu)$ *for all* $\lambda, \mu \in L$, *then the following conditions are equivalent:*

(i) *A is an L-subalgebra;*
(ii) *for each* $\lambda \in L$, $A_{(\lambda)} = \emptyset$ *or it is a subalgebra of E;*
(iii) *for each* $\lambda \in J(L)$, $A_{(\lambda)} = \emptyset$ *or it is a subalgebra of E.*

Let E be an effect algebras and $(2^E)^L$ represents the set of all mapping $H : L \longrightarrow 2^E$. Then we have the following definitions.

Definition 10 ([36]). *Let* $H \in (2^E)^L$.

(i) *If* $\mu \in \beta(\lambda)$ *implies* $H(\lambda) \subset H(\mu)$, *then H is called an* L_β*-nest of E.*
(ii) *If* $\mu \in \alpha(\lambda)$ *implies* $H(\mu) \subset H(\lambda)$, *then H is called an* L_α*-nest of E.*

By Theorem 2.3 in [36], we can prove the following two theorems.

Theorem 3. *Let L be completely distributive and let* $\{A(\lambda) \mid \lambda \in L\})$ $(A(\lambda) \neq \emptyset)$ *be an* L_α*-nest of subalgebras on an effect algebra E. Then there exists an L-subalgebra B such that*

(i) $B^{(\lambda)} \subseteq A(\lambda) \subseteq B^{[\lambda]}$ *for all* $\lambda \in L$;
(ii) $B^{(\lambda)} = \bigcup_{\mu \in \alpha(\lambda)} A(\mu)$ *for all* $\lambda \in P(L)$;
(iii) $B^{[\mu]} = \bigcap_{\mu \in \alpha(\lambda)} A(\lambda)$ *for all* $\mu \in P(L)$.

Theorem 4. *Let L be completely distributive and let* $\{A(\lambda) \mid \lambda \in L\}$ $(A(\lambda) \neq \emptyset)$ *be an* L_β*-nest of subalgebras on an effect algebra E. Then there exists an L-subalgebra B such that*

(i) $B_{(\lambda)} \subseteq A(\lambda) \subseteq B_{[\lambda]}$ *for all* $\lambda \in L$;
(ii) $B_{(\lambda)} = \bigcup_{\lambda \in \beta(\mu)} A(\mu)$ *for all* $\lambda \in L$;
(iii) $B_{[\mu]} = \bigcap_{\lambda \in \beta(\mu)} A(\lambda)$ *for all* $\mu \in L$.

In particular, when $L = [0, 1]$, we have the following result.

Corollary 4. *Let* $\{A(\lambda) \mid \lambda \in (0, 1]\}$ $(A(\lambda) \neq \emptyset)$ *be a family of subalgebras on an effect algebra E. If* $\lambda < \mu \Rightarrow A(\mu) \subseteq A(\lambda)$, *then there exists a fuzzy subalgebra B satisfying*

(i) $B_{(\lambda)} \subseteq A(\lambda) \subseteq B_{[\lambda]}$;
(ii) $B_{(\lambda)} = \bigcup_{\lambda < \mu} A(\mu)$ *for all* $\lambda \in (0, 1]$;
(iii) $B_{[\mu]} = \bigcap_{\lambda < \mu} A(\lambda)$ *for all* $\mu \in (0, 1]$.

4. *L*-Fuzzy Convexity on Effect Algebras

In this section, we study the relationship between the L-fuzzy subalgebra degree and L-fuzzy convexity on an effect algebra. Further, we prove that a morphism between two effect algebras is an L-fuzzy convexity preserving mapping and a monomorphism is an L-fuzzy convex-to-convex mapping. Finally, we prove that the set of all L-subalgebras on an effect algebra is an L-convexity. For instance, we give its L-convex hull formula.

For each $A \in L^E$, $\mathcal{E}$ can be naturally considered as a mapping $\mathcal{E} : L^E \longrightarrow L$ defined by $A \longmapsto \mathcal{E}(A)$. The following theorem will show that $\mathcal{E}$ is an L-fuzzy convexity on an effect algebra E.

Theorem 5. *Let E be an effect algebra. Then $\mathcal{E} : L^E \longrightarrow L$ is an L-fuzzy convexity on E.*

Proof. (LMC1). It is clear that $\mathcal{E}(\underline{\perp}) = \mathcal{E}(\overline{\top}) = \top_L$ by Definition 8.

(LMC2). Let $\{A_i\}_{i\in I} \subseteq L^E$ be nonempty. Now we show $\mathcal{E}(\bigwedge_{i\in I} A_i) \geq \bigwedge_{i\in I} \mathcal{E}(A_i)$. Let $\lambda \in L$ with $\lambda \leq \bigwedge_{i\in I} \mathcal{E}(A_i)$. Then for any $i \in I$, we have $\lambda \leq \mathcal{E}(A_i)$, which implies $A_i(x) \wedge A_i(y) \wedge \lambda \leq A_i(x \oplus y)$ and $A_i(x) \wedge \lambda \leq A_i(x') \wedge A_i(0)$ for all $x, y \in E$ satisfying $x \perp y$. It follows that

$$\bigwedge_{i\in I} A_i(x) \wedge \bigwedge_{i\in I} A_i(y) \wedge \lambda \leq \bigwedge_{i\in I} A_i(x \oplus y),$$

$$\bigwedge_{i\in I} A_i(x) \wedge \lambda \leq \bigwedge_{i\in I} A_i(x') \wedge \bigwedge_{i\in I} A_i(0).$$

This gives that $\lambda \leq \mathcal{E}(\bigwedge_{i\in I} A_i)$. Hence $\mathcal{E}(\bigwedge_{i\in I} A_i) \geq \bigwedge_{i\in I} \mathcal{E}(A_i)$.

(LMC3) Let $\{A_i\}_{i\in I} \subseteq L^E$ be nonempty and directed. Now, we show $\mathcal{E}(\bigvee_{i\in I} A_i) \geq \bigwedge_{i\in I} \mathcal{E}(A_i)$. Let $\lambda \in L$ with $\lambda \leq \bigwedge_{i\in I} \mathcal{E}(A_i)$. Then for any $i \in I$, $\lambda \leq \mathcal{E}(A_i)$ implies $A_i(x) \wedge A_i(y) \wedge \lambda \leq A_i(x \oplus y)$ and $A_i(x) \wedge \lambda \leq A_i(x') \wedge A_i(0)$ for all $x, y \in E$ satisfying $x \perp y$. Now, we prove

$$\bigvee_{i\in I} A_i(x) \wedge \bigvee_{i\in I} A_i(y) \wedge \lambda \leq \bigvee_{i\in I} A_i(x \oplus y).$$

Take $\mu \in L$ with

$$\mu \prec \bigvee_{i\in I} A_i(x) \wedge \bigvee_{i\in I} A_i(y) \wedge \lambda,$$

then there exists $i, j \in I$ such that $\mu \leq A_i(x)$, $\mu \leq A_j(y)$ and $\mu \leq \lambda$. Since $\{A_i\}_{i\in I}$ is directed, there exists $i_0 \in I$ such that $A_i \leq A_{i_0}$ and $A_j \leq A_{i_0}$, we thus have $\mu \leq A_{i_0}(x) \wedge A_{i_0}(y) \wedge \lambda$. By the fact that $A_{i_0}(x) \wedge A_{i_0}(y) \wedge \lambda \leq A_{i_0}(x \oplus y)$, so $\mu \leq A_{i_0}(x \oplus y) \leq \bigvee_{i\in I} A_i(x \oplus y)$. Hence

$$\bigvee_{i\in I} A_i(x) \wedge \bigvee_{i\in I} A_i(y) \wedge \lambda \leq \bigvee_{i\in I} A_i(x \oplus y).$$

Similarly, we can prove

$$\bigvee_{i\in I} A_i(x) \wedge \lambda \leq \bigvee_{i\in I} A_i(x') \wedge \bigvee_{i\in I} A_i(0).$$

Hence $\lambda \leq \mathcal{E}(\bigvee_{i\in I} A_i)$, it implies $\mathcal{E}(\bigvee_{i\in I} A_i) \geq \bigwedge_{i\in I} \mathcal{E}(A_i)$.

Therefore, $\mathcal{E} : L^E \longrightarrow L$ is an L-fuzzy convexity. □

In order to investigate L-fuzzy convexity preserving mappings and L-fuzzy convex-to-convex mappings, we first give following definition and lemma.

Definition 11 ([40]). *Let E and F be effect algebras and g: $E \longrightarrow F$ is called*

(i) *a morphism if* $g(1_E) = 1_F$, *and* $x \perp y, x, y \in E$ *implies* $g(x) \perp g(y)$ *in* F, *and* $g(x \oplus y) = g(x) \oplus g(y)$;
(ii) *a monomorphism if* g *is a morphism and* $g(x) \perp g(y)$ *iff* $x \perp y$.

Lemma 2. *Let E and F be effect algebras and* $g\colon E \longrightarrow F$ *is a morphism. Then* $g(0_E) = 0_F$ *and* $g(x') = g(x)'$ *for all* $x \in E$

Proof. Since $g(0_E \oplus 1_E) = g(0_E) \oplus g(1_E) = g(0_E) \oplus 1_F$ by Definition 11, it follows from Definition 1 (E3) that $g(0_E) = 0_F$. Since $x \perp x'$ for any $x \in E$, we have $g(1_E) = g(x \oplus x') = g(x) \oplus g(x')$, thus $g(x') = g(x)'$. □

Theorem 6. *Let* $g : E \longrightarrow F$ *be a morphism between two effect algebras E and F. Then*
(i) $g\colon (E, \mathcal{E}_E) \longrightarrow (F, \mathcal{E}_F)$ *is L-fuzzy convexity preserving.*
(ii) *If* $g : E \longrightarrow F$ *is a monomorphism, then* $g\colon (E, \mathcal{E}_E) \longrightarrow (F, \mathcal{E}_F)$ *is L-fuzzy convex-to-convex.*

Proof. (i) In order to prove $g\colon (E, \mathcal{E}_E) \longrightarrow (M, \mathcal{E}_F)$ is L-fuzzy convexity preserving, we just need to prove for any $A \in L^F$, $\mathcal{E}_E(g_L^{\leftarrow}(A)) \geq \mathcal{E}_F(A)$.

Let $\lambda \in L$ with $\lambda \leq \mathcal{E}_F(A)$. Then by Lemma 1 we obtain $A(x) \wedge A(y) \wedge \lambda \leq A(x \oplus y)$ and $A(x) \wedge \lambda \leq A(x') \wedge A(0)$ for all $x, y \in F$ satisfying $x \perp y$. So for any $a, b \in E$ with $a \perp b$,

$$\begin{aligned} \lambda \wedge g_L^{\leftarrow}(A)(a) \wedge g_L^{\leftarrow}(A)(b) &= \lambda \wedge A(g(a)) \wedge A(g(b)) \\ &\leq A(g(a) \oplus g(b)) \\ &= A(g(a \oplus b)) \\ &= g_L^{\leftarrow}(A)(a \oplus b). \end{aligned}$$

Similarly, we can obtain $\lambda \wedge g_L^{\leftarrow}(A)(a) \leq g_L^{\leftarrow}(A)(a')$ and $\lambda \wedge g_L^{\leftarrow}(A)(a) \leq g_L^{\leftarrow}(A)(0_E)$, which implies $\lambda \leq \mathcal{E}_E(g_L^{\leftarrow}(A))$. Hence $\mathcal{E}_E(g_L^{\leftarrow}(A)) \geq \mathcal{E}_F(A)$.

(ii) In order to prove $g\colon (E, \mathcal{E}_E) \longrightarrow (F, \mathcal{E}_F)$ is L-fuzzy convex-to-convex, we just need to prove for any $A \in L^E$, $\mathcal{E}_E(A) \leq \mathcal{E}_F(g_L^{\rightarrow}(A))$.

Let $\lambda \in L$ with $\lambda \leq \mathcal{E}_E(A)$, then $A(a) \wedge A(b) \wedge \lambda \leq A(a \oplus b)$ and $A(a) \wedge \lambda \leq A(a') \wedge A(0)$ for all $a, b \in E$ satisfying $a \perp b$. It follows that for $x, y \in F$ satisfying $x \perp y$,

$$\begin{aligned} (g_L^{\rightarrow}(A))(x) \wedge (g_L^{\rightarrow}(A))(y) \wedge \lambda &= \lambda \wedge \bigvee_{g(a)=x} A(a) \wedge \bigvee_{g(b)=y} A(b) \\ &= \bigvee\{\lambda \wedge A(a) \wedge A(b) \mid g(a) = x, g(b) = y\} \\ &\leq \bigvee\{A(a \oplus b) \mid g(a) = x, g(b) = y\} \\ &\leq \bigvee\{A(a \oplus b) \mid g(a \oplus b) = x \oplus y\} \\ &\leq \bigvee\{A(c) \mid g(c) = x \oplus y\} \\ &= (g_L^{\rightarrow}(A))(x \oplus y). \end{aligned}$$

Similarly, we can prove $(g_L^{\rightarrow}(A))(x) \wedge \lambda \leq (g_L^{\rightarrow}(A))(x') \wedge (g_L^{\rightarrow}(A))(0)$. This gives that $\lambda \leq \mathcal{E}_F(g_L^{\rightarrow}(A))$. Hence $\mathcal{E}_E(A) \leq \mathcal{E}_F(g_L^{\rightarrow}(A))$. Therefore, $g : (E, \mathcal{E}_E) \longrightarrow (F, \mathcal{E}_F)$ is L-fuzzy convex-to-convex. □

Corollary 5. *Let* $\{E_i\}_{i\in\Lambda}$ *be a family of effect algebras and* $\prod_{i\in\Lambda} A_i$ *be the direct product of* $\{A_i\}_{i\in\Lambda}$, *where* $A_i \in L^{E_i}$. *If* $p_i : \prod_{i\in\Lambda} E_i \longrightarrow E_i$ *is the projection, then* $p_i\colon (\prod_{i\in\Lambda} E_i, \mathcal{E}_{\prod_{i\in\Lambda} E_i}) \longrightarrow (E_i, \mathcal{E}_{E_i})$ *is L-fuzzy convexity preserving for all* $i \in \Lambda$.

Proof. It is easy to check that $p_i : \prod_{i\in\Lambda} E_i \longrightarrow E_i$ is a morphism between effects algebras, so by Theorem 6, we know p_i is L-fuzzy convexity preserving for all $i \in \Lambda$. □

Theorem 7. *Let $\{E_i\}_{i\in\Lambda}$ be a family of effect algebras and $\prod_{i\in\Lambda} A_i$ be the direct product of $\{A_i\}_{i\in\Lambda}$, where $A_i \in L^{E_i}$. Then $\mathcal{E}_{\prod_{i\in\Lambda} E_i}\Big(\prod_{i\in\Lambda} A_i\Big) \geq \bigwedge_{i\in\Lambda} \mathcal{E}_{E_i}(A_i)$.*

Proof. Let $p_i : \prod_{i\in\Lambda} E_i \longrightarrow E_i$ be the projection. It can be easily proved that $\prod_{i\in\Lambda} A_i = \bigwedge_{i\in\Lambda} p_i^{\leftarrow}(A_i)$. So by the proof of **(LMC2)** in Theorem 8, we have

$$\mathcal{E}_{\prod_{i\in\Lambda} E_i}\Big(\prod_{i\in\Lambda} A_i\Big) = \mathcal{E}_{\prod_{i\in\Lambda} E_i}\Big(\bigwedge_{i\in\Lambda} p_i^{\leftarrow}(A_i)\Big) \geq \bigwedge_{i\in\Lambda} \mathcal{E}_{\prod_{i\in\Lambda} E_i}(p_i^{\leftarrow}(A_i)) \geq \bigwedge_{i\in\Lambda} \mathcal{E}_{E_i}(A_i).$$

The last inequality holds because p_i is L-fuzzy convexity preserving by Corollary 4. □

From paper [27], we know $(E, \mathcal{E}_\lambda)$ is an L-convex space for all $\lambda \in L\backslash\bot_L$. In particular, $(E, \mathcal{E}_{\top_L})$ is an L-convex space. We should note that $\mathcal{E}_{\top_L} = \{A \in L^E \mid \mathcal{E}(A) = \top_L\}$, by Definition 9, we obtain that $\mathcal{E}_{\top_L}$ is the set of all L-subalgebras on an effect algebra E.

The L-convex hull operator with respect to the L-convex space $(E, \mathcal{E}_{\top_L})$ is as follows: for $A \in L^E$, $co(A) = \bigwedge\{B \in L^E \mid A \leq B \in \mathcal{E}_{\top_L}\}$ [17], i.e., $co(A)$ is the least L-subalgebra containing A. In the following, we will give the corresponding L-convex hull formula.

Theorem 8. *Let A be an L-subset of an effect algebra E. Define for any $x \in E$,*

$$\sigma_0 = A, \sigma_1(x) = \left(\bigvee_{x=x_1\oplus x_2} [A(x_1) \vee A(x_1')] \wedge [A(x_2) \vee A(x_2')]\right) \vee A(x),$$

$$when\ n \geq 2,\ \sigma_n(x) = \bigvee_{x=x_1\oplus x_2} [\sigma_{n-1}(x_1) \vee \sigma_{n-1}(x_1')] \wedge [\sigma_{n-1}(x_2) \vee \sigma_{n-1}(x_2')].$$

Then $co(A) = \bigvee_{n=0}^{\infty} \sigma_n$.

Proof. We first show $\bigvee_{n=0}^{\infty} \sigma_n$ is an L-subalgebra of E. To achieve this, we give the following statements.

(i). For any $x \in E$ and $n \geq 1$, $\sigma_n(1) \geq \sigma_n(x)$.

Let $n = 1$. For any $x \in E$, it follows from $x \oplus x' = 1$ that

$$\sigma_1(1) \geq A(x) \vee A(x') \geq A(x) = \sigma_0(x).$$

Let $n \geq 2$. For any $x \in E$, since $x \oplus x' = 1$, we have that for each $x \in E$.

$$\begin{aligned}\sigma_n(1) &= \bigvee_{1=y\oplus z} [\sigma_{n-1}(y) \vee \sigma_{n-1}(y')] \wedge [\sigma_{n-1}(z) \vee \sigma_{n-1}(z')] \\ &\geq [\sigma_{n-1}(x) \vee \sigma_{n-1}(x')] \wedge [\sigma_{n-1}(x) \vee \sigma_{n-1}(x')] \\ &= \sigma_{n-1}(x) \vee \sigma_{n-1}(x').\end{aligned}$$

Since for any $x \in E$,

$$\sigma_n(x) = \bigvee_{x=x_1\oplus x_2} [\sigma_{n-1}(x_1) \vee \sigma_{n-1}(x_1')] \wedge [\sigma_{n-1}(x_2) \vee \sigma_{n-1}(x_2')],$$

then for every pair (x_1, x_2) satisfying $x = x_1 \oplus x_2$, it holds that $\sigma_{n-1}(x_1) \vee \sigma_{n-1}(x_1') \leq \sigma_n(1)$ and $\sigma_{n-1}(x_2) \vee \sigma_{n-1}(x_2') \leq \sigma_n(1)$, we thus obtain $\sigma_n(x) \leq \sigma_n(1)$.

(ii). $\bigvee_{n=0}^{\infty} \sigma_n(x) \leq \bigvee_{n=0}^{\infty} \sigma_n(1)$.

Since $\sigma_0 = A$ and $1 = 1 \oplus 0$, we have $\sigma_1(1) \geq A(1) \vee A(0) \geq A(1)$. By the above proof, we obtain $\sigma_1(1) \geq A(x) \vee A(x') \geq A(x)$. This implies

$$\bigvee_{n=0}^{\infty} \sigma_n(x) = A(x) \vee \bigvee_{n=1}^{\infty} \sigma_n(x) \leq \sigma_1(1) \vee \bigvee_{n=1}^{\infty} \sigma_n(1) \leq \bigvee_{n=0}^{\infty} \sigma_n(1).$$

(iii). $\sigma_0(x) \leq \sigma_1(x) \leq \sigma_2(x) \leq \cdots$

For any $x \in E$, from the construction of $\sigma_1(x)$, it is clear that $\sigma_1(x) \geq A(x) = \sigma_0(x)$. Let $n \geq 2$. For any $x \in E$, by the fact that $x = x \oplus 0$ and (i), we have

$$\begin{aligned} \sigma_n(x) &= \bigvee_{x=x_1 \oplus x_2} [\sigma_{n-1}(x_1) \vee \sigma_{n-1}(x_1')] \wedge [\sigma_{n-1}(x_2) \vee \sigma_{n-1}(x_2')] \\ &\geq [\sigma_{n-1}(x) \vee \sigma_{n-1}(x')] \wedge [\sigma_{n-1}(0) \vee \sigma_{n-1}(1)] \\ &\geq \sigma_{n-1}(x). \end{aligned}$$

(iv). $\bigvee_{n=0}^{\infty} \sigma_n(x) \leq \bigvee_{n=0}^{\infty} \sigma_n(x')$.

For any $x \in E$ and $n \geq 2$, since $x' = x' \oplus 0$, we have

$$\sigma_n(x') \geq [\sigma_{n-1}(x) \vee \sigma_{n-1}(x')] \wedge [\sigma_{n-1}(0) \vee \sigma_{n-1}(1)].$$

By (i) we know that $\sigma_{n-1}(x), \sigma_{n-1}(x') \leq \sigma_{n-1}(1)$, it follows that $\sigma_n(x') \geq \sigma_{n-1}(x)$. Thus $\bigvee_{n=1}^{\infty} \sigma_n(x) \leq \bigvee_{n=2}^{\infty} \sigma_n(x')$. By (iii) we obtain

$$\bigvee_{n=0}^{\infty} \sigma_n(x) = \bigvee_{n=1}^{\infty} \sigma_n(x) \leq \bigvee_{n=2}^{\infty} \sigma_n(x') = \bigvee_{n=0}^{\infty} \sigma_n(x').$$

(v). For $x, y \in E$ and $x \perp y$,

$$\bigvee_{n=0}^{\infty} \sigma_n(x \oplus y) \geq \left(\bigvee_{n=0}^{\infty} \sigma_n(x) \right) \wedge \left(\bigvee_{n=0}^{\infty} \sigma_n(y) \right).$$

Let $t \in L$ such that $t \prec \left(\bigvee_{n=0}^{\infty} \sigma_n(x) \right) \wedge \left(\bigvee_{n=0}^{\infty} \sigma_n(y) \right)$, then we have $t \prec \bigvee_{n=0}^{\infty} \sigma_n(x)$ and $t \prec \bigvee_{n=0}^{\infty} \sigma_n(y)$. So there exist $i, k \geq 0$ such that $t \leq \sigma_i(x)$ and $t \leq \sigma_j(y)$. Put $j = \max\{i, k\}$, then by (iii) we obtain $t \leq \sigma_j(x)$ and $t \leq \sigma_j(y)$. It results that

$$t \leq \sigma_j(x) \wedge \sigma_j(y) \leq [\sigma_j(x) \vee \sigma_j(x')] \wedge [\sigma_j(y) \vee \sigma_j(y')] \leq \sigma_{j+1}(x \oplus y) \leq \bigvee_{n=0}^{\infty} \sigma_n(x \oplus y).$$

Hence $\bigvee_{n=0}^{\infty} \sigma_n(x \oplus y) \geq \left(\bigvee_{n=0}^{\infty} \sigma_n(x) \right) \wedge \left(\bigvee_{n=0}^{\infty} \sigma_n(y) \right)$.

(vi). $\bigvee_{n=0}^{\infty} \sigma_n(x) \leq \bigvee_{n=0}^{\infty} \sigma_n(0)$.

By (ii) and (iv), we have,

$$\bigvee_{n=0}^{\infty} \sigma_n(0) = \bigvee_{n=0}^{\infty} \sigma_n(1') \geq \bigvee_{n=0}^{\infty} \sigma_n(1) \geq \bigvee_{n=0}^{\infty} \sigma_n(x).$$

Therefore, by (iii), (iv), (v), (vi) and Proposition 1, we can obtain that $\bigvee_{n=0}^{\infty} \sigma_n$ is an L-subalgebra containing A. Next, we prove $\bigvee_{n=0}^{\infty} \sigma_n$ is the least L-subalgebra containing A.

Assume $B \in L^E$ is an L-subalgebra containing A, so $A(x) \leq B(x)$ for all $x \in E$. It is clear that $\sigma_0 \leq B$. Since

$$\begin{aligned}
\sigma_1(x) &= \Big(\bigvee_{x=x_1\oplus x_2}[A(x_1)\vee A(x_1')]\wedge[A(x_2)\vee A(x_2')]\Big)\vee A(x)\\
&\le \Big(\bigvee_{x=x_1\oplus x_2}[B(x_1)\vee B(x_1')]\wedge[B(x_2)\vee B(x_2')]\Big)\vee B(x)\\
&= \Big(\bigvee_{x=x_1\oplus x_2}B(x_1)\wedge B(x_2)\Big)\vee B(x) \qquad \text{(by Corrollary 1 and } B \text{ is an } L\text{-subalgebra)}\\
&\le \Big(\bigvee_{x=x_1\oplus x_2}B(x_1\oplus x_2)\Big)\vee B(x) = B(x)\vee B(x) = B(x).
\end{aligned}$$

Assume $\sigma_k \le B$ $(k \ge 2)$ holds. Now,

$$\begin{aligned}
\sigma_{k+1}(x) &= \bigvee_{x=x_1\oplus x_2}[\sigma_k(x_1)\vee\sigma_k(x_1')]\wedge[\sigma_k(x_2)\vee\sigma_k(x_2')]\\
&\le \bigvee_{x=x_1\oplus x_2}[B(x_1)\vee B(x_1')]\wedge[B(x_2)\vee B(x_2')]\\
&= \bigvee_{x=x_1\oplus x_2}B(x_1)\wedge B(x_2)\\
&\le \bigvee_{x=x_1\oplus x_2}B(x_1\oplus x_2) = B(x).
\end{aligned}$$

Thus for any $n \ge 0$, it holds that $\sigma_n \le B$, which implies $\bigvee_{n=0}^{\infty}\sigma_n \le B$. This implies $\bigvee_{n=0}^{\infty}\sigma_n$ is the least L-subalgebra containing A. Therefore, $co(A) = \bigvee_{n=0}^{\infty}\sigma_n$. □

5. Conclusions

In this paper, we proposed the notion of L-fuzzy subalgebra degree on an effect algebra based on a completely distributive lattice L. We gave their characterizations in terms of four kinds of cut sets of L-subsets. We say an L-subset is an L-subalgebra if its L-fuzzy subalgebra degree is equal to $\top_L$. In fact, the L-fuzzy subalgebra degree can describe the degree to which an L-subset is an L-fuzzy subalgebra. An L-fuzzy convex structure on an effect algebra was naturally constructed and some of its properties were studied. For instance, an L-convex structure was induced by the set of all L-subalgebras and its L-convex hull formula was given.

It should be noted that the same thought can be applied to different algebraic structure such as MV-algebras, BL-algebras, residuated lattices and so on. Thus, there exist L-fuzzy convexities in many logical algebras, which enrich the convexity theory in logical algebras.

Author Contributions: Conceptualization, Y.-Y.D.; formal analysis, Y.-Y.D. and F.-G.S.; funding acquisition, F.-G.S.; investigation, Y.-Y.D. and F.-G.S.; methodology, Y.-Y.D. and F.-G.S.; project administration, Y.-Y.D. and F.-G.S.; supervision, F.-G.S.; validation, Y.-Y.D. and F.-G.S.; visualization, F.-G.S.; writing—original draft, Y.-Y.D.; writing—review and editing, Y.-Y.D. and F.-G.S. Both authors have read and agreed to the published version of the manuscript.

Funding: The project is funded by the National Natural Science Foundation of China (11871097, 12071033), Beijing Institute of Technology Science and Technology Innovation Plan Cultivation Project (No. 2021CX01030) and Chinese Postdoctoral Science Foundation (No. 2020M670142).

Institutional Review Board Statement: Not applicable.

Informed Consent Statement: Not applicable.

Data Availability Statement: Not applicable.

Acknowledgments: The authors express thanks to the handling Editor and the reviewers for their careful reading and useful suggestions.

Conflicts of Interest: The authors declare no conflict of interest.

References

1. Foulis, D.J.; Bennett, M.K. Effect algebras and unsharp quantum logics. *Found. Phys.* **1994**, *24*, 1331–1352. [CrossRef]
2. Zadeh, L.A. Fuzzy sets. *Inf. Control* **1965**, *8*, 338–353. [CrossRef]
3. Ma, Z. Note on ideals of effect algebras. *Inf. Sci.* **2009**, *179*, 505–507. [CrossRef]
4. Jenča, G.; Pulmannová, S. Ideals and quotients in lattice ordered effect algebras. *Soft Comput.* **2001**, *5*, 376–380. [CrossRef]
5. Foulis, D.J.; Greechie, R.J.; Rüttimann, G.T. Filters and supports in orthoalgebras. *Int. J. Theor. Phys.* **1992**, *31*, 789–807. [CrossRef]
6. Liu, D.L.; Wang, G.J. Fuzzy filters in effect algebras. *Fuzzy Syst. Math.* **2009**, *23*, 6–16.
7. Berger, M. Convexity. *Am. Math. Mon.* **1990**, *97*, 650–678. [CrossRef]
8. Van de Vel, M. *Theory of Convex Structures*; Elsevier: Amesterdam, The Netherlands, 1993.
9. Fan, C.Z.; Shi, F.-G.; Mehmood, F. M-Hazy Γ-Semigroups. *J. Nonlinear Convex Anal.* **2020**, *21*, 2659–2669.
10. Mehmood, F.; Shi, F.-G.; Hayat, K.; Yang, X.P. The homomorphism theorems of M-Hazy rings and their induced fuzzifying convexities. *Mathematics* **2020**, *8*, 411. [CrossRef]
11. Mehmood, F.; Shi, F.-G.; Hayat, K. A new approach to the fuzzification of rings. *J. Nonlinear Convex Anal.* **2020**, *21*, 2637–2646.
12. Changat, M.; Mulder, H.M.G.; Sierksma, G. Convexities related to path properties on graphs. *Discrete Math.* **2005**, *290*, 117–131. [CrossRef]
13. Pang, B. Categorical properties of L-fuzzifying convergence spaces. *Filomat* **2018**, *32*, 4021–4036. [CrossRef]
14. Rosa, M.V. On fuzzy topology fuzzy convexity spaces and fuzzy local convexity. *Fuzzy Sets Syst.* **1994**, *62*, 97–100. [CrossRef]
15. Maruyama, Y. Lattice-valued fuzzy convex geometry. *RIMS Kokyuroku* **2009**, *164*, 22–37.
16. Dong, Y.Y.; Li, J. Fuzzy convex structure and prime fuzzy ideal space on residuated lattices. *J. Nonlinear Convex Anal.* **2020**, *21*, 2725–2735.
17. Shen, C.; Shi, F.-G. Characterizations of L-convex spaces via domain theory. *Fuzzy Sets Syst.* **2020**, *380*, 44–63. [CrossRef]
18. Wei, X.W.; Wang, B. Fuzzy (restricted) hull operators and fuzzy convex structures on L-sets. *J. Nonlinear Convex Anal.* **2020**, *21*, 2805–2815.
19. Li, H.Y.; Wang, K. L-ordered neighborhood systems of stratified L-concave structures. *J. Nonlinear Convex Anal.* **2020**, *21*, 2783–2793.
20. Zhang, L.; Pang, B. Strong L-concave structures and L-convergence structures. *J. Nonlinear Convex Anal.* **2020**, *21*, 2759–2769.
21. Xiu, Z.-Y. Convergence structures in L-concave spaces. *J. Nonlinear Convex Anal.* **2020**, *21*, 2693–2703.
22. Shi, F.-G.; Xiu, Z.-Y. A new approach to the fuzzification of convex structures. *J. Appl. Math.* **2014**, *2014*, 1–12. [CrossRef]
23. Pang, B. Convergence structures in M-fuzzifying convex spaces. *Quaest. Math.* **2020**, *43*, 1541–1561. [CrossRef]
24. Pang, B. L-fuzzifing convex structures as L-convex structures. *J. Nonlinear Convex Anal.* **2020**, *21*, 2831–2841.
25. Xiu, Z.-Y.; Pang, B. Base axioms and subbase axioms in M-fuzzifying convex spaces. *Iran. J. Fuzzy Syst.* **2018**, *15*, 75–87.
26. Zhou, X.-W.; Wang, L. Four kinds of special mappings between M-fuzzifying convex spaces. *J. Nonlinear Convex Anal.* **2020**, *21*, 2683–2692.
27. Shi, F.-G.; Xiu, Z.-Y. (L, M)-fuzzy convex structures. *J. Nonlinear Sci. Appl.* **2017**, *10*, 3655–3669. [CrossRef]
28. Li, L.Q. On the category of enriched (L, M)-convex spaces. *J. Intell. Fuzzy Syst.* **2017**, *33*, 3209–3216. [CrossRef]
29. Li, J.; Shi, F.-G. L-fuzzy convex induced by L-convex fuzzy sublattice degree. *Iran. J. Fuzzy Syst.* **2017**, *14*, 83–102.
30. Pang, B. Bases and subbases in (L, M)-fuzzy convex spaces. *Comput. Appl. Math.* **2020**, *39*. [CrossRef]
31. Pang, B. Hull operators and interval operators in (L, M)-fuzzy convex spaces. *Fuzzy Sets Syst.* **2021**, *405*, 106–127. [CrossRef]
32. Xiu, Z.-Y.; Li, L.; Zhu, Y. A degree approach to special mappings between (L, M)-fuzzy convex spaces. *J. Nonlinear Convex Anal.* **2020**, *21*, 2625–2635.
33. Dvurečenskij, A.; Pulmannová, S. *New Trends in Quantum Structures*; Kluwer: Dordrecht, The Netherlands, 2000.
34. Wang, G.J. Theory of topological molecular lattices. *Fuzzy Sets Syst.* **1992**, *47*, 351–376.
35. Shi, F.-G. A new approach to the fuzzification of matroids. *Fuzzy Sets Syst.* **2009**, *160*, 696–705. [CrossRef]
36. Shi, F.-G. Theory of L_β-nested sets and L_α-nested sets and application. *Fuzzy Syst. Math.* **1995**, *4*, 65–72.
37. War, M.; Dilworth, R.P. Residuated lattices. *Trans. Am. Math. Soc.* **1939**, *45*, 335–354.
38. Galatos, N.; Jipsen, P.; Kowalski, T.; Ono, H. *Residuated Lattices: An Algebraic Glimpse at Substructural Logics*; Elsevier: Amsterdam, The Netherlands, 2007.
39. Shi, F.-G.; Xin, X. L-fuzzy subgroup degrees and L-fuzzy normal subgroup degrees. *J. Adv. Res. Pure Math.* **2011**, *3*, 92–108.
40. Jenča, G.; Pulmannová, S. Effect algebras with state operator. *Fuzzy Sets Syst.* **2015**, *260*, 43–61. [CrossRef]

Σ mathematics MDPI

Review

Soft Computing for Decision-Making in Fuzzy Environments: A Tribute to Professor Ioan Dzitac

Simona Dzitac [1,†] and Sorin Nădăban [2,*,†]

1 Department of Energy Engineering, Faculty of Energy Engineering and Industrial Management, University of Oradea, Universitatii 1, RO-485620 Oradea, Romania; simona.dzitac@gmail.com
2 Department of Mathematics and Computer Science, Faculty of Exact Sciences, Aurel Vlaicu University of Arad, Elena Drăgoi 2, RO-310330 Arad, Romania
* Correspondence: snadaban@gmail.com
† These authors contributed equally to this work.

Abstract: This paper is dedicated to Professor Ioan Dzitac (1953–2021). Therefore, his life has been briefly presented as well as a comprehensive overview of his major contributions in the domain of soft computing methods in a fuzzy environment. This paper is part of a special reverential volume, dedicated to the Centenary of the Birth of Lotfi A. Zadeh, whom Ioan Dzitac considered to be is his mentor, and to whom he showed his gratitude many times and in innumerable ways, including by being the Guest Editor of this Special Issue. Professor Ioan Dzitac had many important achievements throughout his career: he was co-founder and Editor-in-Chief of an ISI Expanded quoted journal, International Journal of Computers Communications & Control; together with L.A. Zadeh, D. Tufis and F.G. Filip he edited the volume "From Natural Language to Soft Computing: New Paradigms in Artificial Intelligence"; his scientific interest focused on different sub-fields: fuzzy logic applications, soft computing in a fuzzy environment, artificial intelligence, learning platform, distributed systems in internet. He had the most important contributions in soft computing in a fuzzy environment. Some of them will be presented in this paper. Finally, some future trends are discussed.

Keywords: soft computing; decision-making; fuzzy sets; fuzzy environment; Dzitac

Citation: Dzitac, S.; Nădăban, S. Soft Computing for Decision-Making in Fuzzy Environments: A Tribute to Professor Ioan Dzitac. *Mathematics* **2021**, *9*, 1701. https://doi.org/10.3390/math9141701

Academic Editor: Daniel Gómez Gonzalez

Received: 20 May 2021
Accepted: 15 July 2021
Published: 20 July 2021

1. Introduction

This paper is dedicated to Professor Ioan Dzitac (14 February 1953–6 February 2021). Therefore the paper begins with a short presentation of his life and work.

Ioan Dzitac was born in the village of Poienile de sub Munte, in the County of Maramures, Transylvania, Romania. He graduated from the Faculty of Mathematics of Babeş-Bolyai University of Cluj-Napoca in 1977 and continued as a high school math teacher in Bihor (Aleşd and Oradea), Romania. In 2002, Prof. Dzitac obtained his PhD Thesis at Babeş-Bolyai University of Cluj-Napoca and in the next few years he published several works in field of distributed and information systems.

In 2007, Dzitac had the great privilege of meeting the world-renowned scientist Lotfi A. Zadeh and since then, up to the end of his career, his scientific interest focused on different sub-fields:

1. Fuzzy logic applications. For major achievement see: [1,2] etc.
2. Soft computing in a fuzzy environment. See: [3–8] etc.
3. Artificial intelligence: [9,10] etc.
4. E-learning platform: [11,12] etc.
5. Distributed systems in internet: [9,13] etc.

He had the most important contributions in soft computing in a fuzzy environment. Some of them will be presented in this paper.

His collaboration with Lotfi A. Zadeh started in 2008. Zadeh was invited as speaker at the International Conference on Computers Communications and Control (ICCCC) (see Figure 1), an ISI indexed conference, founded and chaired by Dzitac.

Figure 1. Ioan Dzitac and Lotfi A. Zadeh at ICCCC 2008.

Just a week before passing away, I. Dzitac remembered: "I waited for him (Lofti A. Zadeh) at the airport in Budapest. At 87 years old, he was traveling unattended from San Francisco, where he lived and was the active director of the research institute BISC at the University of California, Berkeley (position that he held until his death in 2017, at 96 years old). He took a nap in the car. However, when he woke up, he started to tell me about his first visit in Romania and the encouragements offered by Grigore C. Moisil, in 1967, two years later, with great courage, he published "Fuzzy Sets". He really needed those pats on the back because many mathematicians, logicians and engineers met his theories with skepticism and sometimes even with mockery [14]".

L.A. Zadeh had a major influence on Ioan Dzitac's career, because, after their encounter, Ioan Dzitac had a very prosperous period from a scientific point of view, as he published many articles in well-known journals, either as a unique author or in cooperation with: F.G. Filip, M.J. Manolescu, S. Negulescu, A.E. Lascu, C. Butaci, S. Dzitac, G. Bologa, D. Benta, S. Nadaban, B. Barbat, I. Moisil, I. Felea, T. Vesselenyi, C. Secui, V. Lupse and abroad: B. Stanojevic (Serbia), H. Liu (China), S. Gao (China), R. Andonie (USA), A.M. Brasoveanu (Austria), Y. Shi (China), G. Kou (China), F. Cordova (Chile), H. Lee (Korea), etc.

One volume, very dear to Professor I. Dzitac, cannot be omitted. It is "From Natural Language to Soft Computing: New Paradigms in Artificial Intelligence", volume co-edited by L.A. Zadeh (University of California), D. Tufis (Romanian Academy), F.G. Filip (Romanian Academy), I. Dzitac (Agora University), Romanian Academy Ed. House, 2008 [15].

The recognition of his results appeared very soon, his papers being quoted by authors from Romania, Chile, India, USA, Iran, Malaysia, Serbia, Canada, France, Russia, Turkey, Australia, Hungary, Lithuania, Morocco, Spain, Tunisia, Algeria, Czech Republic, in some prestigious journals.

To highlight the impact of his research it should be mentioned that Ioan Dzitac has in Web of Science h-index = 11 and 472 citations, of which 64 in the first 5 months of 2021. The paper [6] has 10 citations in the period January 2021–May 2021, while the paper [2] has 16 citations in the same period. The last citations from May 2021 are:

1. Liang, H.; Cai, R. A new correlation coefficient of BPA based on generalized information quality. *Int. J. Intell. Syst.* **2021**, doi:10.1002/int.22490 [16];
2. Roszkowska, E.; Kusterka-Jefmanska, M.; Jefmanski, B. Intuitionistic Fuzzy TOPSIS as a Method for Assessing Socioeconomic Phenomena on the Basis of Survey Data. *Entropy* **2021**, *23*, 563, doi:10.3390/e23050563 [17];

3. Hamzelou, N.; Ashtiani, M; Sadeghi, R. A propagation trust model in social networks based on the A* algorithm and multi-criteria decision-making. *Computing* **2021**, *103*, 827–867, doi:10.1007/s00607-021-00918-w [18].

I. Dzitac was co-founder and Editor-in-Chief of an ISI Expanded quoted journal, *International Journal of Computers Communications & Control* (nominee by Elsevier for Journal Excellence Award - Scopus Awards Romania 2015) and member in Editorial Board of 12 scientific journals. Additionally, he is co-founder and General Chair of International Conference on Computers Communications and Control. He was member of the Program Committee of more than 80 international conferences.

He was Senior Member IEEE (since 2011). He was invited speaker and/or invited special sessions's organizer and chair in China (2013: Beijing, Suzhou, Chengdu, 2015: Dalian, 2016: Beijing), India (2014: Madurai, 2017: Delhi), Russia (2014: Moscow), Brazil (2015: Rio), Lithuania (2015: Druskininkai), South Korea (2016: Asan), USA (2018: Nebraska).

He was included among 100 Romanian computer scientists from all over the past 100 years in the volume "One Hundred Romanian Scientists in Theoretical Computer Science", Romanian Academy Publishing House, 2018.

I. Dzitac was full professor at Aurel Vlaicu University of Arad (since 2009), professor at Agora University of Oradea (since 2017) and Rector at Agora University of Oradea (2012–2020). He was an Adjunct Professor at University of Chinese Academy of Sciences—Beijing, China (2013–2016) and since 2016 he was in Advisory Board Member at Graduate School of Management of Technology, Hoseo University, South Korea.

In 2019 he defended his Habilitation Thesis "Soft Computing for Decision-Making" at "Alexandru Ioan Cuza" University of Iasi, which conferred him the right to conduct doctorates. Thus, he became PhD supervisor at University of Craiova, Romania.

In all those years, I. Dzitac did not cease to show his gratitude towards the one who considered to be his mentor: Lofti A. Zadeh. Thus, in 2011, he edited a Special Issue of IJCCC at 90th Zadeh's birthday and another in 2015 at 50th Fuzzy Sets anniversary. In 2017, at Zadeh's death, I. Dzitac published a survey about his life and his famous contributions in scientific world [2]. In January 2021, just a month before his death, I. Dzitac edited another Special Issue of IJCCC dedicated to the centenary of the birth of Lotfi A. Zadeh (1921–2017).

As already mentioned, beginning with 2007, Ioan Dzitac's interest was in soft computing methods in fuzzy environments. Starting from here, the structure of the paper will continue as follows: Section 2 will present some general considerations regarding soft computing methods, highlighting the fundamental differences between soft computing and hard computing; considering that soft computing methods are numerous, in Section 3 we will resume to presenting some fundamental ideas of fuzzy logic, which is the most used Soft Computing method in a variety of decision-making problems; in Section 4 we will present a survey on I. Dzitac's contributions to this domain. Finally, in Section 5 we will have some conclusions but mostly some future trends will be discussed.

2. Soft Computing Methods

Hard computing (HC) is the conventional calculation and it needs an analytical model well defined and many times a log time for calculating. Many analytical models are valid only in ideal cases. The problems of the real world exist within a non-ideal frame. Thus, many complex systems that are found in engineering, biology, medicine, economy remain unsolved to HC.

Soft computing is a concept introduced for the first time by L.A. Zadeh [19]. According to Zadeh's definition, Soft computing (SC) methods are opposed to HC techniques, consisting of computational techniques in Informatics, machine learning and certain engineering subjects that study, shape and analyze a very complex reality for which the traditional methods prove ineffective. Soft computing can work with ambiguous data, and it is tolerant to vagueness, uncertainty, partial truth and approximation. The model for SC is human mind.

Lotfi A. Zadeh said about Natural Language (NL) Computation: "NL-Computation is of intrinsic importance because much of human knowledge is described in natural language. This is particularly true in such fields as economics, data mining, systems engineering, risk assessment and emergency management. It is safe to predict that as we move further into the age of machine intelligence and mechanized decision-making, NL-Computation will grow in visibility and importance. Computation with information described in natural language cannot be dealt with using machinery of natural language processing. The problem is semantic imprecision of natural languages. More specifically, a natural language is basically a system for describing perceptions. Perceptions are intrinsically imprecise, reflecting the bounded ability of sensory organs, and ultimately the brain, to resolve detail and store information. Semantic imprecision of natural languages is a concomitant of imprecision of perceptions. Our approach to NL-Computation centers on what is referred to as generalized-constraint-based computation, or GC-Computation for short. A fundamental thesis which underlies NL-Computation is that information may be interpreted as a generalized constraint." [20].

Soft Computing includes: (1) Fuzzy Logic; (2) Neural Computing: Perceptions, Artificial Neural Networks, Neuro-Fuzzy Systems; (3) Evolutionary Computation: Genetic Algorithms (GA), Ant Colony Optimization (ACO), Particle Swarm Optimization (PSO), Artificial Life(AL); (4) Machine Learning: Intelligent Agents, Expert Systems, Data Mining; (5) Probabilistic Reasoning: Bayesian Networks, Markov Networks, Belief Networks.

To discuss all these methods would exceed the space for an article. Therefore, the next section will be limited to the presentation of only a few fundamental ideas of fuzzy logic, which is the most used Soft Computing method in a variety of decision-making problems.

The next table (see Table 1) is adapted from [9] and it presents the conclusions of the HC paradigm versus SC paradigm.

Table 1. Hard Computing vs. Soft Computing.

Hard Computing	Soft Computing
Well-posed problem solving	*Ill-posed problem* solving
Bivalent logic-based (tertium non datur)	*Fuzzy logic*-based (tertium included)
Deterministic environment (closed, static, known)	*Nondeterministic* environment (open, dynamic, uncertain)
Well-defined *problem* (quantity, precision, certainty)	Fuzzy-defined *Situation* (quality, imprecision, uncertainty)
Solving accurately problems (imperative, firm, reliable)	*Managing* "Just In Time" situations (descriptive, flexible, robust)
Optimal, lasting, *solution* (algorithmic, apodictic, general)	Suboptimal, temporary, *answer* (non-algorithmic, revisable, local)
Technocentric design	*Anthropocentric* design
Software entity: *PROGRAM* (*object devised as tool*)	Software entity: *AGENT* (*process devised as interactant*)
Client-Server paradigm (object-oriented, sequential)	*"Computing as Interaction"* paradigm (agent-oriented, parallel)

3. Fuzzy Logic in Decision-Making

In classical logic the sentences are bivalent, this meaning that all sentences that describe the state of an event are either true or false. With this bivalent logic computers were endowed, they can make massive computations, which are very difficult for people. On the other hand, computers cannot imitate the intuitive human mind and this because people can operate with vague linguistic information. It is difficult though for these to be modelled in classical logic. At the same time, in many real-world situations there are uncertainties and vague information.

To be able to deal with these uncertainties and ambiguities, L. A. Zadeh [21] introduced in 1965 the concept of fuzzy sets.

If X is an arbitrary set. A fuzzy set in X is a function $\mu : X \to [0,1]$. The function μ is called the membership function and $\mu(x)$ represents the value of truth as x belongs to the fuzzy set.

In 1975, L.A. Zadeh [22] introduced the concept of linguistic variable. This is a variable whose values are words or sentences. For example, an important criterion in location problems is "accessibility". This is a linguistic variable. It can take values linguistic terms. For the linguistic variable "accessibility" the linguistic terms set is

$$T(\text{"}accessibility\text{"}) = \{Very\ good, Good, Medium, Poor, Very\ poor\}.$$

The structure of a fuzzy logic system is presented in Figure 2. We notice that a fuzzy logic system is made of four components: fuzzyfication, fuzzy rules, inference engine, defuzzification.

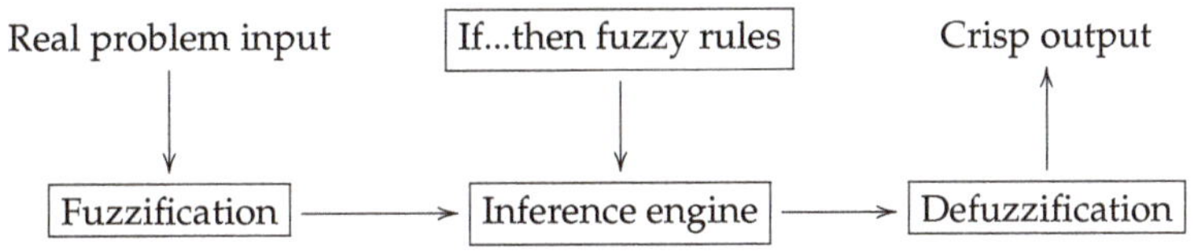

Figure 2. The structure of a fuzzy logic system.

3.1. *Fuzzification*

At this stage , first, each linguistic input A is mapped to the linguistic terms set $T(A) = \{T_A^1, T_A^2, \cdots, T_A^k\}$, and then, the meaning of each linguistic term $T_A^j (j = \overline{1,k})$ is represented by the membership function $\mu_A^j (j = \overline{1,k})$.

According to the needs, considering the computational efficiency, different types of membership functions can be used. The most used are *fuzzy numbers* (FNs).

By a FN we understand a mapping $f : \mathbb{R} \to [0,1]$, such that:

1. $f(y) \geq \min\{f(x), f(z)\}$, for $x \leq y \leq z$;
2. $(\exists)x_0 \in \mathbb{R} \, : \, f(x_0) = 1$;
3. f is upper semicontinuous, i.e.

$$(\forall)x \in \mathbb{R}, (\forall)\alpha \in (0,1] \, : \, f(x) < \alpha, (\exists)\delta > 0 \text{ such that } |y - x| < \delta \Rightarrow f(y) < \alpha\,.$$

Among the various types of fuzzy sets or their generalizations, the most common (see [6,23,24]) are:

(1) *Triangular FNs* which have membership function

$$f(x) = \begin{cases} 0 & \text{if} \quad x < a \\ \frac{x-a}{b-a} & \text{if} \quad a \leq x < b \\ \frac{c-x}{c-b} & \text{if} \quad b \leq x < c \\ 0 & \text{if} \quad x > c \end{cases} \text{, where } a \leq b \leq c\,,$$

and they are denoted by $\tilde{f} = (a,b,c)$.

(2) *Trapezoidal FNs* defined by membership function

$$f(x) = \begin{cases} 0 & \text{if} \quad x < a \\ \frac{x-a}{b-a} & \text{if} \quad a \leq x \leq b \\ 1 & \text{if} \quad b < x < c \\ \frac{d-x}{d-c} & \text{if} \quad c \leq x \leq d \\ 0 & \text{if} \quad x > d \end{cases} \text{, where } a \leq b \leq c \leq d\,,$$

and expressed as $\tilde{f} = (a, b, c, d)$.

(3) *Gaussian FNs* defined by $f(x) = e^{-\frac{(x-m)^2}{2\sigma^2}}$.

(4) *Interval-valued fuzzy sets*, defined by the membership mapping $f : X \to \mathcal{I}([0,1])$, where $\mathcal{I}([0,1])$ represents the set of all closed subintervals of $[0,1]$.

(5) *Intuitionistic fuzzy sets*, defined by a membership function f and also by a non-membership function g such that $0 \le f(x) + g(x) \le 1, (\forall)x \in X$.

(6) *Interval-valued intuitionistic fuzzy sets*, defined by two functions $f, g : X \to \mathcal{I}([0,1])$ such that $0 \le \sup_{x\in X} f(x) + \sup_{x\in X} g(x) \le 1$.

Our basic references for interval-valued fuzzy sets, intuitionistic fuzzy sets and interval-valued intuitionistic fuzzy sets are [25–32].

(7) F. Smarandache [33] proposed in 1999 the concept of neutrosophic set. A *Neutrosophic set* in X is defined as

$$A = \{< x, T_A(x), I_A(x), F_A(x) >: x \in X\}$$

where $T_A(x), I_A(x), F_A(x)$ are subsets of $]0^-, 1^+[$ and represent the truth-membership function, indeterminacy-membership function and falsity-membership function such that

$$0^- \le \sup_{x\in X} T_A(x) + \sup_{x\in X} I_A(x) + \sup_{x\in X} F_A(x) \le 3^+,$$

where

$$a^- = \{a - \epsilon \ : \ \epsilon \in \mathbb{R}^*, \epsilon \text{ is infinitesimal}\}$$
$$b^+ = \{b + \epsilon \ : \ \epsilon \in \mathbb{R}^*, \epsilon \text{ is infinitesimal}\}.$$

For applications we can consider that $T_A(x), I_A(x), F_A(x)$ are subsets of $[0,1]$.

(8) *Pythagorean fuzzy sets* were proposed by R.R. Yager in 2013 [34]. A *Pythagorean fuzzy set* is defined by the functions $f, g : X \to [0,1]$ which give us the degree of membership and degree of non-membership, respectively such that

$$0 \le (f(x))^2 + (g(x))^2 \le 1, (\forall)x \in X .$$

The function $h : X \to [0,1]$ defined by

$$h(x) = \sqrt{1 - [(f(x))^2 + (g(x))^2]}$$

is called the degree of indeterminacy.

3.2. Fuzzy Rules

Fuzzy rules allow the logic fuzzy system to take rational decisions. Fuzzy rules offer a frame in which human knowledge is integrated. In the fuzzy process the relation between input variables and output variables are described through schemes like: "If ... then". Generally we have a number N of fuzzy rules, each of them with the form [24,35]:

If A_1 is $T^j_{A_1}$ and A_2 is $T^j_{A_2}$, then B_0 is $T^j_{B_0}$, where B_0 is the consequent (output) linguistic variable of the rule.

3.3. Inference Engine

The mission of the inference engine is to obtain output variables from input variables, based on fuzzy logic rules.

For example, let us assume there are two rules:

If A_1 is $T^1_{A_1}$ and A_2 is $T^1_{A_2}$, then B_0 is $T^1_{B_0}$,

If A_1 is $T^2_{A_1}$ and A_2 is $T^2_{A_2}$, then B_0 is $T^2_{B_0}$.

The firing strengths of the two rules are: $f_1 = \mu^1_{A_1} \wedge \mu^1_{A_2}$ and $f_2 = \mu^2_{A_1} \wedge \mu^2_{A_2}$, where $\wedge$ represents AND operation in fuzzy logic, this being the minimum most often, namely

$\mu^1_{A_1} \wedge \mu^1_{A_2} = \min\{\mu^1_{A_1}, \mu^1_{A_2}\}$ and $\mu^2_{A_1} \wedge \mu^2_{A_2} = \min\{\mu^2_{A_1}, \mu^2_{A_2}\}$, but there can be used other t-norms as well.

We also consider $\hat{\mu}^1_{B_0}$ and $\hat{\mu}^2_{B_0}$ defined by: $\hat{\mu}^1_{B_0} = f_1 \wedge \mu^1_{B_0}$ and $\hat{\mu}^2_{B_0} = f_2 \wedge \mu^2_{B_0}$.

Finally, the membership degree of the output is obtained using OR operation (denoted $\vee$), namely $\mu = \hat{\mu}^1_{B_0} \vee \hat{\mu}^2_{B_0}$.

For OR operation most often we meet: $\hat{\mu}^1_{B_0} \vee \hat{\mu}^2_{B_0} = \max\{\hat{\mu}^1_{B_0}, \hat{\mu}^2_{B_0}\}$, but other t-conorms can be used.

3.4. Defuzzification

In this phase the fuzzy sets obtained of the inference engine are defuzzied into crisp outputs.

There are several methods that can be used. Center of gravity is a method used in many studies. Let us assume that the result is the membership function $\mu : \mathbb{R} \to [0,1]$ with the support the interval $[t_0, t_k]$. We consider an equidistant division $\Delta = \{t_0, t_1, t_2, \cdots, t_{k-1}, t_k\}$ of this interval. The crisp value is obtained by

$$\hat{t}_j = \frac{\sum\limits_i t_i \mu(t_i)}{\sum\limits_i \mu(t_i)}.$$

Other defuzzification methods are: max membership principle, weighted average method, mean max membership method, center of largest area etc.

4. On Some Results Obtained by Ioan Dzitac

4.1. Fuzzy Method for Multiple Criteria Fractional Programming

In paper [8], B. Stanojević, I. Dzitac and S. Dzitac proposed a new method of solving a full fuzzy linear fractional programming (FFLFP) problems using α-cut representations for the triangular fuzzy coefficients involved in both the objective function and constraints. The decision variables were considered fuzzy valued, but their shapes were determined a posteriori.

The authors first discussed the issues arising when a ratio of fuzzy numbers had to be evaluated: (i) the error induced by approximating the exact membership function of the ratio with the membership function of the triangular fuzzy number that had the same support and the same value with maximal amplitude as the exact (non-triangular) fuzzy number representing the ratio; (ii) the misleading computation of the ratio of two functions of fuzzy numbers, in the case that the same fuzzy number appeared on both numerator and denominator of the ratio, and in the global α-cut evaluation distinct endpoints of the α-cut were used for that same fuzzy number on numerator and denominator respectively. The authors established the formula for computing the area between the graphic representations of the exact membership function of the ratio of fuzzy numbers and its above-mentioned approximation. The difference between the correct and a misleading evaluation of a rational function of one triangular fuzzy number valued variable, and the effect of the translation of the triangular fuzzy numbers on the rational function correctly evaluated were both illustrated in the paper.

Further on, the authors introduced a new methodology to solve the FFLFP problems by making use of two models to yield the left and the right endpoints of an arbitrary α-cut. Their approach properly handled the multiple occurrences of variables in the fractional objective function and in constraints as well. Discrete representations of the inverses of the exact membership functions of the products and ratios of fuzzy numbers were used to derive the fuzzy solutions expressed by (non-triangular) fuzzy numbers. The main advantage of the solution approach was in its ability to work with the exact membership functions of the ratios, and to avoid the usage of any special definition for fuzzy inequality evaluation.

By their approach, the authors disclosed a direction that can be useful in managing decision-making processes involving fractional objective functions under uncertainty. The illustrative example provided in the paper analyzed and solved a decision-making problem in production planning.

4.2. Prudent Decision to Estimate the Risk in Insurance

In paper [3], the authors investigated the issue of capital allocation with risk exposure starting from the regulations of the new solvency system "Solvency II" built by European Commission to regulate all insurance activities in European Union member countries.

As stated in the Solvency II Directive, 101 Article, the allocation of capital in insurance is done either using a standard formula or by developing an internal model taking into account the particularities of the insurance company.

Making prudent risk assessment decisions to which the insurance company is exposed ultimately means a capital allocation that better corresponds to the company's risk exposure. The risk measure advanced by the new Solvency II system under the standard formula is Value-at-Risk (VaR) or Tail Value-at-Risk (TVaR). The authors proposed for determination of VaR or TVaR, the use of the theory of extreme values. This decision proves more prudent in the sense of correlating the values of VaR and TVaR with the amplitude of risk exposure.

To argue quantitatively the usefulness of applying the theory of extreme events to set the level of prudence regarding the determination of the solvency capital requirement (SCR), the authors built a hypothetical portfolio composed of shares and bonds. Although the standard formula in the Solvency II system was built on the assumption of the normal distribution of the risk to which the company was exposed, this hypothesis was verified too few times, especially when it came to the occurrence of rare events. Under these conditions, the use of distributions better adapted to the company's risk exposure may prove to be a more prudent decision.

In the case of the built-up portfolio, compared with the use of normal law, the authors proved that more prudent decisions had been shown to be the use of specific distribution laws for determination of extreme quantiles such as: Chi2 (from the attraction domain of Fréchet—laws with heavy tails); Exponential, Gamma and Weibull (from the attraction domain of Gumbel—laws with exponential decreasing tails).

4.3. Decision Support Model for Production Disturbance Estimation

There are several models in the literature for the study of disturbance in manufacturing systems to optimize them. Unfortunately, these models do not give us a dynamic evaluation of multiple consequences of the disturbance. Many production bottlenecks have the effect of severe economic consequences, and they are caused by disturbances in technological lines. A good estimate of disturbance dynamics is needed to reduce losses.

In paper [4] the authors introduced a model which allows a dynamic evaluation of consequences of disturbance. The model used several indicators: time, energy and costs. Model testing was done using MATLAB.

4.4. Manufacture System Control Using ANN Software

In paper [5] the authors presented a technique for improving the autonomy of the flexible manufacturing systems. The paper presented a system capable of monitoring the tool flank and widening the manufacturing system whiteout input from the users.

The entire structure of the proposed system was presented as well as the different connections between the system itself and other components of the flexible manufacturing systems. The concept behind the system was the continuous monitoring of tools in the machine tool storage systems. From the hardware point of view the paper presented an experimental system designed to acquire the tool flank images.

The paper also presented the software component. The first step taken by the authors was the realization of an application able to extract significant data from the pictures obtained. All the steps taken were presented in the paper. Several parameters were identified and compared to determine their efficiency in determining the tool ware.

An alternative decisional process regarding the tool ware is based on artificial neural networks. The paper presented the realization and setting of such a network realized in MATLAB and performed an in-depth analysis of the network. Methods of Image Classification Using One-Hidden-Layer ANN on Image Data and Image Classification with Autoencoders on Image Data were presented.

The conclusion section of the paper makes an exhaustive presentation of the results for each method of image classification using ANN, a comparative network training parameters table is presented. Als the paper includes the statistical data for misclassification of images.

The paper offers significant theoretical and experimental data on the hardware component of the system, and it is also focusing on the software component. Several methods for image classification are presented and the results are compared to determine optimal solutions both from accuracy and computational time.

4.5. Ant Colony Optimization for a Practical Economic Problem

The source of inspiration for ant colony optimization (ACO) algorithm is the behavior of ants in search of food, which is a problem of optimization, more precisely of determining the optimal path between food and nest.

Several theoretical aspects of artificial ants were presented in papers [10,36].

In paper [7] the authors presented an algorithm based on Ant Colony Optimization (ACO) applied in a practical economic problem, more precisely, in finding an optimal solution in an electricity distribution network. A nonlinear function was used in the mathematical model of optimization. The mathematical model contains both equality constraints and inequality constraints. The results are validated on a network with 35 nodes.

5. Conclusions and Future Trends

We wrote this paper for a Special Issue dedicated to the centenary of birth of Lotfi A. Zadeh published by Mathematics. Therefore, this paper is an homage to Lofti A. Zadeh as a sign of respect for his great contributions to scientific knowledge. At the same time, this paper is dedicated to I. Dzitac, who passed away on 6 February 2021 as he stood out as one of the followers of Zadeh's work by publishing, between 2007 and 2020, tens of articles in the domain of soft computing.

I. Dzitac's most important contributions have just been presented, wishing and hoping that either his former partners or other researchers will continue his work.

We will finish with some challenges and future research directions, because the rapid development of the economy and of the modern industrial systems make the process of decision-making more complicated, leading to difficulties in dealing with vague information. In this environment, soft computing methods have been successfully applied. Nevertheless, as the human thinking process is very complex, raised attention and new research are necessary for the applications of soft computing methods in intelligent decision-making. We will raise a few problems, more general and larger, but that deserve the attention of researchers.

1. In Section 3.1 various types of fuzzy sets used for the representation of the linguistic terms are presented. The development of new types of fuzzy sets deserves to be solved in order to obtain better results in problems of decision-making. Thus, it is necessary to develop new algorithms for generalized fuzzy sets. Choosing membership functions has an important role in fuzzification process. To increase the accuracy of soft computing methods, membership functions must be improved to finally reach a better representation of linguistic variables.
2. Because of the complexity of the economic problems and of the modern industries, we must operate with cu massive input items. In such a situation the number of fuzzy rules increases exponentially as related to the number of input items [37]. As a result, some fuzzy rules can become incomplete or unsubstantial and there can appear conflicts among these numerous rules. There are only a few studies in this direction

and it is necessary that in the future more attention to be granted and greater efforts to be made.

3. Fuzzy inference methods are largely used to model the experts' behavior. Despite this, fuzzy inference methods need standard procedures for a quantity analyze and to optimize the process of transformation of experience and experts' knowledge (see [24]). The development of artificial intelligence can be of great help as fuzzy inference systems can be combined with mining techniques.
4. Another great challenge is to improve the linguistic operators and to increase in this way, the flexibility outcomes as a result of the defuzzification process.
5. Consensus reaching is a subject of great importance in the group decision-making field. The obtaining of a high level of consensus or agreement among decision-makers is of crucial importance in any group decision-making problem (see [38,39]). This consensus can be obtained by integrating the opinions of decision-makers according to some rules, because it is natural and normal for the decision-makers to have different point of views upon the same issue. Therefore, making adequate rules to solve conflicts in a group is a subject that still needs to be studied.
   In the literature of genre can be found many models of consensus with linguistic information. Nevertheless, some questions remain open: the development of new models of consensus that include feedback mechanisms, the creation of new technologies to implement consensus models etc.

In conclusion, as the problems of decision-making become more and more difficult when vague linguistic information interferes, soft computing methods have gained more attention and a great interest in the latest years. In this paper, a comprehensive review upon some of I. Dzitac's results has been realized. The structure of a fuzzy system has been presented and its components have been analyzed. Finally, some challenges and future trends have been presented.

Author Contributions: Conceptualization, S.N.; formal analysis, S.D. and S.N.; methodology, S.D. and S.N.; supervision, S.D.; validation, S.D. and S.N.; writing-original draft, S.N.; writing-review and editing, S.D. and S.N. All authors have read and agreed to the published version of the manuscript.

Funding: This research received no external funding.

Institutional Review Board Statement: Not applicable.

Informed Consent Statement: Not applicable.

Data Availability Statement: Not applicable.

Conflicts of Interest: The authors declare no conflict of interest.

References

1. Dzitac, I. The Fuzzification of Classical Structures: A General View. *Int. J. Comput. Commun. Control* **2015**, *10*, 12–28. [CrossRef]
2. Dzitac, I.; Filip, F.G.; Manolescu, M.J. Fuzzy logic is not fuzzy: World renowned computer scientist Lotfi A. Zadeh. *Int. J. Comput. Commun. Control* **2017**, *12*, 748–789. [CrossRef]
3. Butaci, C.; Dzitac, S.; Dzitac, I.; Bologa G. Prudent decisions to estimate the risk of loss in insurance. *Technol. Econ. Dev. Econ.* **2017**, *23*, 428–440. [CrossRef]
4. Felea, I.; Dzitac, S.; Vesselenyi, T.; Dzitac, I. Decision support model for production disturbance estimation. *Int. J. Inf. Technol. Decis. Mak.* **2014**, *13*, 623–647. [CrossRef]
5. Moldovan, O.G.; Dzitac, S.; Moga, I.; Vesselenyi, T.; Dzitac, I. Tool-Wear Analysis Using Image Processing of the Tool Flank. *Symmetry* **2017**, *9*, 296. [CrossRef]
6. Nădăban, S.; Dzitac, S.; Dzitac, I. Fuzzy TOPSIS: A general view. *Procedia Comput. Sci.* **2016**, *91*, 823–831. [CrossRef]
7. Secui D.C.; Dzitac S.; Bendea G.V.; Dzitac I. An ACO Algorithm for Optimal Capacitor Banks Placement in Power Distribution Networks. *Stud. Inform. Control* **2009**, *18*, 305–314.
8. Stanojević, B.; Dzitac, I.; Dzitac, S. On the ratio of fuzzy numbers—Exact membership function computation and applications to decision making. *Technol. Econ. Dev. Econ.* **2015**, *21*, 815–832. [CrossRef]
9. Dzitac, I.; Barbat, B.E. Artificial Intelligence + Distributed Systems = Agents. *Int. J. Comput. Commun. Control* **2009**, *4*, 17–26. [CrossRef]

10. Negulescu S.C.; Dzitac I.; Lascu A.E. Synthetic Genes for Artificial Ants. Diversity in Ant Colony Optimization Algorithms. *Int. Comput. Commun. Control* **2010**, *5*, 216–223. [CrossRef]
11. Benta, D.; Bologa, G.; Dzitac, I. E-learning Platforms in Higher Education. Case Study. *Procedia Comput. Sci.* **2014** *31*, 1170–1176. [CrossRef]
12. Benta, D.; Bologa, G.; Dzitac, S.; Dzitac, I. University Level Learning and Teaching via E-Learning Platforms. *Procedia Comput. Sci.* **2015**, *55*, 1366–1373. [CrossRef]
13. Lupse, V.; Dzitac, I.; Dzitac, S.; Manolescu, A.; Manolescu, M.J. CRM Kernel-based Integrated Information System for a SME: An Object-oriented Design. *Int. J. Comput. Commun. Control* **2008**, *3*, 375–380.
14. Dzitac, I. Zadeh's Centenary. *Int. J. Comput. Commun. Control* **2021**, *16*, 4102. [CrossRef]
15. Zadeh, L.A.; Tufis, D.; Filip, F.G.; Dzitac I. *From Natural Language to Soft Computing: New Paradigms in Artificial Intelligence*; Editing House of Romanian Academy: Bucharest, Romania, 2008.
16. Liang, H.; Cai, R. A new correlation coefficient of BPA based on generalized information quality. *Int. J. Intell. Syst.* **2021**. [CrossRef]
17. Roszkowska, E.; Kusterka-Jefmanska, M.; Jefmanski, B. Intuitionistic Fuzzy TOPSIS as a Method for Assessing Socioeconomic Phenomena on the Basis of Survey Data. *Entropy* **2021**, *23*, 563. [CrossRef] [PubMed]
18. Hamzelou, N.; Ashtiani, M.; Sadeghi, R. A propagation trust model in social networks based on the A* algorithm and multi-criteria decision making. *Computing* **2021**, *103*, 827–867. [CrossRef]
19. Zadeh, L.A. Soft computing and fuzzy logic. *IEEE Softw.* **1994**, *11*, 48–56. [CrossRef]
20. Zadeh, L.A. A new frontier in Computation—Computation with information described in natural language. In Proceedings of the IEEE International Sumposium on Intelligent Sigmal Processing, 1st International North American Simulation Technology Conference, Nastec 2008, Montreal, QC, Canada, 13–15 August 2008; pp. 73–74.
21. Zadeh, L.A. Fuzzy Sets. *Inf. Control* **1965**, *8*, 338–353. [CrossRef]
22. Zadeh, L.A. The concept of a linguistiv variable and its application to approximate reasoning-I. *Inf. Sci.* **1975** *8*, 199–249. [CrossRef]
23. Nădăban, S. From classical logic to fuzzy logic and quantum logic: A general view. *Int. J. Comput. Commun. Control* **2021**, *16*, 4125. [CrossRef]
24. Wu, H.; Xu, Z.S. Fuzzy logic in decision support: Methods, applications and future trends. *Int. J. Comput. Commun. Control* **2021**, *16*, 4044.
25. Atanassov, K. *Intuitionistic Fuzzy Sets: Theory and Applications*; Physica: Heidelberg, Germany; New York, NY, USA, 1999.
26. Atanassov, K. *Interval-Valued Intuitionistic Fuzzy Sets*; Springer International Publishing: Berlin, Germany, 2020; Volume 388.
27. Bustince, H.; Burillo, P. Vague sets are intuitionistic fuzzy sets. *Fuzzy Sets Syst.* **1996**, *79*, 403–405. [CrossRef]
28. Burillo, P.; Bustince, H. Entropy on intuitionistic fuzzy sets and on interval-valued fuzzy sets. *Fuzzy Sets Syst.* **1996**, *78*, 305–316. [CrossRef]
29. Deschrijver, G.; Kerre, E. On the relationship between some extensions of fuzzy set theory. *Fuzzy Sets Syst.* **2003**, *133*, 227–235. [CrossRef]
30. Turksen, I.B. Interval valued fuzzy sets based on normal forms. *Fuzzy Sets Syst.* **1986** *20*, 191–210. [CrossRef]
31. Wang, G.J.; He, Y.Y. Intuitionistic fuzzy sets and L-fuzzy sets. *Fuzzy Sets Syst.* **2000** *110*, 271–274. [CrossRef]
32. Zhang, H.Y.; Zhang, W.X.; Mei, C.L. Entropy of interval-valued fuzzy sets based on distance and its relationship with similarity measure. *Knowl. Based Syst.* **2009**, *22*, 449–454. [CrossRef]
33. Smarandache, F. A unifying field in logics. In *Neutrosophy: Neutrosophic Probability, Set and Logic*; American Research Press: Rehoboth, DE, USA, 1999.
34. Yager, R.R. Pythagorean fuzzy subsets. In Proceedings of the 2013 Joint IFSA World Congress and NAFIPS Annual Meeting (IFSA/NAFIPS), Edmonton, AB, Canada, 24–28 June 2013.
35. Wu, Z.; Liao, H.; Lu, K.; Zavadskas, E.K. Soft computing techniques and their applications in intelligent industrial control system: A survey. *Int. J. Comput. Commun. Control* **2021**, *16*, 4142. [CrossRef]
36. Negulescu A.E.; Negulescu S.C.; Dzitac I. Balancing Between Exploration and Exploitation in ACO. *Int. J. Comput. Control* **2017**, *12*, 265–275. [CrossRef]
37. Wu, D.R.; Lin, C.T.; Huang, J.; Zeng, Z.G. On the functional equivalence of TSK fuzzy systems to neural networks, mixture of experts, CART, and stacking ensemble regression. *IEEE Trans. Fuzzy Syst.* **2020**, *28*, 2570–2580. [CrossRef]
38. Herrera, F.; Alonso, S.; Chiclana, F.; Herrera-Viedma, E. Computing with words in decision making: foundations, trends and prospects. *Fuzzy Optim Decis Mak.* **2009**, *8*, 337–364. [CrossRef]
39. Saint, S.; Lawson, J.R. *Rules for Reaching Consensus: A Modern Approach to Decision Making*; Jossey-Bass: San-Francisco, CA, USA, 1994.

mathematics MDPI

Article

New Applications of Sălăgean and Ruscheweyh Operators for Obtaining Fuzzy Differential Subordinations

Alina Alb Lupaş *,† and Georgia Irina Oros †

Department of Mathematics and Computer Science, University of Oradea, 1 Universitatii Street, 410087 Oradea, Romania; georgia_oros_ro@yahoo.co.uk
* Correspondence: alblupas@gmail.com
† These authors contributed equally to this work.

Abstract: The present paper deals with notions from the field of complex analysis which have been adapted to fuzzy sets theory, namely, the part dealing with geometric function theory. Several fuzzy differential subordinations are established regarding the operator L_α^m, given by $L_\alpha^m : \mathcal{A}_n \to \mathcal{A}_n$, $L_\alpha^m f(z) = (1-\alpha)R^m f(z) + \alpha S^m f(z)$, where $\mathcal{A}_n = \{f \in \mathcal{H}(U),\ f(z) = z + a_{n+1}z^{n+1} + \ldots,\ z \in U\}$ is the subclass of normalized holomorphic functions and the operators $R^m f(z)$ and $S^m f(z)$ are Ruscheweyh and Sălăgean differential operator, respectively. Using the operator L_α^m, a certain fuzzy class of analytic functions denoted by $SL_{\mathcal{F}}^m(\delta,\alpha)$ is defined in the open unit disc. Interesting results related to this class are obtained using the concept of fuzzy differential subordination. Examples are also given for pointing out applications of the theoretical results contained in the original theorems and corollaries.

Keywords: fuzzy differential subordination; convex function; fuzzy best dominant; differential operator

Citation: Alb Lupaş, A.; Oros, G.I. New Applications of Sălăgean and Ruscheweyh Operators for Obtaining Fuzzy Differential Subordinations. *Mathematics* **2021**, *9*, 2000. https://doi.org/10.3390/math9162000

Academic Editor: Alfonso Mateos Caballero

Received: 22 July 2021
Accepted: 16 August 2021
Published: 21 August 2021

1. Introduction

Fuzzy sets theory epic started in 1965 when Lotfi A. Zadeh published the paper "Fuzzy Sets" [1], received with distrust at first but currently cited by over 95,000 papers. Mathematicians have been constantly concerned with adapting fuzzy sets theory to different branches of mathematics, and many such connections have been made. The beautiful review paper published in 2017 [2] is a tribute to Lotfi A. Zadeh's contribution to the scientific world and shows the evolution of the notion of fuzzy set in time and its numerous connections with different topics of mathematics, science, and technique. Another great review article published as part of this Special Issue dedicated to the Centenary of the Birth of Lotfi A. Zadeh [3] gives further details on the development of fuzzy sets theory and highlights the contributions of Professor I. Dzitac who has had Lotfi A. Zadeh as mentor. In 2008, he edited a volume [4], tying his name to that of Lotfi A. Zadeh for posterity.

The first applications of fuzzy sets theory in the part of complex analysis studying analytic functions of one complex variable were marked by the introduction of the concept of fuzzy subordination in 2011 [5]. The study was continued, and the notion of fuzzy differential subordination was introduced in 2012 [6]. All the aspects of the classical theory of differential subordination which are synthesized in the monograph published in 2000 [7] by the same authors who have introduced the notion in 1978 [8] and 1981 [9] were then adapted in light of the connection to fuzzy sets theory. At some point, fuzzy differential subordinations began to be studied in connection with different operators with many applications in geometric function theory as it can be seen in the first papers published starting with 2013 [10–12]. The topic is of obvious interest at this time, a fact proved by the numerous papers published in the last 2 years, of which we mention only a few [13–16].

In this paper, fuzzy differential subordinations will be obtained using the differential operator defined and studied in several aspects in [17,18].

The basic notions used for conducting the study are denoted as previously established in literature.

Let $U = \{z \in \mathbb{C} : |z| < 1\}$ and denote by $\mathcal{H}(U)$ the class of holomorphic functions in the unit disc U. Let $\mathcal{A}_n = \{f \in \mathcal{H}(U) : f(z) = z + a_{n+1}z^{n+1} + \dots,\ z \in U\}$ be the subclass of normalized holomorphic functions writing $\mathcal{A}_1$ as $\mathcal{A}$. When $a \in \mathbb{C}$ and $n \in \mathbb{N}^*$, denote by $\mathcal{H}[a,n] = \{f \in \mathcal{H}(U) : f(z) = a + a_n z^n + a_{n+1}z^{n+1} + \dots,\ z \in U\}$ writing $\mathcal{H}_0 = \mathcal{H}[0,1]$. The class of convex functions is obtained for $\alpha = 0$ when $0 < \alpha < 1$ the class denoted by $\mathcal{K}(\alpha) = \left\{f \in \mathcal{A} : \mathrm{Re}\,\frac{zf''(z)}{f'(z)} + 1 > \alpha,\ z \in U\right\}$ contains convex functions of order α.

The definitions necessary for using the concept of fuzzy differential subordinations introduced in previously published cited papers are next reminded.

Definition 1 ([19]). *A pair (A, F_A), where $F_A : X \to [0,1]$ and $A = \{x \in X : 0 < F_A(x) \le 1\}$ is called fuzzy subset of X. The set A is called the support of the fuzzy set (A, F_A) and F_A is called the membership function of the fuzzy set (A, F_A). One can also denote $A = \mathrm{supp}(A, F_A)$.*

Remark 1. *If $A \subset X$, then $F_A(x) = \begin{cases} 1, \text{ if } x \in A \\ 0, \text{ if } x \notin A \end{cases}$.*

For a fuzzy subset, the real number 0 represents the smallest membership degree of a certain $x \in X$ to A and the real number 1 represents the biggest membership degree of a certain $x \in X$ to A.

The empty set $\emptyset \subset X$ is characterized by $F_\emptyset(x) = 0,\ x \in X$, and the total set X is characterized by $F_X(x) = 1,\ x \in X$.

Definition 2 ([5]). *Let $D \subset \mathbb{C}$, $z_0 \in D$ be a fixed point and let the functions $f, g \in \mathcal{H}(D)$. The function f is said to be fuzzy subordinate to g and write $f \prec_{\mathcal{F}} g$ or $f(z) \prec_{\mathcal{F}} g(z)$, if the conditions are satisfied:*

(1) $f(z_0) = g(z_0)$,

(2) $F_{f(D)}f(z) \le F_{g(D)}g(z),\ z \in D$.

Definition 3 ([6] (Definition 2.2)). *Let $\psi : \mathbb{C}^3 \times U \to \mathbb{C}$ and h univalent in U, with $\psi(a,0;0) = h(0) = a$. If p is analytic in U, with $p(0) = a$ and satisfies the (second-order) fuzzy differential subordination*

$$F_{\psi(\mathbb{C}^3 \times U)}\psi(p(z), zp'(z), z^2p''(z); z) \le F_{h(U)}h(z), \quad z \in U, \tag{1}$$

then p is called a fuzzy solution of the fuzzy differential subordination. The univalent function q is called a fuzzy dominant of the fuzzy solutions of the fuzzy differential subordination, or more simple a fuzzy dominant, if $F_{p(U)}p(z) \le F_{q(U)}q(z),\ z \in U$, for all p satisfying (1). A fuzzy dominant $\tilde{q}$ that satisfies $F_{\tilde{q}(U)}\tilde{q}(z) \le F_{q(U)}q(z),\ z \in U$, for all fuzzy dominants q of (1) is said to be the fuzzy best dominant of (1).

Lemma 1 ([7] (Corollary 2.6g.2, p. 66)). *Let $h \in \mathcal{A}_n$ and*

$$L[f](z) = F(z) = \frac{1}{nz^{\frac{1}{n}}} \int_0^z h(t)t^{\frac{1}{n}-1}dt,\ z \in U.$$

If

$$Re\left(\frac{zh''(z)}{h'(z)} + 1\right) > -\frac{1}{2},\ z \in U,$$

then $L(f) = F \in \mathcal{K}$.

Lemma 2 ([20]). *Let h be a convex function with $h(0) = a$, and let $\gamma \in \mathbb{C}^*$ be a complex number with $Re\,\gamma \geq 0$. If $p \in \mathcal{H}[a,n]$ with $p(0) = a$, $\psi : \mathbb{C}^2 \times U \to \mathbb{C}$, $\psi(p(z), zp'(z); z) = p(z) + \frac{1}{\gamma}zp'(z)$ an analytic function in U and*

$$F_{\psi(\mathbb{C}^2\times U)}\left(p(z) + \frac{1}{\gamma}zp'(z)\right) \leq F_{h(U)}h(z), \text{ i.e., } p(z) + \frac{1}{\gamma}zp'(z) \prec_{\mathcal{F}} h(z), \quad z \in U, \tag{2}$$

then

$$F_{p(U)}p(z) \leq F_{g(U)}g(z) \leq F_{h(U)}h(z), \text{ i.e., } p(z) \prec_{\mathcal{F}} g(z) \prec_{\mathcal{F}} h(z), \quad z \in U,$$

where $g(z) = \frac{\gamma}{nz^{\gamma/n}}\int_0^z h(t)t^{\gamma/n-1}dt$, $z \in U$. The function q is convex and is the fuzzy best dominant.

Lemma 3 ([20]). *Let g be a convex function in U and let $h(z) = g(z) + n\alpha zg'(z)$, $z \in U$, where $\alpha > 0$ and n is a positive integer.*

If $p(z) = g(0) + p_n z^n + p_{n+1}z^{n+1} + \ldots$, $z \in U$, is holomorphic in U and

$$F_{p(U)}(p(z) + \alpha zp'(z)) \leq F_{h(U)}h(z), \text{ i.e., } p(z) + \alpha zp'(z) \prec_{\mathcal{F}} h(z), \quad z \in U,$$

then

$$F_{p(U)}p(z) \leq F_{g(U)}g(z), \text{ i.e., } p(z) \prec_{\mathcal{F}} g(z), \quad z \in U,$$

and this result is sharp.

Sălăgean and Ruscheweyh differential operators are well known in geometric function theory for the nice results obtained by implementing them in the studies. Their definitions and basic properties are given in the next two definitions and remarks.

Definition 4 (Sălăgean [21]). *For $f \in \mathcal{A}_n$, $m, n \in \mathbb{N}$, the operator S^m is defined by $S^m : \mathcal{A}_n \to \mathcal{A}_n$,*

$$\begin{aligned} S^0 f(z) &= f(z) \\ S^1 f(z) &= zf'(z) \\ &\ldots \\ S^{m+1}f(z) &= z(S^m f(z))', \; z \in U. \end{aligned}$$

Remark 2. *If $f \in \mathcal{A}_n$, $f(z) = z + \sum_{j=n+1}^{\infty} a_j z^j$, then $S^m f(z) = z + \sum_{j=n+1}^{\infty} j^m a_j z^j$, $z \in U$.*

Definition 5 (Ruscheweyh [22]). *For $f \in \mathcal{A}_n$, $m, n \in \mathbb{N}$, the operator R^m is defined by $R^m : \mathcal{A}_n \to \mathcal{A}_n$,*

$$\begin{aligned} R^0 f(z) &= f(z) \\ R^1 f(z) &= zf'(z) \\ &\ldots \\ (m+1)R^{m+1}f(z) &= z(R^m f(z))' + mR^m f(z), \; z \in U. \end{aligned}$$

Remark 3. *If $f \in \mathcal{A}_n$, $f(z) = z + \sum_{j=n+1}^{\infty} a_j z^j$, then $R^m f(z) = z + \sum_{j=n+1}^{\infty} C_{m+j-1}^m a_j z^j$, $z \in U$.*

The next definition shows the operator used for obtaining the original results of this paper, defined in a previously published paper. Two remarks regarding it are also listed.

Definition 6 ([17]). *Let $\alpha \geq 0$, $m, n \in \mathbb{N}$. Denote by L_α^m the operator given by $L_\alpha^m : \mathcal{A}_n \to \mathcal{A}_n$,*

$$L_\alpha^m f(z) = (1-\alpha)R^m f(z) + \alpha S^m f(z), \quad z \in U.$$

Remark 4. *L_α^m is a linear operator and if $f \in \mathcal{A}_n$, $f(z) = z + \sum_{j=n+1}^{\infty} a_j z^j$, then $L_\alpha^m f(z) = z + \sum_{j=n+1}^{\infty}\left(\alpha j^m + (1-\alpha)C_{m+j-1}^m\right)a_j z^j$, $z \in U$.*

Remark 5. *For $\alpha = 0$, $L_0^m f(z) = R^m f(z)$, $z \in U$, and for $\alpha = 1$, $L_1^m f(z) = S^m f(z)$, $z \in U$.*

For $m = 0$, $L_\alpha^0 f(z) = (1-\alpha)R^0 f(z) + \alpha S^0 f(z) = f(z) = R^0 f(z) = S^0 f(z)$, $z \in U$, and for $m = 1$, $L_\alpha^1 f(z) = (1-\alpha)R^1 f(z) + \alpha S^1 f(z) = zf'(z) = R^1 f(z) = S^1 f(z)$, $z \in U$.

Definition 7 ([11])**.** *Let $f(D) = \sup p\left(f(D), F_{f(D)}\right) = \{z \in D : 0 < F_{f(D)}f(z) \leq 1\}$, where $F_{f(D)}\cdot$ is the membership function of the fuzzy set $f(D)$ associated to the function f.*

The membership function of the fuzzy set $(\mu f)(D)$ associated to the function μf coincides with the membership function of the fuzzy set $f(D)$ associated to the function f, i.e., $F_{(\mu f)(D)}((\mu f)(z)) = F_{f(D)}f(z)$, $z \in D$.

The membership function of the fuzzy set $(f+g)(D)$ associated to the function $f+g$ coincide with the half of the sum of the membership functions of the fuzzy sets $f(D)$, respectively $g(D)$, associated to the function f, respectively g, i.e., $F_{(f+g)(D)}((f+g)(z)) = \frac{F_{f(D)}f(z)+F_{g(D)}g(z)}{2}$, $z \in D$.

Remark 6. *As $0 < F_{f(D)}f(z) \leq 1$ and $0 < F_{g(D)}g(z) \leq 1$, it is evident that $0 < F_{(f+g)(D)}((f+g)(z)) \leq 1$, $z \in D$.*

2. Main Results

First, a new fuzzy class of analytic functions is defined using the operator given by Definition 6.

Definition 8. *The fuzzy class denoted $SL_{\mathcal{F}}^m(\delta,\alpha)$ contains all functions $f \in \mathcal{A}_n$ which satisfy the fuzzy inequality*

$$F_{(L_\alpha^m f)'(U)}(L_\alpha^m f(z))' > \delta, \quad z \in U, \tag{3}$$

when $\delta \in (0,1]$, $\alpha \geq 0$ and $m, n \in \mathbb{N}$.

The first result for this class is related to its convexity.

Theorem 1. *The set $SL_{\mathcal{F}}^m(\delta,\alpha)$ is convex.*

Proof. Consider the functions

$$f_j(z) = z + \sum_{j=n+1}^{\infty} a_{jk} z^j \in SL_{\mathcal{F}}^m(\delta,\alpha), \quad k = 1,2, \quad z \in U.$$

For obtaining the required conclusion, the function

$$h(z) = \eta_1 f_1(z) + \eta_2 f_2(z)$$

must be part of the class $SL_{\mathcal{F}}^m(\delta,\alpha)$, with η_1 and η_2 non-negative such that $\eta_1 + \eta_2 = 1$.

We next show that $h \in SL_{\mathcal{F}}^m(\delta,\alpha)$,

$h'(z) = (\mu_1 f_1 + \mu_2 f_2)'(z) = \mu_1 f_1'(z) + \mu_2 f_2'(z)$, $z \in U$, and
$(L_\alpha^m h(z))' = (L_\alpha^m(\mu_1 f_1 + \mu_2 f_2)(z))' = \mu_1(L_\alpha^m f_1(z))' + \mu_2(L_\alpha^m f_2(z))'$.

From Definition 7 we obtain that

$$F_{(L_\alpha^m h)'(U)}(L_\alpha^m h(z))' = F_{(L_\alpha^m(\mu_1 f_1+\mu_2 f_2))'(U)}(L_\alpha^m(\mu_1 f_1 + \mu_2 f_2)(z))' =$$
$$F_{(L_\alpha^m(\mu_1 f_1+\mu_2 f_2))'(U)}\left(\mu_1(L_\alpha^m f_1(z))' + \mu_2(L_\alpha^m f_2(z))'\right) =$$
$$\frac{F_{(\mu_1 L_\alpha^m f_1)'(U)}(\mu_1(L_\alpha^m f_1(z))')+F_{(\mu_2 L_\alpha^m f_2)'(U)}(\mu_2(L_\alpha^m f_2(z))')}{2} = \frac{F_{(L_\alpha^m f_1)'(U)}(L_\alpha^m f_1(z))'+F_{(L_\alpha^m f_2)'(U)}(L_\alpha^m f_2(z))'}{2}.$$

As $f_1, f_2 \in SL^m_{\mathcal{F}}(\delta, \alpha)$ we have $\delta < F_{(L^m_\alpha f_1)'(U)}(L^m_\alpha f_1(z))' \leq 1$ and $\delta < F_{(L^m_\alpha f_2)'(U)}$ $(L^m_\alpha f_2(z))' \leq 1$, $z \in U$. Therefore, $\delta < \frac{F_{(L^m_\alpha f_1)'(U)}(L^m_\alpha f_1(z))' + F_{(L^m_\alpha f_2)'(U)}(L^m_\alpha f_2(z))'}{2} \leq 1$ and we obtain that $\delta < F_{(L^m_\alpha h)'(U)}(L^m_\alpha h(z))' \leq 1$, which means that $h \in SL^m_{\mathcal{F}}(\delta, \alpha)$ and $SL^m_{\mathcal{F}}(\delta, \alpha)$ is convex. □

A fuzzy subordination result is given in the next theorem and a related example follows.

Theorem 2. *Considering the convex function in U denoted by g and defining $h(z) = g(z) + \frac{1}{c+2} z g'(z)$, with $c > 0$, $z \in U$, if $f \in SL^m_{\mathcal{F}}(\delta, \alpha)$ and $G(z) = I_c(f)(z) = \frac{c+2}{z^{c+1}} \int_0^z t^c f(t) dt$, $z \in U$, then the fuzzy differential subordination*

$$F_{(L^m_\alpha f)'(U)}(L^m_\alpha f(z))' \leq F_{h(U)} h(z), \text{ i.e., } (L^m_\alpha f(z))' \prec_{\mathcal{F}} h(z), \quad z \in U, \tag{4}$$

implies

$$F_{(L^m_\alpha G)'(U)}(L^m_\alpha G(z))' \leq F_{g(U)} g(z), \text{ i.e., } (L^m_\alpha G(z))' \prec_{\mathcal{F}} g(z), \quad z \in U,$$

and this result is sharp.

Proof. Using the definition of function $G(z)$, we obtain

$$z^{c+1} G(z) = (c+2) \int_0^z t^c f(t) dt. \tag{5}$$

Differentiating (5) with respect to z, we have $(c+1)G(z) + zG'(z) = (c+2)f(z)$ and

$$(c+1) L^m_\alpha G(z) + z (L^m_\alpha G(z))' = (c+2) L^m_\alpha f(z), \quad z \in U. \tag{6}$$

Differentiating (6) we have

$$(L^m_\alpha G(z))' + \frac{1}{c+2} z (L^m_\alpha G(z))'' = (L^m_\alpha f(z))', \quad z \in U. \tag{7}$$

Using (7), the fuzzy differential subordination (4) becomes

$$F_{L^m_\alpha G(U)}\left((L^m_\alpha G(z))' + \frac{1}{c+2} z (L^m_\alpha G(z))'' \right) \leq F_{g(U)}\left(g(z) + \frac{1}{c+2} z g'(z) \right). \tag{8}$$

If we denote

$$p(z) = (L^m_\alpha G(z))', \quad z \in U, \tag{9}$$

then $p \in \mathcal{H}[1, n]$.

Replacing (9) in (8) we obtain

$$F_{p(U)}\left(p(z) + \frac{1}{c+2} z p'(z) \right) \leq F_{g(U)}\left(g(z) + \frac{1}{c+2} z g'(z) \right), \quad z \in U.$$

Using Lemma 3, we have

$$F_{p(U)} p(z) \leq F_{g(U)} g(z), \quad z \in U, \quad \text{i.e.,} \quad F_{(L^m_\alpha G)'(U)}(L^m_\alpha G(z))' \leq F_{g(U)} g(z), \quad z \in U,$$

and g is the best dominant. We have obtained

$$(L^m_\alpha G(z))' \prec_{\mathcal{F}} g(z), \quad z \in U.$$

□

Example 1. *If $f \in SL^1_{\mathcal{F}}\left(1, \frac{1}{2}\right)$, then $f'(z) + z f''(z) \prec_{\mathcal{F}} \frac{3-2z}{3(1-z)^2}$ implies $G'(z) + zG''(z) \prec_{\mathcal{F}} \frac{1}{1-z}$, where $G(z) = \frac{3}{z^2} \int_0^z t f(t) dt$.*

Several fuzzy subordination results are contained in the next theorems and corollaries. Some are followed by examples.

Theorem 3. *Let* $h(z) = \frac{1+(2\beta-1)z}{1+z}$, $\beta \in [0,1)$ *and* $c > 0$. *If* $\alpha \geq 0$, $m \in \mathbb{N}$ *and* $I_c(f)(z) = \frac{c+2}{z^{c+1}} \int_0^z t^c f(t)dt$, $z \in U$, *then*

$$I_c[SL_{\mathcal{F}}^m(\beta, \alpha)] \subset SL_{\mathcal{F}}^m(\beta^*, \alpha), \tag{10}$$

where $\beta^* = 2\delta - 1 + \frac{(c+2)(2-2\delta)}{n} \int_0^1 \frac{t^{\frac{c+2}{n}-1}}{1+t} dt$.

Proof. As function h given in the theorem is convex, we can use the same arguments as in the proof of Theorem 2. Interpreting the hypothesis of Theorem 3, we read that

$$F_{p(U)}\left(p(z) + \frac{1}{c+2} zp'(z)\right) \leq f_{h(U)}h(z),$$

where $p(z)$ is given by (9).

By applying Lemma 2, the following fuzzy inequality is obtained:

$$F_{p(U)}p(z) \leq F_{g(U)}g(z) \leq F_{h(U)}h(z), \quad \text{i.e.,} \quad F_{(L_\alpha^m G)'(U)}(L_\alpha^m G(z))' \leq F_{g(U)}g(z) \leq F_{h(U)}h(z),$$

where

$$g(z) = \frac{c+2}{nz^{\frac{c+2}{n}}} \int_0^z t^{\frac{c+2}{n}-1} \frac{1+(2\delta-1)t}{1+t} dt = (2\delta - 1) + \frac{(c+2)(2-2\delta)}{nz^{\frac{c+2}{n}}} \int_0^z \frac{t^{\frac{c+2}{n}-1}}{1+t} dt.$$

Using the hypothesis of convexity for function g, it is known that $g(U)$ is symmetric with respect to the real axis and we can write

$$F_{L_\alpha^m G(U)}(L_\alpha^m G(z))' \geq \min_{|z|=1} F_{g(U)}g(z) = F_{g(U)}g(1) \tag{11}$$

and $\beta^* = g(1) = 2\delta - 1 + \frac{(c+2)(2-2\delta)}{n} \int_0^1 \frac{t^{\frac{c+2}{n}-1}}{1+t} dt$.

From (11) we deduce inclusion (10). □

Theorem 4. *Let* g *be a convex function with* $g(0) = 1$ *and define the function* $h(z) = g(z) + zg'(z)$, $z \in U$.

If a function $f \in \mathcal{A}_n$ *satisfies*

$$F_{(L_\alpha^m f)'(U)}(L_\alpha^m f(z))' \leq F_{h(U)}h(z), \text{ i.e., } (L_\alpha^m f(z))' \prec_{\mathcal{F}} h(z), \ z \in U, \tag{12}$$

for $\alpha \geq 0$ *and* $m, n \in \mathbb{N}$, *then we obtain the fuzzy differential subordination*

$$F_{L_\alpha^m f(U)} \frac{L_\alpha^m f(z)}{z} \leq F_{g(U)}g(z), \text{ i.e., } \frac{L_\alpha^m f(z)}{z} \prec_{\mathcal{F}} g(z), \quad z \in U,$$

and this result is sharp.

Proof. Using Remark 4 concerning the operator L_α^m, we can write

$$L_\alpha^m f(z) = z + \sum_{j=n+1}^{\infty} \left(\alpha j^m + (1-\alpha) C_{m+j-1}^m\right) a_j z^j, \quad z \in U.$$

Consider $p(z) = \frac{L_\alpha^m f(z)}{z} = \frac{z + \sum_{j=n+1}^{\infty} \left(\alpha j^m + (1-\alpha) C_{m+j-1}^m\right) a_j z^j}{z} = 1 + p_n z^n + p_{n+1} z^{n+1} + \ldots,$ $z \in U$.

We deduce that $p \in \mathcal{H}[1, n]$.

Let $L_\alpha^m f(z) = zp(z)$, for $z \in U$. Differentiating the expression we obtain $(L_\alpha^m f(z))' = p(z) + zp'(z)$, $z \in U$.

Using this result in (12), we can write

$$F_{p(U)}(p(z) + zp'(z)) \le F_{h(U)}h(z) = F_{g(U)}(g(z) + zg'(z)), \quad z \in U.$$

We can now apply Lemma 3 and obtain

$$F_{p(U)}p(z) \le F_{g(U)}g(z), \quad z \in U, \quad \text{i.e.,} \quad F_{(L_\alpha^m f)'(U)} \frac{L_\alpha^m f(z)}{z} \le F_{g(U)}g(z), \quad z \in U.$$

Therefore,

$$\frac{L_\alpha^m f(z)}{z} \prec_{\mathcal{F}} g(z), \quad z \in U,$$

and this result is sharp. □

Theorem 5. *Let h be a convex function of order $-\frac{1}{2}$ with $h(0) = 1$. If a function $f \in \mathcal{A}_n$ satisfies*

$$F_{(L_\alpha^m f)'(U)}(L_\alpha^m f(z))' \le F_{h(U)}h(z), \ \text{i.e.,} \ (L_\alpha^m f(z))' \prec_{\mathcal{F}} h(z), \quad z \in U, \tag{13}$$

for $\alpha \ge 0$ and $m, n \in \mathbb{N}$, then

$$F_{L_\alpha^m f(U)} \frac{L_\alpha^m f(z)}{z} \le F_{q(U)}q(z), \ \text{i.e.,} \ \frac{L_\alpha^m f(z)}{z} \prec_{\mathcal{F}} q(z), \quad z \in U,$$

where $q(z) = \frac{1}{nz^{\frac{1}{n}}} \int_0^z h(t)t^{\frac{1}{n}-1}dt$ is convex and is the fuzzy best dominant.

Proof. Let

$$\begin{aligned} p(z) &= \frac{L_\alpha^m f(z)}{z} = \frac{z + \sum_{j=n+1}^{\infty}\left(\alpha j^m + (1-\alpha)C_{m+j-1}^m\right)a_j z^j}{z} \\ &= 1 + \sum_{j=n+1}^{\infty}\left(\alpha j^m + (1-\alpha)C_{m+j-1}^m\right)a_j z^{j-1} = 1 + \sum_{j=n+1}^{\infty} p_j z^{j-1}, \quad z \in U, \ p \in \mathcal{H}[1,n]. \end{aligned}$$

As $\mathrm{Re}\left(1 + \frac{zh''(z)}{h'(z)}\right) > -\frac{1}{2}$, $z \in U$, from Lemma 1, we obtain that $q(z) = \frac{1}{nz^{\frac{1}{n}}} \int_0^z h(t)t^{\frac{1}{n}-1}dt$ is a convex function and verifies the differential equation associated to the fuzzy differential subordination (13) $q(z) + zq'(z) = h(z)$, therefore it is the fuzzy best dominant.

Differentiating, we obtain $(L_\alpha^m f(z))' = p(z) + zp'(z)$, $z \in U$ and (13) becomes

$$F_{p(U)}(p(z) + zp'(z)) \le F_{h(U)}h(z), \quad z \in U.$$

Using Lemma 3, we have

$$F_{p(U)}p(z) \le F_{q(U)}q(z), \ z \in U, \quad \text{i.e.,} \quad F_{L_\alpha^m f(U)} \frac{L_\alpha^m f(z)}{z} \le F_{q(U)}q(z), \ z \in U.$$

We have obtained that

$$\frac{L_\alpha^m f(z)}{z} \prec_{\mathcal{F}} q(z), \quad z \in U.$$

□

Corollary 1. *Let $h(z) = \frac{1+(2\beta-1)z}{1+z}$ a convex function in U, $0 \le \beta < 1$. If $\alpha \ge 0$, $m, n \in \mathbb{N}$, $f \in \mathcal{A}_n$ and verifies the fuzzy differential subordination*

$$F_{(L_\alpha^m f)'(U)}(L_\alpha^m f(z))' \le F_{h(U)}h(z), \text{i.e.,} \ (L_\alpha^m f(z))' \prec_{\mathcal{F}} h(z), \quad z \in U, \tag{14}$$

then

$$F_{L_\alpha^m f(U)} \frac{L_\alpha^m f(z)}{z} \le F_{q(U)} q(z), \text{ i.e., } \frac{L_\alpha^m f(z)}{z} \prec_{\mathcal{F}} q(z), \quad z \in U,$$

where q *is given by* $q(z) = 2\beta - 1 + \frac{2(1-\beta)}{nz^{\frac{1}{n}}} \int_0^z \frac{t^{\frac{1}{n}-1}}{1+t} dt$, $z \in U$. *The function* q *is convex and it is the fuzzy best dominant.*

Proof. We have $h(z) = \frac{1+(2\beta-1)z}{1+z}$ with $h(0) = 1$, $h'(z) = \frac{-2(1-\beta)}{(1+z)^2}$ and $h''(z) = \frac{4(1-\beta)}{(1+z)^3}$, therefore $Re\left(\frac{zh''(z)}{h'(z)} + 1\right) = Re\left(\frac{1-z}{1+z}\right) = Re\left(\frac{1-\rho\cos\theta - i\rho\sin\theta}{1+\rho\cos\theta + i\rho\sin\theta}\right) = \frac{1-\rho^2}{1+2\rho\cos\theta+\rho^2} > 0 > -\frac{1}{2}$.

Following the same steps as in the proof of Theorem 5 and considering $p(z) = \frac{L_\alpha^m f(z)}{z}$, the fuzzy differential subordination (14) becomes

$$F_{L_\alpha^n f(U)}(p(z) + zp'(z)) \le F_{h(U)} h(z), \quad z \in U.$$

By using Lemma 2 for $\gamma = 1$, we have $F_{p(U)} p(z) \le F_{q(U)} q(z)$, i.e.,

$$F_{L_\alpha^n f(U)} \frac{L_\alpha^m f(z)}{z} \le F_{q(U)} q(z)$$

and

$$q(z) = \frac{1}{nz^{\frac{1}{n}}} \int_0^z h(t) t^{\frac{1}{n}-1} dt =$$

$$\frac{1}{nz^{\frac{1}{n}}} \int_0^z t^{\frac{1}{n}-1} \frac{1+(2\beta-1)t}{1+t} dt = 2\beta - 1 + \frac{2(1-\beta)}{nz^{\frac{1}{n}}} \int_0^z \frac{t^{\frac{1}{n}-1}}{1+t} dt, \quad z \in U.$$

□

Example 2. *Let* $h(z) = \frac{1-z}{1+z}$ *with* $h(0) = 1$, $h'(z) = \frac{-2}{(1+z)^2}$ *and* $h''(z) = \frac{4}{(1+z)^3}$.

As $Re\left(\frac{zh''(z)}{h'(z)} + 1\right) = Re\left(\frac{1-z}{1+z}\right) = Re\left(\frac{1-\rho\cos\theta - i\rho\sin\theta}{1+\rho\cos\theta + i\rho\sin\theta}\right) = \frac{1-\rho^2}{1+2\rho\cos\theta+\rho^2} > 0 > -\frac{1}{2}$, *the function* h *is convex in* U.

Let $f(z) = z + z^2$, $z \in U$. *For* $n = 1$, $m = 1$, $\alpha = 2$, *we obtain* $L_2^1 f(z) = -R^1 f(z) + 2S^1 f(z) = -zf'(z) + 2zf'(z) = zf'(z) = z + 2z^2$. *Then,* $(L_2^1 f(z))' = 1 + 4z$ *and* $\frac{L_2^1 f(z)}{z} = 1 + 2z$.

We have $q(z) = \frac{1}{z} \int_0^z \frac{1-t}{1+t} dt = -1 + \frac{2\ln(1+z)}{z}$.

Using Theorem 5 we obtain

$$1 + 4z \prec_{\mathcal{F}} \frac{1-z}{1+z}, \quad z \in U,$$

induces

$$1 + 2z \prec_{\mathcal{F}} -1 + \frac{2\ln(1+z)}{z}, \quad z \in U.$$

Theorem 6. *Define the function* $h(z) = g(z) + zg'(z)$, $z \in U$, *using* g *a convex function in* U *with* $g(0) = 1$. *If a function* $f \in \mathcal{A}_n$ *satisfies*

$$F_{L_\alpha^m f(U)} \left(\frac{zL_\alpha^{m+1} f(z)}{L_\alpha^m f(z)}\right)' \le F_{h(U)} h(z), \text{ i.e., } \left(\frac{zL_\alpha^{m+1} f(z)}{L_\alpha^m f(z)}\right)' \prec_{\mathcal{F}} h(z), \quad z \in U \tag{15}$$

for $\alpha \ge 0$ *and* $m, n \in \mathbb{N}$, *then we obtain the sharp fuzzy differential subordination*

$$F_{L_\alpha^m f(U)} \frac{L_\alpha^{m+1} f(z)}{L_\alpha^m f(z)} \le F_{g(U)} g(z), \text{ i.e., } \frac{L_\alpha^{m+1} f(z)}{L_\alpha^m f(z)} \prec_{\mathcal{F}} g(z), \quad z \in U.$$

Proof. For $f \in \mathcal{A}_n$, $f(z) = z + \sum_{j=n+1}^{\infty} a_j z^j$ we have

$L_\alpha^m f(z) = z + \sum_{j=n+1}^{\infty} \left(\alpha j^m + (1-\alpha) C_{m+j-1}^m\right) a_j z^j,\ z \in U.$

Consider

$$p(z) = \frac{L_\alpha^{m+1} f(z)}{L_\alpha^m f(z)} = \frac{z + \sum_{j=n+1}^{\infty} \left(\alpha j^{m+1} + (1-\alpha) C_{m+j}^{m+1}\right) a_j z^j}{z + \sum_{j=n+1}^{\infty} \left(\alpha j^m + (1-\alpha) C_{m+j-1}^m\right) a_j z^j}.$$

We have $p'(z) = \frac{\left(L_\alpha^{m+1} f(z)\right)'}{L_\alpha^m f(z)} - p(z) \cdot \frac{\left(L_\alpha^m f(z)\right)'}{L_\alpha^m f(z)}$ and we obtain $p(z) + z \cdot p'(z) = \left(\frac{z L_\alpha^{m+1} f(z)}{L_\alpha^m f(z)}\right)'$.

Relation (15) becomes

$$F_{p(U)}(p(z) + zp'(z)) \le F_{h(U)} h(z) = F_{g(U)}(g(z) + zg'(z)),\ \ z \in U.$$

By using Lemma 3, we have

$$F_{p(U)} p(z) \le F_{g(U)} g(z),\ \ z \in U, \quad \text{i.e.,} \quad F_{L_\alpha^m f(U)} \frac{L_\alpha^{m+1} f(z)}{L_\alpha^m f(z)} \le F_{g(U)} g(z), \quad z \in U.$$

We obtain

$$\frac{L_\alpha^{m+1} f(z)}{L_\alpha^m f(z)} \prec_{\mathcal{F}} g(z),\ \ z \in U.$$

□

Theorem 7. *Given a convex function g with $g(0) = 1$, define function $h(z) = g(z) + zg'(z)$, $z \in U$.*

If we take $\alpha \ge 0$ and $m, n \in \mathbb{N}$ and a function $f \in \mathcal{A}_n$ satisfying

$$F_{L_\alpha^m f(U)}\left(\left(L_\alpha^{m+1} f(z)\right)' + \frac{(1-\alpha) m z (R^m f(z))''}{m+1}\right) \le F_{h(U)} h(z),\ \textit{i.e.,}$$

$$\left(L_\alpha^{m+1} f(z)\right)' + \frac{(1-\alpha) m z (R^m f(z))''}{m+1} \prec_{\mathcal{F}} h(z),\ \ z \in U, \tag{16}$$

then the sharp fuzzy differential subordination results

$$F_{(L_\alpha^m f)'(U)} [L_\alpha^m f(z)]' \le F_{g(U)} g(z),\ \textit{i.e.,}\ [L_\alpha^m f(z)]' \prec_{\mathcal{F}} g(z), \quad z \in U.$$

Proof. Using the definition of operator L_α^m, we get

$$L_\alpha^{m+1} f(z) = (1-\alpha) R^{m+1} f(z) + \alpha S^{m+1} f(z), \quad z \in U. \tag{17}$$

Using this result in (16), we obtain

$$F_{L_\alpha^m f(U)}\left(\left((1-\alpha) R^{m+1} f(z) + \alpha S^{m+1} f(z)\right)' + \frac{(1-\alpha) m z (R^m f(z))''}{m+1}\right) \le F_{h(U)} h(z),\ z \in U,$$

which can be easily transformed into

$$F_{L_\alpha^m f(U)}\left((1-\alpha)(R^m f(z))' + \alpha (S^m f(z))' + z\left((1-\alpha)(R^m f(z))'' + \alpha (S^m f(z))''\right)\right) \le F_{h(U)} h(z), z \in U.$$

Let

$$p(z) = (1-\alpha)(R^m f(z))' + \alpha (S^m f(z))' = (L_\alpha^m f(z))' \tag{18}$$

$$= 1 + \sum_{j=n+1}^{\infty} \left(\alpha j^{m+1} + (1-\alpha) j C_{m+j-1}^m\right) a_j z^{j-1} = 1 + p_n z^n + p_{n+1} z^{n+1} + \ldots.$$

We deduce that $p \in \mathcal{H}[1, n]$.

Using the notation in (18), the fuzzy differential subordination becomes

$$F_{p(U)}(p(z)+zp'(z)) \leq F_{h(U)}h(z) = F_{g(U)}(g(z)+zg'(z)).$$

By using Lemma 3, we have

$$F_{p(U)}p(z) \leq F_{g(U)}g(z), \ z \in U, \quad \text{i.e.,} \quad F_{L_\alpha^m f(U)}(L_\alpha^m f(z))' \leq F_{g(U)}g(z), \ z \in U,$$

and this result is sharp. □

Theorem 8. *Let h be a convex function of order $-\frac{1}{2}$ which satisfies $h(0)=1$. If a function $f \in \mathcal{A}_n$ satisfies*

$$F_{L_\alpha^m f(U)}\left(\left(L_\alpha^{m+1}f(z)\right)' + \frac{(1-\alpha)mz(R^m f(z))''}{m+1}\right) \leq F_{h(U)}h(z), \text{ i.e.,}$$

$$\left(L_\alpha^{m+1}f(z)\right)' + \frac{(1-\alpha)mz(R^m f(z))''}{m+1} \prec_{\mathcal{F}} h(z), \ z \in U, \tag{19}$$

for $\alpha \geq 0$ and $m, n \in \mathbb{N}$, then the fuzzy differential subordination can be written as

$$F_{L_\alpha^m f(U)}(L_\alpha^m f(z))' \leq F_{q(U)}q(z), \text{ i.e., } (L_\alpha^m f(z))' \prec_{\mathcal{F}} q(z), \ z \in U,$$

with $q(z) = \frac{1}{nz^{\frac{1}{n}}}\int_0^z h(t)t^{\frac{1}{n}-1}dt$ being convex and the best fuzzy dominant.

Proof. As h is a convex function of order $-\frac{1}{2}$, Lemma 1 can be applied and we have that $q(z) = \frac{1}{nz^{\frac{1}{n}}}\int_0^z h(t)t^{\frac{1}{n}-1}dt$ is a convex function and verifies the differential equation associated to the fuzzy differential subordination (19) $q(z)+zq'(z) = h(z)$, therefore it is the fuzzy best dominant.

Using the properties of operator L_α^m and considering $p(z) = (L_\alpha^m f(z))'$, we obtain

$$\left(L_\alpha^{m+1}f(z)\right)' + \frac{(1-\alpha)mz(R^m f(z))''}{m+1} = p(z)+zp'(z), \ z \in U.$$

Then, (19) becomes

$$F_{p(U)}(p(z)+zp'(z)) \leq F_{h(U)}h(z), \ z \in U.$$

As $p \in \mathcal{H}[1,n]$, using Lemma 3, we deduce

$$F_{p(U)}p(z) \leq F_{q(U)}q(z), \ z \in U, \quad \text{i.e.,} \quad F_{L_\alpha^m f(U)}(L_\alpha^m f(z))' \leq F_{q(U)}q(z), \ z \in U.$$

We have obtained

$$(L_\alpha^m f(z))' \prec_{\mathcal{F}} q(z), \ z \in U.$$

□

Corollary 2. *Consider the special case when using the convex function $h(z) = \frac{1+(2\beta-1)z}{1+z}$, where $0 \leq \beta < 1$.*

If $\alpha \geq 0$, $m, n \in \mathbb{N}$, $f \in \mathcal{A}_n$ and satisfies the differential subordination

$$F_{L_\alpha^m f(U)}\left([L_\alpha^{m+1}f(z)]' + \frac{(1-\alpha)mz(R^m f(z))''}{m+1}\right) \leq F_{h(U)}h(z), \text{ i.e.,}$$

$$[L_\alpha^{m+1}f(z)]' + \frac{(1-\alpha)mz(R^m f(z))''}{m+1} \prec_{\mathcal{F}} h(z), \quad z \in U, \tag{20}$$

then

$$F_{L_\alpha^m f(U)}(L_\alpha^m f(z))' \le F_{q(U)}q(z), \text{ i.e., } (L_\alpha^m f(z))' \prec_{\mathcal{F}} q(z), \quad z \in U,$$

where q is given by $q(z) = 2\beta - 1 + \frac{2(1-\beta)}{nz^{\frac{1}{n}}} \int_0^z \frac{t^{\frac{1}{n}-1}}{1+t}dt$, *for* $z \in U$. *The function q is convex and it is the fuzzy best dominant.*

Proof. Following the same argumentation as for the proof of Theorem 7 and taking $p(z) = (L_\alpha^m f(z))'$, the fuzzy differential subordination (20) becomes

$$F_{p(U)}(p(z) + zp'(z)) \le F_{h(U)}h(z), \quad z \in U.$$

By using Lemma 2 for $\gamma = 1$, we have $F_{p(U)}p(z) \le F_{q(U)}q(z)$, i.e.,

$$F_{(L_\alpha^m f)'(U)}(L_\alpha^m f(z))' \le F_{q(U)}q(z), \text{ i.e., } (L_\alpha^m f(z))' \prec_{\mathcal{F}} q(z), \ z \in U,$$

and $q(z) = \frac{1}{nz^{\frac{1}{n}}} \int_0^z h(t)t^{\frac{1}{n}-1}dt = \frac{1}{nz^{\frac{1}{n}}} \int_0^z t^{\frac{1}{n}-1}\frac{1+(2\beta-1)t}{1+t}dt = 2\beta - 1 + \frac{2(1-\beta)}{nz^{\frac{1}{n}}} \int_0^z \frac{t^{\frac{1}{n}-1}}{1+t}dt$, $z \in U$. □

Example 3. *Let* $h(z) = \frac{1-z}{1+z}$ *a convex function in U with* $h(0) = 1$ *and* $Re\left(\frac{zh''(z)}{h'(z)} + 1\right) > -\frac{1}{2}$ *(see Example 2).*

Let $f(z) = z + z^2$, $z \in U$. *For* $n = 1$, $m = 1$, *and* $\alpha = 2$, *we obtain* $L_2^1 f(z) = zf'(z) = z + 2z^2$ *and* $(L_2^1 f(z))' = 1 + 4z$. *We also obtain* $(L_\alpha^{m+1} f(z))' + \frac{(1-\alpha)mz(R^m f(z))''}{m+1} = (L_2^2 f(z))' - \frac{z(R^1 f(z))''}{2} = (z^2 + z)' - \frac{z(z+2z^2)''}{2} = 1.$

We have $q(z) = \frac{1}{z}\int_0^z \frac{1-t}{1+t}dt = -1 + \frac{2\ln(1+z)}{z}$.

Using Theorem 8 we obtain

$$1 \prec_{\mathcal{F}} \frac{1-z}{1+z}, \quad z \in U,$$

induce

$$1 + 4z \prec_{\mathcal{F}} -1 + \frac{2\ln(1+z)}{z}, \quad z \in U.$$

3. Conclusions

Further studies on the newly introduced class can be conducted for obtaining results that give coefficient estimates, distortion theorems, or closure theorems, as it is usual in geometric function theory. Furthermore, the way this class is introduced can inspire research for introducing other interesting fuzzy classes and studying their properties. The limit imposed on $\delta \in (0, 1]$ could be further investigated so that other possible values of δ for correct definitions of fuzzy classes could be found.

Author Contributions: Conceptualization, A.A.L. and G.I.O.; methodology, G.I.O.; software, A.A.L.; validation, A.A.L. and G.I.O.; formal analysis, A.A.L. and G.I.O.; investigation, A.A.L.; resources, G.I.O.; data curation, G.I.O.; writing—original draft preparation, A.A.L.; writing—review and editing, A.A.L. and G.I.O.; visualization, A.A.L.; supervision, G.I.O.; project administration, A.A.L.; funding acquisition, G.I.O. All authors have read and agreed to the published version of the manuscript.

Funding: This research received no external funding.

Institutional Review Board Statement: Not applicable.

Informed Consent Statement: Not applicable.

Data Availability Statement: Not applicable.

Conflicts of Interest: The authors declare no conflict of interest.

References

1. Zadeh, L.A. Fuzzy Sets. *Inf. Control* **1965**, *8*, 338–353. [CrossRef]
2. Dzitac, I.; Filip, F.G.; Manolescu, M.J. Fuzzy Logic Is Not Fuzzy: World-renowned Computer Scientist Lotfi A. Zadeh. *Int. J. Computers Commun. Control.* **2017**, *12*, 748–789. [CrossRef]
3. Dzitac, S.; Nădăban, S. Soft Computing for Decision-Making in Fuzzy Environments: A Tribute to Professor Ioan Dzitac. *Mathematics* **2021**, *9*, 1701. [CrossRef]
4. Zadeh, L.A.; Tufis, D.; Filip, F.G.; Dzitac, I. *From Natural Language to Soft Computing: New Paradigms in Artificial Intelligence*; Editing House of Romanian Academy: Bucharest, Romania, 2008.
5. Oros, G.I.; Oros, G. The notion of subordination in fuzzy sets theory. *Gen. Math.* **2011**, *19*, 97–103.
6. Oros, G.I.; Oros, G. Fuzzy differential subordination. *Acta Univ. Apulensis* **2012**, *3*, 55–64.
7. Miller, S.S.; Mocanu, P.T. *Differential Subordinations. Theory and Applications*; Marcel Dekker, Inc.: New York, NY, USA; Basel, Switzerland, 2000.
8. Miller, S.S.; Mocanu, P.T. Second order-differential inequalities in the complex plane. *J. Math. Anal. Appl.* **1978**, *65*, 298–305. [CrossRef]
9. Miller, S.S.; Mocanu, P.T. Differential subordinations and univalent functions. *Michig. Math. J.* **1981**, *28*, 157–171. [CrossRef]
10. Alb Lupaş, A. A note on special fuzzy differential subordinations using generalized Sălăgean operator and Ruscheweyh derivative. *J. Comput. Anal. Appl.* **2013**, *15*, 1476–1483.
11. Alb Lupaş, A.; Oros, G. On special fuzzy differential subordinations using Sălăgean and Ruscheweyh operators. *Appl. Math. Comput.* **2015**, *261*, 119–127.
12. Venter, A.O. On special fuzzy differential subordination using Ruscheweyh operator. *Analele Univ. Oradea Fasc. Mat.* **2015**, *XXII*, 167–176.
13. El-Deeb, S.M.; Alb Lupaş, A. Fuzzy differential subordinations associated with an integral operator. *An. Univ. Oradea Fasc. Mat.* **2020**, *XXVII*, 133–140.
14. El-Deeb, S.M.; Oros, G.I. Fuzzy differential subordinations connected with the linear operator. *Math. Bohem.* **2021**, 1–10. [CrossRef]
15. Oros, G.I. New fuzzy differential subordinations. *Commun. Fac. Sci. Univ. Ank. Ser. A1 Math. Stat.* **2021**, *70*, 229–240.
16. Srivastava, H.M.; El-Deeb, S.M. Fuzzy Differential Subordinations Based upon the Mittag-Leffler Type Borel Distribution. *Symmetry* **2021**, *13*, 1023. [CrossRef]
17. Alb Lupaş, A. On special differential subordinations using Sălăgean and Ruscheweyh operators. *Math. Inequal. Appl.* **2009**, *12*, 781–790. [CrossRef]
18. Alb Lupaş, A. On a certain subclass of analytic functions defined by Sălăgean and Ruscheweyh operators. *J. Math. Appl.* **2009**, *31*, 67–76.
19. Gal, S.G.; Ban, A.I. *Elemente de Matematică Fuzzy*; University of Oradea: Oradea, Romania, 1996.
20. Oros, G.I.; Oros, G. Dominants and best dominants in fuzzy differential subordinations. *Stud. Univ. Babeş-Bolyai Math.* **2012**, *57*, 239–248.
21. Sălăgean, G.S. Subclasses of univalent functions. In *Lecture Notes in Math*; Springer: Berlin, Germany, 1983; Volume 1013; pp. 362–372.
22. Ruscheweyh, S. New criteria for univalent functions. *Proc. Amet. Math. Soc.* **1975**, *49*, 109–115. [CrossRef]

mathematics MDPI

Article

Regular and Intra-Regular Semigroups in Terms of m-Polar Fuzzy Environment

Shahida Bashir [1,*], Sundas Shahzadi [1], Ahmad N. Al-Kenani [2] and Muhammad Shabir [3]

[1] Department of Mathematics, University of Gujrat, Gujrat 50700, Pakistan; 19011709-020@uog.edu.pk
[2] Department of Mathematics, Faculty of Science, King Abdulaziz University, P.O. Box 80219, Jeddah 21589, Saudi Arabia; analkenani@kau.edu.sa
[3] Department of Mathematics, Quaid-I-Azam University, Islamabad 44000, Pakistan; mshabirbhatti@qau.edu.pk
* Correspondence: shahida.bashir@uog.edu.pk

Abstract: The central objective of the proposed work in this research is to introduce the innovative concept of an m-polar fuzzy set (m-PFS) in semigroups, that is, the expansion of bipolar fuzzy set (BFS). Our main focus in this study is the generalization of some important results of BFSs to the results of m-PFSs. This paper provides some important results related to m-polar fuzzy subsemigroups (m-PFSSs), m-polar fuzzy ideals (m-PFIs), m-polar fuzzy generalized bi-ideals (m-PFGBIs), m-polar fuzzy bi-ideals (m-PFBIs), m-polar fuzzy quasi-ideals (m-PFQIs) and m-polar fuzzy interior ideals (m-PFIIs) in semigroups. This research paper shows that every m-PFBI of semigroups is the m-PFGBI of semigroups, but the converse may not be true. Furthermore this paper deals with several important properties of m-PFIs and characterizes regular and intra-regular semigroups by the properties of m-PFIs and m-PFBIs.

Keywords: m-PF subsemigroups; m-PF generalized bi-ideals; m-PF bi-ideals; m-PF quasi-ideals; m-PF interior ideals

MSC: 03E72; 18B40

Citation: Bashir, S.; Shahzadi, S.; Al-Kenani, A.N.; Shabir, M. Regular and Intra-Regular Semigroups in Terms of m-Polar Fuzzy Environment. *Mathematics* **2021**, *9*, 2031. https://doi.org/10.3390/math9172031

Academic Editor: Sorin Nadaban

Received: 30 July 2021
Accepted: 20 August 2021
Published: 24 August 2021

1. Introduction

In 2014, Chen et al. [1] presented the m-PFS as an expansion of the BFS. The mathematical theories of a 2-polar fuzzy set and BFS are equivalent, and we can see that one connected to the other. The BFS is expanded into an m-PFS by applying the notion of one-to-one correspondence. Sometimes, different things are monitored in different ways. The m-PFS is effective in assigning degrees of membership to various objects in multi-polar data. The m-PFS gives only a positive degree of membership to each element. The m-PFS has an extensive range of implementations in real world problems related to the multi-agent, multi-objects, multi-polar information, multi-index and multi-attributes. This theory is applicable when a company decides to construct an item, a country elects its political leaders, or a group of friends wants to visit a country, with various options. It can be used in decision making, co-operative games, disease diagnosis, to discuss the confusions and conflicts of communication signals in wireless communications and as a model to define clusters or categorization and multi-relationships. In sum, an m-PFS on H is a mapping $I : H \to [0,1]^m$.

Here, we will make a model-based example on m-PFS, and use it to conveniently select an appropriate employee in a company. Here, the selection of an employee is based on 4-PFS with their four qualities, which are honesty, punctuality, communication skills, and being hardworking. Let $H = \{a_1, a_2, a_3, a_4, a_5\}$ be the set of five employees in a company. We shall characterize them according to four qualities in the form of 4-PFS, given in Table 1:

Table 1. Table of qualities in persons with their membership values.

	Honesty	Punctual	Communication	Hardworking
a_1	0.6	0.5	0.8	1
a_2	1	0.8	0.5	0.4
a_3	0.5	1	1	0.8
a_4	0.8	0.5	1	0.7
a_5	1	0.5	0	0.6

Therefore, we attain a 4-PFS $\eta : H \rightarrow [0,1]^4$ of H such that

$$\begin{aligned}\eta(a_1) &= (0.6, 0.5, 0.8, 1)\\ \eta(a_2) &= (1, 0.8, 0.5, 0.4)\\ \eta(a_3) &= (0.5, 1, 1, 0.8)\\ \eta(a_4) &= (0.8, 0.5, 1, 0.7)\\ \eta(a_5) &= (1, 0.5, 0, 0.6).\end{aligned}$$

Figure 1 is the graphical representation of 4-PFS:

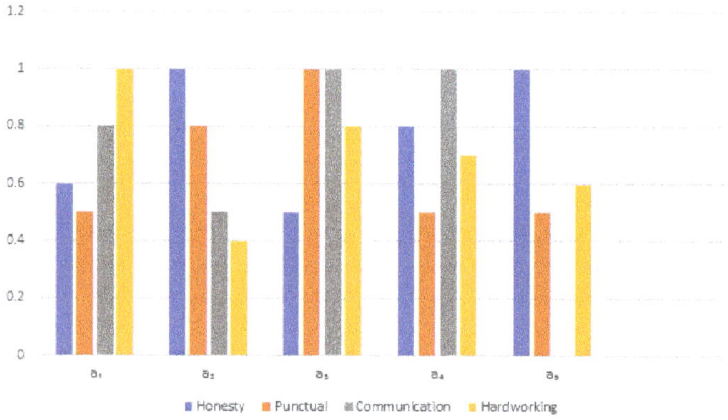

Figure 1. Graphical representation of 4-polar fuzzy subset.

Here, 1 represents good remarks, 0.5 represents average and 0 represents bad remarks. Similarly, we can solve any other problem with uncertainty in multiple directions.

Zhang [2] proposed that the function is mapped to the interval $[-1, 1]$ rather than $[0, 1]$ in BFS theory. Lee [3] coined the term bipolar fuzzy ideals. BFS is useful for solving uncertain problems with two poles of a situation: positive and negative pole. For more applications of BFS, see [4–9]. In medical science, environmental research, and engineering, we may find data or information that are ambiguous or complicated. All mathematical equations and techniques in classical mathematics are exact, they cannot deal with such problems. Many tools have been developed to deal with such issues. After extensive effort, Zadeh [10] was the first to propose fuzzy set theory as a solution to such complicated issues. This idea is used in a variety of areas, including logic, measure theory, topological space, ring theory, group theory and real analysis. The theory of fuzzy group was first intitated by Rosenfeld [11]. Kuroki [12] and Mordeson [13] have extensively explored fuzzy semigroups.

Semigroups are very useful in many applications containing dynamical systems, control problems, partial differential equations, sociology, stochastic differential equations, biology, etc. Some examples of semigroups are the collection of all mappings of a set, under the composition of functions (termed a full transformation monoid) and the set of natural numbers $\mathbb{N}$ under either addition or multiplication. The word "semigroup" was introduced to provide a title for some structures that were not groups but were created through the expansion of consequences. The proper semigroup theory was initiated by the working of Russian mathematician Anton Kazimirovich Suschkewitsch [14]. Quasi-ideals in semigroups were introduced by Otto Steinfeld [15].

The study of m-PF algebraic structures began with the concept of m-PF Lie subalgebras [16]. After that, the m-PF Lie ideals were studied in Lie algebras [17]. In 2017, Sarwar and Akram worked on new applications of m-PF matroids [18]. In 2019, Ahmad and Al-Masarwah introduced the concept of m-PF (commutative) ideals and m-polar (α, β)-fuzzy ideals in BCK/BCI-algebras [19,20]. To continue their work, they introduced a new aspect of generalized m-PF ideals and studied the normalization of m-PF subalgebras in [21,22]. Recently, Muhiuddin and Al-Kadi presented interval-valued m-PF BCK/BCI-Algebras [23]. Shabir et al. [24] studied regular and intra-regular semirings in terms of BFIs. Then, Bashir et al. [25,26] studied regular ordered ternary semigroups and semirings in terms of BFIs. Shabir et al. extended the work of [24], initiated the concept of m-PFIs in LA-semigroups and characterized the regular LA-semigroups according to the properties of these m-PFIs [27]. By extending the work of [24,27], the concept of m-PFIs in semigroups was introduced and characterizations of regular and intra-regular semigroups according to the properties of m-PFIs are given in this paper.

This paper is charaterized as follows: We present some basic concepts related to m-PFS in Section 2. The major part of this paper is Section 3, the m-PFSSs, m-PFIs (left, right), m-PFBIs, m-PFGBIs, m-PFQIs, m-PFIIs of semigroups are discussed with examples. In Section 4, the regular and intra-regular semigroups are characterized by the properties of m-PFIs. A comparison between this research and previous work is shown in Section 5. In Section 6, we also talk about the conclusions and future work.

The list of acronyms used in the research article is given in Table 2.

Table 2. List of acronyms.

Acronyms	Representation
BFS	Bipolar fuzzy set
BFIs	Bipolar fuzzy ideals
m-PFS	m-Polar fuzzy set
m-PFSs	m-Polar fuzzy subsets
m-PFSSs	m-Polar fuzzy subsemigroups
m-PFIs	m-Polar fuzzy ideals
m-PFGBIs	m-Polar fuzzy generalized bi-ideals
m-PFBIs	m-Polar fuzzy bi-ideals
m-PFQIs	m-Polar fuzzy quasi-ideals
m-PFIIs	m-Polar fuzzy interior ideals

2. Preliminaries

In this Section, we have studied the fundamental but essential definitions and preliminary findings based on semigroups that are important in their own right. These are necessary for later Sections.

If a groupoid $(P, \cdot)$ satisfies the associative property, then it is called a semigroup. Throughout this paper, P will denote the semigroup, unless specified otherwise. A non-empty subset H of P is called a subsemigroup of P if $ab \in H$ for every $a, b \in H$. In this paper, subsets mean non-empty subsets. A subset H of P is called a left ideal (resp. right ideal) of P if $PH \subseteq H$ (resp. $HP \subseteq H$). If H is left and right ideal, then H is called a two-sided ideal or ideal of P [28].

A subset H of P is called a generalized bi-ideal of P if $HPH \subseteq H$. The subsemigroup H of P is called a bi-ideal of P if $HPH \subseteq H$. A subset H of P is called a quasi-ideal of P if $HP \cap PH \subseteq H$. The subsemigroup H of P is called an interior ideal of P if $(PH)P \subseteq H$ [28].

A fuzzy subset η of P is a mapping from P to closed interval $[0,1]$, *that is* $\eta : P \to [0,1]$ [10]. A bipolar fuzzy subset η of P is a mapping from P to closed interval $[-1,1]$ written as $\eta = (P, \eta^-, \eta^+)$, where $\eta^- : P \to [-1,0]$ and $\eta^+ : P \to [0,1]$. It can differentiate between unrelated and contrary components of fuzzy problems. A natural one-to-one correspondence exists among the BFS and 2-polar fuzzy set ($[0,1]^2$-set). When data for real world complex situations come from m factors ($m \geq 2$), then m-PFS is used to deal with such problems. An m-PFS (or a $[0,1]^m$-set) on P is a function $\eta = P \to [0,1]^m$. More generally, the m-PFS is the m-tuple of membership degree function of P that is $\eta = (\eta_1, \eta_2, \ldots, \eta_m)$, where $\eta_\kappa : P \to [0,1]$ is the mapping for every $\kappa \in \{1, 2, \ldots, m\}$. Here, $\mathbf{0} = (0, 0, \ldots, 0)$ is the smallest value in $[0,1]^m$ and $\mathbf{1} = (1, 1, \ldots, 1)$ is the largest value in $[0,1]^m$ [1].

The set of all m-PFSs of P is represented by $m(P)$. We define relation $\leq$ on $m(P)$ as follows: For any m-PFSs $\eta = (\eta_1, \eta_2, \ldots, \eta_m)$ and $m' = (m_1, m_2, \ldots, m_m)$ of P, $\eta \leq m'$ means that $\eta_\kappa(a) \leq m'_\kappa(a)$ for every $a \in P$ and $\kappa \in \{1, 2, \ldots, m\}$. The symbols $\eta \wedge m'$ and $\eta \vee m'$ mean the following m-PFSs of P. $(\eta \wedge m')(a) = \eta(a) \wedge m'(a)$ and $(\eta \vee m')(a) = \eta(a) \vee m'(a)$ that is $(\eta_\kappa \wedge m'_\kappa)(a) = \eta_\kappa(a) \wedge m'_\kappa(a)$ for each $a \in P$ and $\kappa \in \{1, 2, \ldots, m\}$; $(\eta_\kappa \vee m'_\kappa)(a) = \eta_\kappa(a) \vee m'_\kappa(a)$ for each $a \in P$ and $\kappa \in \{1, 2, \ldots, m\}$. For two m-PFSs $\eta = (\eta_1, \eta_2, \ldots, \eta_m)$ and $m' = (m_1, m_2, \ldots, m_m)$, the product of $\eta \circ m' = (\eta_1 \circ m_1, \eta_2 \circ m_2, \ldots, \eta_m \circ m_m)$ is defined as

$$(\eta_\kappa \circ m'_\kappa)(a) = \begin{cases} \bigvee\limits_{a=st} \{\eta_\kappa(s) \wedge m_\kappa(t), & \text{if } a = st \text{ for some } s, t \in P; \\ 0, & \text{otherwise}; \end{cases}$$

for all $\kappa \in \{1, 2, \ldots, m\}$. The next example shows the product of m-PFSs η and m' of P for $m = 4$.

Example 1. *Consider the semigroup* $P = \{\imath, \jmath, \ell, \hbar\}$ *given in Table 3.*

Table 3. Table of multiplication of P.

$\cdot$	$\imath$	$\jmath$	ℓ	$\hbar$
$\imath$	$\imath$	$\imath$	$\imath$	$\imath$
$\jmath$	$\imath$	$\imath$	$\imath$	$\imath$
ℓ	$\imath$	$\imath$	$\jmath$	$\imath$
$\hbar$	$\imath$	$\imath$	$\imath$	$\jmath$

We define 4-PFSs $\eta = (\eta_1, \eta_2, \eta_3, \eta_4)$ and $m' = (m_1, m_2, m_3, m_4)$ as follows:

$\eta(\imath) = (0.2, 0.1, 0, 0.4)$, $\eta(\jmath) = (0.7, 0.5, 0.1, 0)$, $\eta(\ell) = (0.1, 0.3, 0.7, 0.4)$, $\eta(\hbar) = (0, 0, 0, 0.1)$ and

$m'(\imath) = (0.7, 0.3, 0, 0.4)$, $m'(\jmath) = (0.2, 0, 0, 0.1)$, $m'(\ell) = (0.2, 0.2, 0.4, 0)$, $m'(\hbar) = (0.2, 0.3, 0, 0)$.

By defintion, we obtain

$(\eta_1 \circ m_1)(\imath) = 0.7$, $(\eta_1 \circ m_1)(\jmath) = 0.1$, $(\eta_1 \circ m_1)(\ell) = 0$, $(\eta_1 \circ m_1)(\hbar) = 0$;
$(\eta_2 \circ m_2)(\imath) = 0.3$, $(\eta_2 \circ m_2)(\jmath) = 0.1$, $(\eta_2 \circ m_2)(\ell) = 0$, $(\eta_2 \circ m_2)(\hbar) = 0$;
$(\eta_3 \circ m_3)(\imath) = 0.1$, $(\eta_3 \circ m_3)(\jmath) = 0.4$, $(\eta_3 \circ m_3)(\ell) = 0$, $(\eta_3 \circ m_3)(\hbar) = 0$;
$(\eta_4 \circ m_4)(\imath) = 0.4$, $(\eta_4 \circ m_4)(\jmath) = 0.0$, $(\eta_4 \circ m_4)(\ell) = 0$, $(\eta_4 \circ m_4)(\hbar) = 0$.

Hence, the product of $\eta = (\eta_1, \eta_2, \eta_3, \eta_4)$ and $m' = (m_1, m_2, m_3, m_4)$ is defined by

$(\eta \circ m')(\imath) = (0.7, 0.3, 0.1, 0.4)$, $(\eta \circ m')(\jmath) = (0.1, 0.1, 0.4, 0)$, $(\eta \circ m')(\ell) = (0, 0, 0, 0)$, $(\eta \circ m')(\hbar) = (0, 0, 0, 0)$.

Definition 1. *Let* $\eta = (\eta_1, \eta_2, \ldots, \eta_m)$ *be an m-PFS of P.*

1. *Define* $\eta_t = \{a \in P | \eta(a) \geq t\}$ *for all* t, *where* $t = (t_1, t_2, \ldots, t_m) \in (0,1]^m$, *that is,* $\eta_\kappa(a) \geq t_\kappa$ *for all* $\kappa \in \{1, 2, \ldots, m\}$. *Then,* η_t *is called t-cut or a level set.*
2. *The support of* $\eta : P \to [0,1]^m$ *is defined as the set* $Supp(\eta) = \{a \in P | \eta(a) > (0, 0, \ldots, 0)$ *m-tuple*$\}$, *that is* $\eta_\kappa(a) > 0$ *for all* $\kappa \in \{1, 2, \ldots, m\}$.

Definition 2. *An m-PFS* $\eta = (\eta_1, \eta_2, \ldots, \eta_m)$ *of P is called an m-PFSS of P if, for all* $a, b \in P$, $\eta(ab) \geq \min\{\eta(a), \eta(b)\}$, *that is,* $\eta_\kappa(ab) \geq \min\{\eta_\kappa(a), \eta_\kappa(b)\}$ *for all* $\kappa \in \{1, 2, \ldots, m\}$.

Definition 3. *An m-PFS* $\eta = (\eta_1, \eta_2, \ldots, \eta_m)$ *of P is called an m-PFI left (resp. right) of P for all* $a, b \in P$, $\eta(ab) \geq \eta(b)$ *(resp.* $\eta(ab) \geq \eta(a)$*), that is* $\eta_\kappa(ab) \geq \eta_\kappa(b)$ *(resp.* $\eta_\kappa(ab) \geq \eta_\kappa(a)$*) for all* $\kappa \in \{1, 2, \ldots, m\}$.

An m-PFS η of P is called an m-PFI of P if η is both an m-PFI (left) and m-PFI (right) of P.

The example given below is of 4-PFI of P.

Example 2. *Let* $P = \{\imath, \jmath, \ell, \hbar\}$ *be a semigroup given in Table 4.*

Table 4. Table of multiplication of P.

$\cdot$	$\imath$	$\jmath$	ℓ	$\hbar$
$\imath$	$\imath$	$\imath$	$\imath$	$\imath$
$\jmath$	$\imath$	$\jmath$	ℓ	$\imath$
ℓ	$\imath$	$\imath$	$\imath$	$\imath$
$\hbar$	$\imath$	$\hbar$	$\imath$	$\imath$

We define a 4-PFS $\eta = (\eta_1, \eta_2, \eta_3, \eta_4)$ of P as follows:

$\eta(\imath) = (0.7, 0.6, 0.6, 0.4)$, $\eta(\jmath) = (0.2, 0, 0, 0.1)$, $\eta(\ell) = (0.5, 0.4, 0.3, 0.1)$, $\eta(\hbar) = (0.5, 0.4, 0.3, 0.1)$.

Clearly, $\eta = (\eta_1, \eta_2, \eta_3, \eta_4)$ is both 4-PFIs (left and right) of P. Hence η is a 4-PFI of P.

Definition 4. *Let a subset H of P. Then, the m-polar characteristic function* $C_H : H \to [0,1]^m$ *is defined as*

$$C_H(h) = \begin{cases} (1, 1, \ldots, 1), \text{ m-tuple if } h \in H; \\ (0, 0, \ldots, 0), \text{ m-tuple if } h \notin H. \end{cases}$$

3. Characterization of Semigroups by *m*-Polar Fuzzy Sets

This is the most essential portion, because here we make our major contributions. With the help of several lemmas, theorems, and examples, the notions of m-PFSSs and m-PFIs of semigroups are explained in this section. We have proved that every m-PFBI of P is m-PFGBI, but the converse does not hold. For LA-semigroups, Shabir et al. [27] has proved this result. We have generalized the results in Shabir et al. [27] for semigroups. In whole paper, δ is an m-PFS of P that maps each element of P on $(1, 1, \ldots, 1)$.

Lemma 1. *Consider two subsets H and I of P. Then*

1. $C_H \wedge C_I = C_{H \cap I}$;
2. $C_H \vee C_I = C_{H \cup I}$;
3. $C_H \circ C_I = C_{HI}$.

Proof. The proof of (1) and (2) are obvious.

(3): Case 1: Let $a \in HI$. This implies that $a = hi$ for some $h \in H$ and $i \in I$. Therefore, $C_{HI}(a) = (1, 1, \ldots, 1)$. Since $h \in H$ and $i \in I$, we have $C_H(h) = (1, 1, \ldots, 1)$ or $C_I(i) = (1, 1, \ldots, 1)$. Now,

$$\begin{aligned}(C_H \circ C_I)(a) &= \bigvee_{a=bc} \{C_H(b) \wedge C_I(c)\} \\ &\geq C_H(h) \wedge C_I(i) \\ &= (1,1,\ldots,1).\end{aligned}$$

Therefore, $C_H \circ C_I = C_{HI}$. Case 2: If $a \notin HI$. This implies that $C_{HI}(a) = (0,0,\ldots,0)$, since $a \notin hi$ for every $h \in H$ and $i \in I$. Therefore

$$\begin{aligned}(C_H \circ C_I)(a) &= \bigvee_{a=hi} \{C_H(h) \wedge C_I(i)\} \\ &= (0,0,\ldots,0).\end{aligned}$$

Hence $C_H \circ C_I = C_{HI}$. □

Lemma 2. *Let H be a subset of P. Then, the given statements hold.*

1. *H is a subsemigroup of P if, and only if, C_H is an m-PFSS of P;*
2. *H is a left ideal (resp. right) of P if and only if C_H is an m-PFI left (resp. right) of P.*

Proof. (1) Consider H as the subsemigroup of P. We have to show that $C_H(ab) \geq C_H(a) \wedge C_H(b)$ for all $a, b \in P$. Now, we consider some cases:

Case 1: Let $a, b \in H$. Then, $C_H(a) = C_H(b) = (1,1,\ldots,1)$. As H is a subsemigroup of P, so $ab \in H$ implies that $C_H(ab) = (1,1,\ldots,1)$. Hence $C_H(ab) \geq C_H(a) \wedge C_H(b)$.

Case 2: Let $a \in H$, $b \notin H$. Then, $C_H(a) = (1,1,\ldots,1), C_H(b) = (0,0,\ldots,0)$. Hence, $C_H(ab) \geq (0,0,\ldots,0) = C_H(a) \wedge C_H(b)$.

Case 3: Let $a, b \notin H$. Then, $C_H(a) = C_H(b) = (0,0,\ldots,0)$. Clearly, $C_H(ab) \geq (0,0,\ldots,0) = C_H(a) \wedge C_H(b)$.

Case 4: Let $a \notin H$, $b \in H$. Then, $C_H(a) = (0,0,\ldots,0)$ and $C_H(b) = (1,1,\ldots,1)$. Clearly, $C_H(ab) \geq (0,0,\ldots,0) = C_H(a) \wedge C_H(b)$.

Conversely, let C_H be an m-PFSS of P. Let $a, b \in H$. Then, $C_H(a) = C_H(b) = (1,1,\ldots,1)$. By definition, $C_H(ab) \geq C_H(a) \wedge C_H(b) = (1,1,\ldots,1) \wedge (1,1,\ldots,1) = (1,1,\ldots,1)$, we have $C_H(ab) = (1,1,\ldots,1)$. This implies that $ab \in H$, that is H is a subsemigroup of P.

(2) Suppose that H is the left ideal of P. We have to show that $C_H(ab) \geq C_H(b)$ for every $a, b \in P$. Now, consider the two cases:

Case 1: Let $b \in H$ and $a \in P$. Then, $C_H(b) = (1,1,\ldots,1)$. Since H is a left ideal of P, $ab \in H$ implies that $C_H(ab) = (1,1,\ldots,1)$. Hence $C_H(ab) \geq C_H(b)$.

Case 2: Let $b \notin H$ and $a \in P$. Then, $C_H(b) = (0,0,\ldots,0)$. Clearly, $C_H(ab) \geq C_H(b)$.

Conversely, let C_H be an m-PFI (left) of P. Let $a \in P$ and $b \in H$. Then, $C_H(b) = (1,1,\ldots,1)$. By definition, $C_H(ab) \geq C_H(b) = (1,1,\ldots,1)$, we have $C_H(ab) = (1,1,\ldots,1)$. This implies that $ab \in H$, that is H is a left ideal of P.

In the same way, we can show that H is right ideal of P if, and only if, C_H is an m-PFI (right) of P. Therefore, H is an ideal of P if, and only if, C_H is an m-PFI of P. □

Lemma 3. *For m-PFS $\eta = (\eta_1, \eta_2, \ldots, \eta_m)$ of P, the following properties hold.*

1. *η is an m-PFSS of P if, and only if, $\eta \circ \eta \leq \eta$;*
2. *η is an m-PFI (left) of P if, and only if, $\delta \circ \eta \leq \eta$;*
3. *η is an m-PFI (right) of P if, and only if, $\eta \circ \delta \leq \eta$;*
4. *η is an m-PFI of P if, and only if, $\delta \circ \eta \leq \eta$ and $\eta \circ \delta \leq \eta$, where δ is the m-PFS of P that maps each element of P on $(1,1,\ldots,1)$.*

Proof. (1) Assume that $\eta = (\eta_1, \eta_2, \ldots, \eta_m)$ is an m-PFSS of P, that is, $\eta_\kappa(ab) \geq \eta_\kappa(a) \wedge \eta_\kappa(b)$ for all $\kappa \in \{1, 2, \ldots, m\}$. Let $p \in P$. If p is not expressible as $p = ab$ for some $a, b \in P$;

then, $(\eta \circ \eta)(p) = 0$. Hence, $(\eta \circ \eta)(p) \leq \eta(p)$. However, if p is expressible as $p = ab$ for some $a, b \in P$, then

$$\begin{aligned} (\eta_\kappa \circ \eta_\kappa)(p) &= \bigvee_{p=ab} \{\eta_\kappa(a) \wedge \eta_\kappa(b)\} \\ &\leq \bigvee_{p=ab} \{\eta_\kappa(ab)\} \\ &= \eta_\kappa(p) \text{ for all } \kappa \in \{1, 2, \ldots, m\}. \end{aligned}$$

Hence, $\eta \circ \eta \leq \eta$. Conversely, let $\eta \circ \eta \leq \eta$ and $a, b \in P$. Then

$$\begin{aligned} \eta_\kappa(ab) &\geq (\eta_\kappa \circ \eta_\kappa)(ab) \\ &= \bigvee_{ab=uv} \{\eta_\kappa(u) \wedge \eta_\kappa(v)\} \\ &\geq \eta_\kappa(a) \wedge \eta_\kappa(b) \text{ for all } \kappa \in \{1, 2, \ldots, m\}. \end{aligned}$$

Hence, $\eta_\kappa(ab) \geq \eta_\kappa(a) \wedge \eta_\kappa(b)$. Thus, η is m-PFSS of P.

(2) Assume that $\eta = (\eta_1, \eta_2, \ldots, \eta_m)$ is m-PFI (left) of P, that is, $\eta_\kappa(ab) \geq \eta_\kappa(b)$ for all $\kappa \in \{1, 2, \ldots, m\}$ and $a, b \in P$. Let $p \in P$. If p is not expressible as $p = ab$ for some $a, b \in P$, then $(\delta \circ \eta)(p) = 0$. Hence, $\delta \circ \eta \leq \eta$. However, if p is expressible as $p = ab$ for some $a, b \in P$, then

$$\begin{aligned} (\delta_\kappa \circ \eta_\kappa)(p) &= \bigvee_{p=ab} \{\delta_\kappa(a) \wedge \eta_\kappa(b)\} \\ &= \bigvee_{p=ab} \{\eta_\kappa(b)\} \\ &\leq \bigvee_{p=ab} \eta_\kappa(ab) \\ &= \eta_\kappa(p) \text{ for all } \kappa \in \{1, 2, \ldots, m\}. \end{aligned}$$

Hence $\delta \circ \eta \leq \eta$. Conversely, let $\delta \circ \eta \leq \eta$ and $a, b \in P$. Then,

$$\begin{aligned} \eta_\kappa(ab) &\geq (\delta_\kappa \circ \eta_\kappa)(ab) \\ &= \bigvee_{ab=uv} \{\delta_\kappa(u) \wedge \eta_\kappa(v)\} \\ &\geq \{\delta_\kappa(a) \wedge \eta_\kappa(b)\} \\ &= \eta_\kappa(b) \text{ for all } \kappa \in \{1, 2, \ldots, m\}. \end{aligned}$$

Hence, $\eta(ab) \geq \eta(b)$. Thus, η is m-PFI (left) of P.

(3) This can be proved similarly to the proof of part (2) of Lemma 3.

(4) The proof of this follows from parts (2) and (3) of Lemma 3. □

Lemma 4. *The given statements are true in P.*

1. Let $\eta = (\eta_1, \eta_2, \ldots, \eta_m)$ and $m' = (m_1, m_2, \ldots, m_m)$ be two m-PFSSs of P. Then, $\eta \wedge m'$ is also an m-PFSS of P;
2. Let $\eta = (\eta_1, \eta_2, \ldots, \eta_m)$ and $m' = (m_1, m_2, \ldots, m_m)$ be two m-PFIs of P. Then, $\eta \wedge m'$ is also an m-PFI of P.

Proof. Straightforward. □

Proposition 1. *Let $\eta = (\eta_1, \eta_2, \ldots, \eta_m)$ be an m-PFS of P. Then, η is an m-PFSS (resp. m-PFI) of P if, and only if, $\eta_t = \{a \in P | \eta(a) \geq t\} \neq \phi$ is a subsemigroup (resp. ideal) of P for all $t \in (t_1, t_2, \ldots, t_m) \in (0, 1]^m$.*

Proof. Let η be an m-PFSS of P. Let $a, b \in \eta_t$. Then, $\eta_\kappa(a) \geq t_\kappa$ and $\eta_\kappa(b) \geq t_\kappa$ for all $\kappa \in \{1, 2, \ldots, m\}$. As η is an m-PFSS of P, this implies $\eta_\kappa(ab) \geq \eta_\kappa(a) \wedge \eta_\kappa(b) \geq t_\kappa \wedge t_\kappa = t_\kappa$ for all $\kappa \in \{1, 2, \ldots, m\}$. Therefore, $ab \in \eta_t$. Then η_t is a subsemigroup of P.

Conversely, let $\eta_t \neq \phi$ be a subsemigroup of P. On the contrary, let us consider that η is not an m-PFSS of P. Suppose $a, b \in P$ such that $\eta_\kappa(ab) < \eta_\kappa(a) \wedge \eta_\kappa(b)$ for some $\kappa \in \{1, 2, \ldots, m\}$. Take $t_\kappa = \eta_\kappa(a) \wedge \eta_\kappa(b)$ for all $\kappa \in \{1, 2, \ldots, m\}$. Then, $a, b \in \eta_t$ but $ab \notin \eta_t$, there is a contradiction. Hence, $\eta_\kappa(ab) \geq \eta_\kappa(a) \wedge \eta_\kappa(b)$. Thus, η is an m-PFSS of P. Other cases can be proved on the same lines. □

Now, we define the m-PFGBI of a semigroup.

Definition 5. *An m-PFS $\eta = (\eta_1, \eta_2, \ldots, \eta_m)$ of P is called an m-PFGBI of P if for all $a, b, c \in P$, $\eta(abc) \geq \eta(a) \wedge \eta(c)$, that is $\eta_\kappa(abc) \geq \eta_\kappa(a) \wedge \eta_\kappa(c)$ for all $\kappa \in \{1, 2, \ldots, m\}$.*

Lemma 5. *A subset H of P is generalized bi-ideal of P if and only if C_H is an m-PFGBI of P.*

Proof. This Lemma 5 can be proved similarly to the proof of Lemma 2. □

Lemma 6. *An m-PFS η of P is m-PFGBI of P if and only if, $\eta \circ \delta \circ \eta \leq \eta$, where δ is the m-PFS of P that maps each element of P on $(1, 1, \ldots, 1)$.*

Proof. Suppose $\eta = (\eta_1, \eta_2, \ldots, \eta_m)$ is the m-PFGBI of P, that is, $\eta_\kappa(abc) \geq \eta_\kappa(a) \wedge \eta_\kappa(c)$ for all $\kappa \in \{1, 2, \ldots, m\}$ and $a, b, c \in P$. Let $p \in P$. If p is not expressible as $p = ab$ for some $a, b \in P$, then $(\eta \circ \delta \circ \eta)(p) = 0$. Hence, $\eta \circ \delta \circ \eta \leq \eta$. However, if p is expressible as $p = ab$ for some $a, b \in P$. Then

$$\begin{aligned}
(\eta_\kappa \circ \delta_\kappa \circ \eta_\kappa)(p) &= \bigvee_{p=ab} \{(\eta_\kappa \circ \delta_\kappa)(a) \wedge \eta_\kappa(b)\} \\
&= \bigvee_{p=ab} \{ \bigvee_{a=uv} \{\eta_\kappa(u) \wedge \delta_\kappa(v)\} \wedge \eta_\kappa(b)\} \\
&= \bigvee_{p=ab} \{ \bigvee_{a=uv} \{\eta_\kappa(u) \wedge \eta_\kappa(b)\}\} \\
&\leq \bigvee_{p=ab} \{ \bigvee_{a=uv} \{\eta_\kappa(uv)b)\}\} \\
&= \bigvee_{p=ab} \{\eta_\kappa(ab)\} \text{ for all } \kappa \in \{1, 2, \ldots, m\}. \\
&= \eta_\kappa(p) \text{ for all } \kappa \in \{1, 2, \ldots, m\}.
\end{aligned}$$

Hence, $\eta \circ \delta \circ \eta \leq \eta$. Conversely, let $\eta \circ \delta \circ \eta \leq \eta$ and $a, b, c \in P$. Then,

$$\begin{aligned}
\eta_\kappa(abc) &\geq (\eta_\kappa \circ \delta_\kappa) \circ \eta_\kappa)((ab)c) \\
&= \bigvee_{(ab)c=uv} \{(\eta_\kappa \circ \delta_\kappa)(u) \wedge \eta_\kappa(v)\} \\
&\geq (\eta_\kappa \circ \delta_\kappa)(ab) \wedge \eta_\kappa(c) \\
&= \bigvee_{(ab)=xy} \{(\eta_\kappa(x) \wedge \delta_\kappa)(y))\} \wedge \eta_\kappa(c) \\
&= \{(\eta_\kappa(a) \wedge \delta_\kappa)(b))\} \wedge \eta_\kappa(c) \\
&= \eta_\kappa(a) \wedge \eta_\kappa(c) \text{ for all } \kappa \in \{1, 2, \ldots, m\}.
\end{aligned}$$

Hence, $\eta(abc) \geq \eta(a) \wedge \eta(c)$. Thus, η is m-PFGBI of P. □

Proposition 2. *Assume that $\eta = (\eta_1, \eta_2, \ldots, \eta_m)$ is an m-PFS of P. Then, η is an m-PFGBI of P if, and only if, $\eta_t = \{a \in P | \eta(a) \geq t\} \neq \phi$ is a generalized bi-ideal of P for all $t = (t_1, t_2, \ldots, t_m) \in (0, 1]^m$.*

Proof. Let η be an m-PFGBI of P. Let $a, c \in \eta_t$ and $b \in P$. Then, $\eta_\kappa(a) \geq t_\kappa$ and $\eta_\kappa(c) \geq t_\kappa$ for all $\kappa \in \{1, 2, \ldots, m\}$. Since η is m-PFGBI of P, we have $\eta_\kappa(abc) \geq \eta_\kappa(a) \wedge \eta_\kappa(c) \geq t_\kappa \wedge t_\kappa = t_\kappa$ for all $\kappa \in \{1, 2, \ldots, m\}$. Therefore, $abc \in \eta_t$. That is η_t is a GBI of P.

Conversely, assuming that $\eta_t \neq \phi$ is a GBI of P. On the contrary, assume that η is not m-PFGBI of P. Suppose $a, b, c \in P$, such that $\eta_\kappa(abc) < \eta_\kappa(a) \wedge \eta_\kappa(c)$ for some $\kappa \in \{1, 2, \ldots, m\}$. Take $t_\kappa = \eta_\kappa(a) \wedge \eta_\kappa(c)$ for all $\kappa \in \{1, 2, \ldots, m\}$. Then, $a, c \in \eta_t$ but $abc \notin \eta_t$, which is a contradiction. Hence, $\eta_\kappa(abc) \geq \eta_\kappa(a) \wedge \eta_\kappa(c)$, that is, η is m-PFGBI of P. □

Next, we define the m-PFBI of a semigroup.

Definition 6. *A subsemigroup $\eta = (\eta_1, \eta_2, \ldots, \eta_m)$ of P is called an m-PFBI of P if for all $a, b, c \in P, \eta(abc) \geq \eta(a) \wedge \eta(c)$ that is, $\eta_\kappa(abc) \geq \eta_\kappa(a) \wedge \eta_\kappa(c)$ for all $\kappa \in \{1, 2, \ldots, m\}$.*

Lemma 7. *A subset H of P is a bi-ideal of P if, and only if, C_H is an m-PFBI of P.*

Proof. Follows from Lemmas 2 and 5. □

Lemma 8. *An m-PFSS η of P is an m-PFBI of P if and only if, $\eta \circ \delta \circ \eta \leq \eta$, where δ is the m-PFS of P, which maps each element of P on $(1, 1, \ldots, 1)$.*

Proof. Follows from Lemma 6. □

Proposition 3. *Let $\eta = (\eta_1, \eta_2, \ldots, \eta_m)$ be a subsemigroup of P. Then η is an m-PFBI of P if and only if, $\eta_t = \{a \in P | \eta(a) \geq t\} \neq \phi$ is a bi-ideal of P for all $t = (t_1, t_2, \ldots, t_m) \in (0, 1]^m$.*

Proof. Follows from Proposition 2. □

Remark 1. *Every m-PFBI of P is an m-PFGBI of P.*

The Example 3 illustrates that the converse of above Remark may not be true.

Example 3. *Let $P = \{\imath, \jmath, \ell, \hbar\}$ be a semigroup given in Table 5.*

Table 5. Table of multiplication of P.

$\cdot$	$\imath$	$\jmath$	ℓ	$\hbar$
$\imath$	$\imath$	$\imath$	$\imath$	$\imath$
$\jmath$	$\imath$	$\imath$	$\imath$	$\imath$
ℓ	$\imath$	$\imath$	$\jmath$	$\imath$
$\hbar$	$\imath$	$\imath$	$\imath$	$\jmath$

We define a 4-PFS $\eta = (\eta_1, \eta_2, \eta_3, \eta_4)$ of P as follows: $\eta(\imath) = (0.1, 0.3, 0.3, 0.4)$, $\eta(\jmath) = (0, 0, 0, 0)$, $\eta(\ell) = (0, 0, 0, 0)$, $\eta(\hbar) = (0.5, 0.6, 0.7, 0.8)$. Then, simple calculations show that the η is a 4-PFGBI of P.

Now, $\eta(\jmath) = \eta(\hbar\hbar) = (0, 0, 0, 0) \ngeq (0.5, 0.6, 0.7, 0.8) = \eta(\hbar) \wedge \eta(\hbar)$. Therefore, η is not a bi-ideal of P. Next, we define the m-PFQI of a semigroup.

Definition 7. *An m-PFS $\eta = (\eta_1, \eta_2, \ldots, \eta_m)$ of P is called an m-PFQI of P if $(\eta \circ \delta) \wedge (\delta \circ \eta) \leq \eta$, that is $(\eta_\kappa \circ \delta_\kappa) \wedge (\delta_\kappa \circ \eta_\kappa) \leq \eta_\kappa$, for all $\kappa \in \{1, 2, \ldots, m\}$.*

Lemma 9. *A subset H of P is a quasi ideal of P if and only if C_H is an m-PFQI of P.*

Proof. Let H be a quasi ideal of P, that is $HP \cap PH \subseteq H$. We show that $(C_H \circ \delta) \wedge (\delta \circ C_H) \leq C_H$, that is, $((C_H \circ \delta) \wedge (\delta \circ C_H))(h) \leq C_H(h)$ for every $h \in P$. We study the following cases:

Case 1: If $h \in H$ then $C_H(h) = (1,1,\ldots,1) \geq ((C_H \circ \delta) \wedge (\delta \circ C_H))(h)$. Hence $(C_H \circ \delta) \wedge (\delta \circ C_H) \leq C_H$.

Case 2: If $h \notin H$ then $h \notin HP \cap PH$. This implies that $h \neq bc$ and $h \neq fe$ for some $b \in H, c \in P, f \in P$ and $e \in H$. Therefore, either $(C_H \circ \delta)(h) = (0,0,\ldots,0)$ or $(\delta \circ C_H)(h) = (0,0,\ldots,0)$ that is $((C_H \circ \delta) \wedge (\delta \circ C_H))(h) = (0,0,\ldots,0) \leq C_H(h)$. Hence $(C_H \circ \delta) \wedge (\delta \circ C_H) \leq C_H$.

Conversely, let $n \in HP \cap PH$. Then $n = be$ and $n = af$, where $a, e \in P$ and $b, f \in H$. Since C_H is an m-PFQI of P, we have

$$\begin{aligned} C_H(n) &\geq ((C_H \circ \delta) \wedge (\delta \circ C_H))(n) \\ &= (C_H \circ \delta)(n) \wedge (\delta \circ C_H)(n) \\ &= \{ \bigvee_{n=wv} \{(C_H(w) \wedge \delta(v)\}\} \wedge \{ \bigvee_{n=pq} \{\delta(p) \wedge C_H(q)\}\} \\ &\geq \{C_H(b) \wedge \delta(e)\} \wedge \{\delta(a) \wedge C_H(f)\} \text{ since } n = be \text{ and } n = af \\ &= (1,1,\ldots,1). \end{aligned}$$

Therefore, $C_H(n) = (1,1,\ldots,1)$. Hence, $n \in H$. □

Proposition 4. *An m-PFS $\eta = (\eta_1, \eta_2, \ldots, \eta_m)$ of P is an m-PFQI of P if, and only if, $\eta_t = \{a \in P | \eta(a) \geq t\} \neq \phi$ is a quasi ideal of P for all $t = (t_1, t_2, \ldots, t_m) \in (0,1]^m$.*

Proof. Let η is an m-PFQI of P. To show that $\eta_t P \cap P\eta_t \subseteq \eta_t$. Let $n \in \eta_t P \cap P\eta_t$. Then, $n \in \eta_t P$ and $n \in P\eta_t$. Therefore, $n = ba$ and $n = sd$ for some $a, s \in P$ and $b, d \in \eta_t$. Therefore, $\eta_\kappa \geq t_\kappa$ for all $\kappa \in \{1,2,\ldots,m\}$.

Now

$$\begin{aligned} (\eta_\kappa \circ \delta_\kappa)(n) &= \bigvee_{n=uv} \{\eta_\kappa(u) \wedge \delta_\kappa(v)\} \\ &\geq \eta_\kappa(b) \wedge \delta_\kappa(a) \text{ because } n = ba \\ &= \eta_\kappa(b) \wedge 1 \\ &= \eta_\kappa(b) \\ &\geq t_\kappa. \end{aligned}$$

So

$$(\eta_\kappa \circ \delta_\kappa)(n) \geq t_\kappa \text{ for all } \kappa \in \{1,2,\ldots,m\}.$$

Now

$$\begin{aligned} (\delta_\kappa \circ \eta_\kappa)(n) &= \bigvee_{n=uv} \{\delta_\kappa(u) \wedge \eta_\kappa(v)\} \\ &\geq \delta_\kappa(s) \wedge \eta_\kappa(d) \text{ because } n = sd \\ &= 1 \wedge \eta_\kappa(d) \\ &= \eta_\kappa(d) \\ &\geq t_\kappa. \end{aligned}$$

So

$$(\delta_\kappa \circ \eta_\kappa)(n) \geq t_\kappa \text{ for all } \kappa \in \{1,2,\ldots,m\}.$$

Therefore, $((\eta_\kappa \circ \delta_\kappa) \wedge (\delta_\kappa \circ \eta_\kappa))(n) = (\eta_\kappa \circ \delta_\kappa)(n) \wedge (\delta_\kappa \circ \eta_\kappa)(n) \geq t_\kappa \wedge t_\kappa = t_\kappa$ for all $\kappa \in \{1,2,\ldots,m\}$. So, $((\eta \circ \delta) \wedge (\delta \circ \eta))(n) \geq t$. Since $\eta(n) \geq ((\eta \circ \delta) \wedge (\delta \circ \eta))(n) \geq t$, so $n \in \eta_t$. Hence, η_t is a quasi ideal of P.

Conversely, consider that η is not quasi ideal of P. Let $n \in P$ be such that $\eta_\kappa(n) < (\eta_\kappa \circ \delta_\kappa)(n) \wedge (\delta_\kappa \circ \eta_\kappa)(n)$ for some $\kappa \in \{1,2,\ldots,m\}$. Choose $t_\kappa \in (0,1]$, such that $t_\kappa = (\eta_\kappa \circ \delta_\kappa)(n) \wedge (\delta_\kappa \circ \eta_\kappa)(n)$ for all $\kappa \in \{1,2,\ldots,m\}$. This implies that $n \in (\eta_\kappa \circ \delta_\kappa)_{t_\kappa}$ and $n \in (\delta_\kappa \circ \eta_\kappa)_{t_\kappa}$ but $n \notin (\eta_\kappa)_{t_\kappa}$ for some κ. Hence, $n \in (\eta \circ P)_t$ and $n \in (P \circ \eta)_t$ but $n \notin (\eta)_t$, which is a contradiction. Hence, $(\eta \circ \delta) \wedge (\delta \circ \eta) \leq \eta$. □

Lemma 10. *Every m-PF one-sided ideal of P is an m-PFQI of P.*

Proof. This proof follows from Lemma 3. □

In the next example, it is shown that the converse of the above Lemma may not be true.

Example 4. *Consider the semigroup $P = \{\imath, \jmath, \ell\}$ given in Table 6.*

Table 6. Table of multiplication of P.

$\cdot$	$\imath$	$\jmath$	ℓ
$\imath$	$\imath$	$\jmath$	ℓ
$\jmath$	$\jmath$	$\jmath$	$\jmath$
ℓ	ℓ	$\jmath$	ℓ

Define a 3-PFS $\eta = (\eta_1, \eta_2, \eta_3)$ of P as follows: $\eta(\imath) = (0.3, 0.3, 0.4), \eta(\jmath) = (0.7, 0.8, 0.9)$, $\eta(\ell) = (0, 0, 0)$.

Then, simple calculations show that $\eta_\jmath$ is QI of P. Therefore, by using Proposition 4, η is 3-PFQI of P. Now,

$\eta(\ell) = \eta(\imath\ell) = (0,0,0) \ngeq \eta(\imath) = (0.3, 0.3, 0.4)$. So η is not 3-PFI (right) of P.

Lemma 11. *Let $\eta = (\eta_1, \eta_2, \ldots, \eta_m)$ and $m = (m_1, m_2, \ldots, m_m)$ be two m-PFI(right) and m-PFI(left) of P, respectively. Then $\eta \wedge m$ is m-PFQI of P.*

Proof. Let $n \in P$. If $n \neq bt$ for some $b, t \in P$. Then, $((\eta \wedge m) \circ \delta) \wedge (\delta \circ (\eta \wedge m)) \leq (\eta \wedge m)$. If $n = ab$ for some $a, b \in P$, then

$$
\begin{aligned}
(((\eta_\kappa \wedge m_\kappa) \circ \delta_\kappa) \wedge (\delta_\kappa \circ (\eta_\kappa \wedge m_\kappa)))(n) &= ((\eta_\kappa \wedge m_\kappa) \circ \delta_\kappa)(n) \wedge (\delta_\kappa \circ (\eta_\kappa \wedge m_\kappa))(n) \\
&= \bigvee_{n=ab} \{(\eta_\kappa \wedge m_\kappa)(a) \wedge \delta_\kappa(b)\} \wedge \bigvee_{n=ab} \{\delta_\kappa(a) \wedge (\eta_\kappa \wedge m_\kappa)(b)\} \\
&= \bigvee_{n=ab} \{(\eta_\kappa \wedge m_\kappa)(a)\} \wedge \bigvee_{n=ab} \{(\eta_\kappa \wedge m_\kappa)(b)\} \\
&= \bigvee_{n=ab} \{(\eta_\kappa \wedge m_\kappa)(a) \wedge (\eta_\kappa \wedge m_\kappa)(b)\} \\
&= \bigvee_{n=ab} \{(\eta_\kappa(a) \wedge m_\kappa(a)) \wedge (\eta_\kappa(b) \wedge m_\kappa(b))\} \\
&\leq \bigvee_{n=ab} \{\eta_\kappa(a) \wedge m_\kappa(b)\} \\
&\leq \bigvee_{n=ab} \{\eta_\kappa(ab) \wedge m_\kappa(ab)\} \\
&= \bigvee_{n=ab} \{(\eta_\kappa \wedge m_\kappa)(ab)\} \\
&= (\eta_\kappa \wedge m_\kappa)(n) \text{ for all } \kappa \in \{1,2,\ldots,m\}.
\end{aligned}
$$

Hence, $((\eta \wedge m) \circ \delta) \wedge (\delta \circ (\eta \wedge m)) \leq (\eta \wedge m)$, that is, $\eta \wedge m$ is m-PFQI of P. □

Now, we define the m-PFII of a semigroup.

Definition 8. *An m-PFSS $\eta = (\eta_1, \eta_2, \ldots, \eta_m)$ of P is called an m-PFII of P if for all $a, b, c \in P$, $\eta(abc) \geq \eta(b)$, that is, $\eta_\kappa(abc) \geq \eta_\kappa(b)$ for all $\kappa \in \{1, 2, \ldots, m\}$.*

Lemma 12. *A subset H of P is an interior ideal of P if, and only if, C_H is an m-PFII of P.*

Proof. Let H be any interior ideal of P. From Lemma 2, C_H is an m-PFSS of P. Now we show that $C_H(abc) \geq C_H(b)$ for every $a, b, c \in P$. We consider the following two cases:

Case 1: Let $b \in H$ and $a, c \in P$. Then $C_H(b) = (1, 1, \ldots, 1)$. Since H is an interior ideal of P, then $abc \in H$. Then, $C_H(abc) = (1, 1, \ldots, 1)$. Hence, $C_H(abc) \geq C_H(b)$.

Case 2: Let $b \notin H$ and $a, c \in P$. Then, $C_H(b) = (0, 0, \ldots, 0)$. Clearly, $C_H(abc) \geq C_H(b)$. Hence, C_H of H is an m-PFII of P.

Conversely, consider C_H of H is an m-PFII of P. Then by Lemma 2, H is a subsemigroup of P. Let $b \in H$ and $a, c \in P$. Then $C_H(b) = (1, 1, \ldots, 1)$. By hypothesis, $C_H(abc) \geq C_H(b) = (1, 1, \ldots, 1)$. Hence $C_H(abc) = (1, 1, \ldots, 1)$. This implies that $abc \in H$, that is H is an interior ideal of P. □

Lemma 13. *An m-PFSS η of P is an m-PFII of P if, and only if, $\delta \circ \eta \circ \delta \leq \eta$.*

Proof. Let $\eta = (\eta_1, \eta_2, \ldots, \eta_m)$ be m-PFII of P. We show that $\delta \circ \eta \circ \delta \leq \eta$. Let $n \in P$. Then, for all $\kappa \in \{1, 2, \ldots, m\}$.

$$
\begin{aligned}
(\delta_\kappa \circ \eta_\kappa \circ \delta_\kappa)(n) &= \bigvee_{n=uv} \{(\delta_\kappa \circ \eta_\kappa)(u) \wedge \delta_\kappa(v)\} \\
&= \bigvee_{n=uv} \{(\delta_\kappa \circ \eta_\kappa)(u)\} \\
&= \bigvee_{n=uv} \{ \bigvee_{u=ab} (\delta_\kappa(a) \wedge \eta_\kappa(b)\} \\
&= \bigvee_{n=(ab)v} \{\eta_\kappa(b)\} \\
&\leq \bigvee_{n=(ab)v} \{\eta_\kappa((ab)v)\} \text{ as } \eta \text{ is an } m\text{-PFII of } P. \\
&= \eta_\kappa(n) \text{ for all } \kappa \in \{1, 2, \ldots, m\}.
\end{aligned}
$$

Therefore $\delta \circ \eta \circ \delta \leq \eta$.

Conversely, let $\delta \circ \eta \circ \delta \leq \eta$. We only show that $\eta_\kappa(abc) \geq \eta_\kappa(b)$ for every $a, b, c \in P$ and for all $\kappa \in \{1, 2, \ldots, m\}$. Let $n = abc$. Now, for all $\kappa \in \{1, 2, \ldots, m\}$.

$$
\begin{aligned}
\eta_\kappa(abc) &\geq ((\delta_\kappa \circ \eta_\kappa) \circ \delta_\kappa)((ab)c) \\
&= \bigvee_{(ab)c=uv} \{((\delta_\kappa \circ \eta_\kappa)(u) \wedge \delta_\kappa(v)\} \\
&\geq (\delta_\kappa \circ \eta_\kappa)(ab) \wedge \delta_\kappa(c) \\
&= (\delta_\kappa \circ \eta_\kappa)(ab) \\
&= \bigvee_{ab=pq} \{\delta_\kappa(p) \wedge \eta_\kappa(q)\} \\
&\geq \delta_\kappa(a) \wedge \eta_\kappa(b) \\
&= \eta_\kappa(b) \text{ for all } \kappa \in \{1, 2, \ldots, m\}.
\end{aligned}
$$

Therefore, $\eta_\kappa(abc) \geq \eta_\kappa(b)$ for all $\kappa \in \{1, 2, \ldots, m\}$. Hence, η is m-PFII of P. □

Proposition 5. *A subset $\eta = (\eta_1, \eta_2, \ldots, \eta_m)$ of P is m-PFII of P if, and only if, $\eta_t = \{a \in P | \eta(a) \geq t\} \neq \phi$ is an interior ideal of P for all $t = (t_1, t_2, \ldots, t_m) \in (0, 1]^m$.*

Proof. This is the same as the proof of Propositions 1 and 2. □

4. Characterization of Regular and Intra-Regular Semigroups by m-Polar Fuzzy Ideals

A semigroup P is called regular if for all $x \in P$, there exists an element $a \in P$ such that $x = xax$. A semigroup P is called an intra-regular semigroup if for all $x \in P$, there exists elements $b, c \in P$ such that $x = bx^2c$. Regular and intra-regular semigroups have been studied by several authors, see [24,28]. The characterizations of the regular and intra-regular semigroups in terms of m-PF ideals and m-PFBI are discussed with the help of many theorems in this section.

Theorem 1 ([28]). *The following results are equivalent in P.*

1. P is regular;
2. $H \cap I = HI$, for every right ideal H and left ideal I of P;
3. $J = JPJ$, for every quasi ideal J of P.

Theorem 2. *Each m-PFQI η of P is an m-PFBI of P.*

Proof. Suppose that $\eta = (\eta_1, \eta_2, \ldots, \eta_m)$ be m-PFQI of P. Let $a, b \in P$. Then,

$$\begin{aligned}
\eta_\kappa(ab) &\geq ((\eta_\kappa \circ \delta_\kappa) \wedge (\delta_\kappa \circ \eta_\kappa))(ab) \\
&= (\eta_\kappa \circ \delta_\kappa)(ab) \wedge (\delta_\kappa \circ \eta_\kappa)(ab) \\
&= [\bigvee_{ab=op} \{(\eta_\kappa(o) \wedge \delta_\kappa)(p)\}] \wedge [\bigvee_{ab=uv} \{(\delta_\kappa(u) \wedge \eta_\kappa(v)\}] \\
&\geq \{(\eta_\kappa(a) \wedge \delta_\kappa)(b)\} \wedge \{(\delta_\kappa(a) \wedge \eta_\kappa(b)\} \\
&= \{(\eta_\kappa(a) \wedge 1\} \wedge \{1 \wedge \eta_\kappa(b)\} \\
&= \eta_\kappa(a) \wedge \eta_\kappa(b) \text{ for all } \kappa \in \{1, 2, \ldots, m\}.
\end{aligned}$$

So, $\eta_\kappa(ab) \geq \eta_\kappa(a) \wedge \eta_\kappa(b)$. Now, let $a, b, c \in P$. Then,

$$\begin{aligned}
(\delta_\kappa \circ \eta_\kappa))((ab)c) &= \bigvee_{(ab)c=uv} \{(\delta_\kappa(u) \wedge \eta_\kappa(v)\} \\
&\geq \delta_\kappa(ab) \wedge \eta_\kappa(c) \\
&= 1 \wedge \eta_\kappa(c) \\
&= \eta_\kappa(c).
\end{aligned}$$

Therefore, $(\delta_\kappa \circ \eta_\kappa)(abc) \geq \eta_\kappa(c)$ for all $\kappa \in \{1, 2, \ldots, m\}$. Since $(ab)c = a(bc) \in aP$, so $(ab)c = ap$ for some $p \in P$. Therefore,

$$\begin{aligned}
(\eta_\kappa \circ \delta_\kappa)(abc) &= \bigvee_{(ab)c=ob} \{(\eta_\kappa(o) \wedge \delta_\kappa)(b)\} \\
&\geq \eta_\kappa(a) \wedge \delta_\kappa(p) \text{ since } (ab)c = ap \\
&= \eta_\kappa(a) \wedge 1 \\
&= \eta_\kappa(a).
\end{aligned}$$

Therefore, $(\eta_\kappa \circ \delta_\kappa)(abc) \geq \eta_\kappa(a)$ for all $\kappa \in \{1, 2, \ldots, m\}$. Now, by our supposition

$$\begin{aligned}
\eta_\kappa(abc) &\geq ((\eta_\kappa \circ \delta_\kappa) \wedge (\delta_\kappa \circ \eta_\kappa))(abc) \\
&= (\eta_\kappa \circ \delta_\kappa)(abc) \wedge (\delta_\kappa \circ \eta_\kappa)(abc) \\
&\geq \eta_\kappa(a) \wedge \eta_\kappa(c) \text{ for all } \kappa \in \{1, 2, \ldots, m\}.
\end{aligned}$$

Therefore, $\eta(abc) \geq \eta(a) \wedge \eta(c)$. Hence, η is m-PFBI of P. □

Theorem 3. *The given statements are equivalent in P.*

1. *P is regular;*

2. *$\eta \wedge m' = \eta \circ m'$ for every m-PFI(right) η and m-PFI(left) m' of P.*

Proof. (1) $\Longrightarrow$ (2) : Let $\eta = (\eta_1, \eta_2, \ldots, \eta_m)$ and $m' = (m_1, m_2, \ldots, m_m)$ be two m-PFI(right) and m-PFI(left) of P. Let $o \in P$, we have

$$\begin{aligned}(\eta_\kappa \circ m'_\kappa)(o) &= \bigvee_{o=bc} \{(\eta_\kappa(b) \wedge m'_\kappa(c)\} \\ &\leq \bigvee_{o=bc} \{(\eta_\kappa(bc) \wedge m'_\kappa(bc)\} \\ &= \eta_\kappa(o) \wedge m'_\kappa(o) \\ &= (\eta_\kappa \wedge m'_\kappa)(o) \text{ for all } \kappa \in \{1, 2, \ldots, m\}.\end{aligned}$$

Therefore, $\eta \circ m' \leq \eta \wedge m'$. As P is regular, then for every, $o \in P$, there exists $a \in P$, such that $o = (oa)o$.

$$\begin{aligned}(\eta_\kappa \wedge m'_\kappa)(o) &= \eta_\kappa(o) \wedge m'_\kappa(o) \\ &\leq \eta_\kappa(oa) \wedge m'_\kappa(o) \text{ as } \eta \text{ is } m\text{-PFRI of } P. \\ &\leq \bigvee_{o=bc} \{(\eta_\kappa(b) \wedge m'_\kappa(c)\} \\ &= (\eta_\kappa \circ m'_\kappa)(o) \text{ for all } \kappa \in \{1, 2, \ldots, m\}.\end{aligned}$$

So, $\eta \wedge m' \leq \eta \circ m'$. Therefore, $\eta \wedge m' = \eta \circ m'$.

(2) $\Longrightarrow$ (1) : Let $o \in P$. Then, $\eta = oP$ is a left ideal of P and $m' = oP \cup Po$ is a right ideal of P generated by o. Then, by using Lemma 2, C_η and $C_{m'}$ the m-polar fuzzy characteristic fuctions of η and m' are m-PFI(left) and m-PFI(right) of P, respectively. Then, we have

$$\begin{aligned}C_{m'\eta} &= (C_{m'} \circ C_\eta) \text{ by Lemma 1} \\ &= (C_{m'} \wedge C_\eta) \text{ by 2} \\ &= C_{m' \cap \eta} \text{ by Lemma 1}.\end{aligned}$$

Therefore, $m' \cap \eta = m'\eta$. As a result, Theorem 1 shows that P is regular. □

Theorem 4. *The following statements are equivalent in P.*

1. *P is regular;*
2. *$\eta = \eta \circ \delta \circ \eta$ for every m-PFGBI η of P;*
3. *$\eta = \eta \circ \delta \circ \eta$ for every m-PFQI η of P.*

Proof. (1) $\Longrightarrow$ (2) : Let $\eta = (\eta_1, \eta_2, \ldots, \eta_m)$ be an m-PFGBI of P and $o \in P$. Since P is regular, there exists $a \in P$ *such that* $o = (oa)o$. Therefore, we have

$$\begin{aligned}(\eta_\kappa \circ \delta_\kappa \circ \eta_\kappa)(o) &= \bigvee_{o=bc} \{(\eta_\kappa \circ \delta_\kappa)(b) \wedge \eta_\kappa(c)\} \text{ for some } b, c \in P \\ &\geq (\eta_\kappa \circ \delta_\kappa)(oa) \wedge \eta_\kappa(o) \text{ since } o = (oa)o \\ &= \bigvee_{oa=pq} \{\eta_\kappa(p) \wedge \delta_\kappa)(q)\} \wedge \eta_\kappa(o) \\ &\geq \{\eta_\kappa(o) \wedge \delta_\kappa)(a)\} \wedge \eta_\kappa(o) \\ &= \eta_\kappa(o) \text{ for all } \kappa \in \{1, 2, \ldots, m\}.\end{aligned}$$

Hence, $\eta \circ \delta \circ \eta \geq \eta$. Since η is an m-PFGBI of P. Therefore, we have

$$
\begin{aligned}
(\eta_\kappa \circ \delta_\kappa \circ \eta_\kappa)(o) &= \bigvee_{o=rs} \{(\eta_\kappa \circ \delta_\kappa)(r) \wedge \eta_\kappa(s)\} \text{ for some } r,s \in P \\
&= \bigvee_{o=rs} \{ \bigvee_{r=uv} \{\eta_\kappa(u) \wedge \delta_\kappa(v)\} \wedge \eta_\kappa(s)\}\} \text{ for some } r,s \in P \\
&= \bigvee_{o=rs} \{ \bigvee_{r=uv} \{\eta_\kappa(u) \wedge \eta_\kappa(s)\}\} \\
&\leq \bigvee_{o=rs} \{ \bigvee_{r=uv} \{\eta_\kappa((uv)(s)\}\} \\
&= \bigvee_{o=rs} \eta_\kappa(rs) \\
&= \eta_\kappa(o) \text{ for all } \kappa \in \{1,2,\ldots,m\}.
\end{aligned}
$$

So, $\eta \circ \delta \circ \eta \leq \eta$. Therefore, $\eta = \eta \circ \delta \circ \eta$.

$(2) \Longrightarrow (3)$: It is obvious.

$(3) \Longrightarrow (1)$: Let η, ρ be m-PFI (right) and m-PFI (left) of P, respectively. Then $\eta \wedge \rho$ is an m-PFQI of P. According to hypothesis

$$
\begin{aligned}
\eta_\kappa \wedge \rho_\kappa &\leq (\eta_\kappa \wedge \rho_\kappa) \circ \delta_\kappa \circ (\eta_\kappa \wedge \rho_\kappa) \\
&\leq \eta_\kappa \circ \delta_\kappa \circ \rho_\kappa \\
&\leq \eta_\kappa \circ \rho_\kappa.
\end{aligned}
$$

However, $\eta_\kappa \circ \rho_\kappa \leq \eta_\kappa \wedge \rho_\kappa$ always hold. Hence, $\eta_\kappa \circ \rho_\kappa = \eta_\kappa \wedge \rho_\kappa$, that is $\eta \circ \rho \leq \eta \wedge \rho$. Therefore by Theorem 3, P is a regular semigroup. Hence, proved. □

Theorem 5. *The following statements are equivalent in P.*

1. *P is regular;*
2. *$\rho \wedge m' \wedge \eta \leq \rho \circ m' \circ \eta$ for every m-PFI(right) ρ, every m-PFGBI m' and every m-PFI(left) η of P;*
3. *$\rho \wedge m' \wedge \eta \leq \rho \circ m' \circ \eta$ for every m-PFI(right) ρ, every m-PFBI m' and every m-PFI(left) η of P;*
4. *$\rho \wedge m' \wedge \eta \leq \rho \circ m' \circ \eta$ for every m-PFI(right) ρ, every m-PFQI m' and every m-PFI(left) η of P.*

Proof. $(1) \Longrightarrow (2)$: Consider b is any element of P. As P is regular, there exists $a \in P$ *such that* $b = bab$. It follows that $b = (ba)b = b(ab)$ for each $a \in P$ and P is semigroup. Hence, we have

$$
\begin{aligned}
(\rho \circ m' \circ \eta)(b) &= \bigvee_{b=ac} \{(\rho \circ m')(a) \wedge \eta(c)\} \\
&\geq (\rho \circ m')(b) \wedge \eta(ab) \text{ since } b = b(ab) \\
&\geq \bigvee_{b=pq} \{\rho(p) \wedge m'(q)\} \wedge \eta(b) \text{ as } \eta \text{ is an } m\text{-PFI(left) of } P. \\
&\geq (\rho(ba) \wedge m'(b)) \wedge \eta(b) \text{ since } b = (ba)b \\
&\geq (\rho(b) \wedge m'(b)) \wedge \eta(b) \\
&= ((\rho \wedge m')(b)) \wedge \eta(b) \\
&= (\rho \wedge m' \wedge \eta)(b).
\end{aligned}
$$

Therefore, $\rho \wedge m' \wedge \eta \leq \rho \circ m' \circ \eta$. So (1) implies (2).

$(2) \Longrightarrow (3) \Longrightarrow (4)$: Straightforward.

$(4) \Longrightarrow (1)$: As δ is an m-PFQI of P, by the supposition, we have

$$\begin{aligned}(\rho\wedge\eta)(b) &= ((\rho\wedge\delta)\wedge\eta)(b)\\ &\leq ((\rho\circ\delta)\circ\eta(b)\\ &= \bigvee_{b=rs}\{(\rho\circ\delta)(r)\circ\eta(s)\}\\ &= \bigvee_{b=rs}\{(\bigvee_{r=cd}\{\rho(c)\wedge\delta(d)\})\wedge\eta(s)\}\\ &= \bigvee_{b=rs}\{(\bigvee_{r=cd}\{\rho(c)\wedge 1\})\wedge\eta(s)\}\\ &= \bigvee_{b=rs}\{(\bigvee_{r=cd}\rho(c))\wedge\eta(s)\}\\ &\leq \bigvee_{b=rs}\{(\bigvee_{r=cd}\rho(cd)\})\wedge\eta(s)\}\\ &= \bigvee_{b=rs}\{\rho(r)\wedge\eta(s)\}\\ &= (\rho\circ\eta)(b).\end{aligned}$$

Therefore $\rho\wedge\eta\leq\rho\circ\eta$. But $\rho\circ\eta\leq\rho\wedge\eta$ always. So, $\rho\circ\eta=\rho\wedge\eta$. Hence, by using Theorem 3, P is regular. □

Theorem 6 ([28]). *The following conditions are equivalent in P.*

1. P is intra-regular;
2. $H\cap I\subseteq HI$ for every right ideal H and every left ideal I of P.

Definition 9 ([24]). *A semigroup P is both regular and intra-regular if and only if $H=H^2$ for every bi-ideal H of P.*

Theorem 7. *A semigroup P is intra-regular if, and only if $\eta\wedge\rho\leq\eta\circ\rho$ for every m-PFI(left) η and, for every, m-PFI(right) ρ of P.*

Proof. Consider a is any element of P. As P is intra-regular, there exists $x,y\in P$ *such that* $a=xa^2y$. Hence, we have

$$\begin{aligned}(\eta\circ\rho)(a) &= \bigvee_{a=bc}\{\eta(b)\wedge\rho(c)\}\\ &\geq \eta(xa)\wedge\rho(ay)\\ &\geq \eta(a)\wedge\rho(a)\\ &= (\eta\wedge\rho)(a).\end{aligned}$$

This implies $\eta\circ\rho\geq\eta\wedge\rho$.

Conversely, assume that $\eta\wedge\rho\leq\eta\circ\rho$ for all m-PFI(left) η and m-PFI(right) ρ of P. Let H be a right ideal and I be a left ideal of P, then C_H is an m-PFI(right) and C_I is an m-PFI(left) of P. By Lemma 1, $C_{H\cap I}=C_H\wedge C_I\leq C_H\circ C_I=C_{HI}$ which implies that $H\cap I\subseteq HI$. Therefore, by Theorem 6, P is intra-regular. □

Theorem 8. *For every m-PFBI η of P, $\eta\circ\eta=\eta$ if and only if, P is both regular and intra-regular.*

Proof. Let P be both regular and intra-regular semigroup. Let η be an m-PFBI of P. Thus, for $x\in P$, there exists $a,b,c\in P$ such that $x=xax$ and $x=bx^2c$. Therefore, $x=xax=xaxax=xa(bx^2c)ax=(xabx)(xcax)$. Hence, we have

$$
\begin{aligned}
(\eta \circ \eta)(x) &= \bigvee_{x=bc} \{\eta(b) \wedge \eta(c)\} \\
&\geq \eta(xabx) \wedge \eta(xcax) \\
&\geq \eta(x) \wedge \eta(x) \\
&= \eta(x).
\end{aligned}
$$

This implies $\eta \circ \eta \geq \eta$. By Lemma 3, $\eta \circ \eta \leq \eta$ holds always. Therefore, $\eta \circ \eta = \eta$.

Conversely, let H be a bi-ideal of P. Since every m-PFBI is m-PFSS of P. Then, Lemma 2, implies that C_H is m-PFBI of P. Hence, by our supposition, $C_H = C_H \circ C_H$. Thus, $H = H^2$. Therefore, by Theorem 9, P is both regular and intra-regular. □

5. Comparative Study and Discussion

This section explains how this paper and the previous one are related to Shabir et al. [27]. Shabir et al. [24] studied regular and intra-regular semiring in terms of BFIs. Shabir et al. extended the work of [24] and initiated the concept of m-PFIs in LA-semigroups and characterized the regular LA-semigroups by the properties of these m-PFIs [27]. By extending the work of [24,27], the concept of m-PFIs in semigroups is introduced, and characterizations of regular and intra-regular semigroups by the properties of m-PFIs are given in this paper. Our approach is superior to that of Shabir et al. [27] because the associative property in LA-semigroups does not hold. There are also numerous structures that are handled by semigroups but not by LA-semigroups. If we take any non-empty set and define the operation on it as $a * b = a$, then it is a semigroup, but not an LA-semigroup. To overcome this problem, we used a semigroup to generalize the whole results of Shabir et al. [27] and, as a result, our methodology offers a broader variety of applications than Shabir et al. [27].

6. Conclusions

When data for real world complex situations come from m factors $(m \geq 2)$, then m-PFS is used to deal such problems. The structure of semigroups is investigated using the idea of m-PFS in this research paper. Shabir et al. [27] used LA-semigroups as the basis for their algebraic structure, which we converted into semigroups. Most importantly, we proved some results related to fuzzy ideals in semigroups in terms of m-PFIs in semigroups. This paper presents a significant number of m-PFS theory applications. We also studied the characterization of regular and intra-regular semigroups by m-PFIs (left) (resp. m-PFIs right) and m-PFBI.

Our future plans are to study the m-PFIs in terms of semirings, ternary semigroups, ternary semirings, near rings and hyperstructures.

Author Contributions: Conceptualization: S.B. and M.S.; Methodology: S.S.; Software: S.B.; Validation: A.N.A.-K.; Formal Analysis: S.S.; Investigation: S.S.; Resources: A.N.A.-K.; Data Curation, M.S.; Writing—Original Draft Preparation: S.S.; Writing—Review and Editing: S.B.; Visualization: S.S.; Supervision: S.B.; Project Administration: S.S.; funding acquisition: A.N.A.-K. All authors have read and agreed to the published version of the manuscript.

Funding: This research received no external funding.

Institutional Review Board Statement: Not applicable.

Informed Consent Statement: Not applicable.

Data Availability Statement: We didn't use any data for this research work.

Conflicts of Interest: The authors declare that they have no conflict of interest.

References

1. Chen, J.; Li, S.; Ma, S.; Wang, X. m-polar fuzzy sets: An extension of bipolar fuzzy sets. *Sci. World J.* **2014**, *2014*, 416530. [CrossRef]
2. Zhang, W.R. Bipolar fuzzy sets and relations: A computational framework for cognitive modeling and multiagent decision analysis. In Proceedings of the NAFIPS/IFIS/NASA'94, First International Joint Conference of the North American Fuzzy Information Processing Society Biannual Conference, San Antonio, TX, USA, 18–21 December 1994; pp. 305–309
3. Lee, K.M. Bipolar-valued fuzzy sets and their operations. In Proceedings of the International Confefence on Intelligent Technologies, Bangkok, Thailand, 12–14 December 2000; pp. 307–312.
4. Saqib, M.; Akram, M.; Bashir, S.;Allahviranloo, T. A Runge–Kutta numerical method to approximate the solution of bipolar fuzzy initial value problems. *Comput. Appl. Math.* **2021**, *40*, 1–43. [CrossRef]
5. Saqib, M.; Akram, M.; Bashir, S. Certain efficient iterative methods for bipolar fuzzy system of linear equations. *J. Intelligent Fuzzy Syst.* **2020**, *39*, 3971–3985. [CrossRef]
6. Saqib, M.; Akram, M.; Bashir, S.; Allahviranloo, T. Numerical solution of bipolar fuzzy initial value problem. *J. Intell. Fuzzy Syst.* **2021**, *40*, 1309–1341. [CrossRef]
7. Mehmood, M.A.; Akram, M.; Alharbi, M.G.; Bashir, S. Optimization of-Type Fully Bipolar Fuzzy Linear Programming Problems. Mathematical Problems in Engineering. *Math. Probl. Eng.* **2021**, *2021*, 1199336.
8. Mehmood, M.A.; Akram, M.; Alharbi, M.G.; Bashir, S. Solution of Fully Bipolar Fuzzy Linear Programming Models. *Math. Eng.* **2021**, *2021*, 9961891.
9. Shabir, M.; Abbas, T.; Bashir, S.; Mazhar, R. Bipolar fuzzy hyperideals in regular and intra-regular semihypergroups. *Comput. Appl. Math.* **2021**, *40*, 1–20. [CrossRef]
10. Zadeh, L.A. Fuzzy sets. *Inf. Control.* **1965**, *8*, 338–353. [CrossRef]
11. Rosenfeld, A. Fuzzy groups. *J. Math. Appl.* **1971**, *35*, 512–517. [CrossRef]
12. Kuroki, N. Fuzzy bi-ideals in semigroups. *Rikkyo Daigaku Sugaku Zasshi* **1980**, *28*, 17–21.
13. Mordeson, J.N.; Malik, D.S.; Kuroki, N. *Fuzzy Semigroups*; Springer: Berlin/Heidelberg, Germany, 2012; p. 131.
14. Hollings, C. The early development of the algebraic theory of semigroups. *Arch. Hist. Exact Sci.* **2009**, *63*, 497–536. [CrossRef]
15. Steinfeld, O. *Quasi-Ideals in Rings and Semigroups;* Akadémiai Kiadó: Budapest, Hungary, 1978.
16. Akram, M.; Farooq, A.; Shum, K.P. On m-polar fuzzy lie subalgebras. *Ital. J. Pure Appl. Math.* **2016**, *36*, 445–454.
17. Akram, M.; Farooq, A. m-polar fuzzy Lie ideals of Lie algebras. *Quasigroups Relat. Syst.* **2016**, *24*, 141–150.
18. Sarwar, M.; Akram, M. New applications of m-polar fuzzy matroids. *Symmetry* **2017**, *9*, 319. [CrossRef]
19. Al-Masarwah, A. m-polar fuzzy ideals of BCK/BCI-algebras. *J. King Saud-Univ.-Sci.* **2019**, *31*, 1220–1226. [CrossRef]
20. Al-Masarwah, A.; Ahmad, A.G. m-polar (α, β)-fuzzy ideals in BCK/BCI-algebras. *Symmetry* **2019**, *11*, 44. [CrossRef]
21. Al-Masarwah, A.N.A.S.; Ahmad, A.G. A new form of generalized m-PF ideals in BCK/BCI-algebras. *Ann. Commun. Math.* **2019**, *2*, 11–16.
22. Al-Masarwah, A. On (complete) normality of m-pF subalgebras in BCK/BCI-algebras. *AIMS Math.* **2019**, *4*, 740–750. [CrossRef]
23. Muhiuddin, G.; Al-Kadi, D. Interval Valued m-polar Fuzzy BCK/BCI-Algebras. *Int. J. Comput. Intell.* **2021**, *14*, 1014–1021. [CrossRef]
24. Shabir, M.; Liaquat, S.; Bashir, S. Regular and intra-regular semirings in terms of bipolar fuzzy ideals. *Comput. Appl. Math.* **2019**, *38*, 1–19. [CrossRef]
25. Bashir, S.; Fatima, M.; Shabir, M. Regular ordered ternary semigroups in terms of bipolar fuzzy ideals. *Mathematics* **2019**, *7*, 233. [CrossRef]
26. Bashir, S.; Mazhar, R.; Abbas, H.; Shabir, M. Regular ternary semirings in terms of bipolar fuzzy ideals. *Comput. Appl. Math.* **2020**, *39*, 1–18. [CrossRef]
27. Shabir, M.; Aslam, A.; Pervaiz, F. m-polar fuzzy ideals in terms of LA-semigroups. *Pak. Acad. Sci.* (submitted)
28. Shabir, M.; Nawaz, Y.; Aslam, M. Semigroups characterized by the properties of their fuzzy ideals with thresholds. *World Appl. Sci. J.* **2011**, *14*, 1851–1865.

Σ mathematics

MDPI

Article

Marks: A New Interval Tool for Uncertainty, Vagueness and Indiscernibility

Miguel A. Sainz [1], Remei Calm [1,2], Lambert Jorba [3], Ivan Contreras [1] and Josep Vehi [1,2,*]

[1] Modeling, Identification and Control Engineering (MICELab) Research Group, Institut d'Informatica i Aplicacions, Universitat de Girona, 17003 Girona, Spain; sainz@ima.udg.edu (M.A.S.); remei.calm@udg.edu (R.C.); ivancontrerasfd@gmail.com (I.C.)

[2] Centro de Investigación Biomédica en Red de Diabetes y Enfermedades Metabólicas Asociadas (CIBERDEM), 28029 Madrid, Spain

[3] Department de Matemàtica Econòmica, Financera i Actuarial, University of Barcelona, 08034 Barcelona, Spain; lambert.jorba@ub.edu

* Correspondence: josep.vehi@udg.edu

Abstract: The system of marks created by Dr. Ernest Gardenyes and Dr. Lambert Jorba was first published as a doctoral thesis in 2003 and then as a chapter in the book Modal Interval Analysis in 2014. Marks are presented as a tool to deal with uncertainties in physical quantities from measurements or calculations. When working with iterative processes, the slow convergence or the high number of simulation steps means that measurement errors and successive calculation errors can lead to a lack of significance in the results. In the system of marks, the validity of any computation results is explicit in their calculation. Thus, the mark acts as a safeguard, warning of such situations. Despite the obvious contribution that marks can make in the simulation, identification, and control of dynamical systems, some improvements are necessary for their practical application. This paper aims to present these improvements. In particular, a new, more efficient characterization of the difference operator and a new implementation of the marks library is presented. Examples in dynamical systems simulation, fault detection and control are also included to exemplify the practical use of the marks.

Keywords: modal interval analysis; interval arithmetic; marks; uncertainty modeling; indiscernibility

Citation: Sainz, M.A.; Calm, R.; Jorba, L.; Contreras, I.; Vehi, J. Marks: A New Interval Tool for Uncertainty, Vagueness and Indiscernibility. *Mathematics* **2021**, *9*, 2116. https://doi.org/10.3390/math9172116

Academic Editor: Ioannis Konstantinos Argyros

Received: 12 July 2021
Accepted: 27 August 2021
Published: 1 September 2021

Publisher's Note: MDPI stays neutral with regard to jurisdictional claims in published maps and institutional affiliations.

1. Introduction

Measurements of a variable are made using numerical scales. Usually, the value of a measurement is associated with a real number, but this association is not exact because of imperfect or partial knowledge due to uncertainty, vagueness or indiscernibility.

Incomplete knowledge comes from limited reliability of technical devices, partial knowledge, an insufficient number of observations, or other causes [1]. Among the different types of uncertainty, we find imprecision, vagueness, or indiscernibility. Vagueness, in the colloquial sense of the term, refers to ambiguity, which remains in a datum due to lack of precision, although its meaning is understood. An example could be the measurement of a person's weight using a scale, which provides a value within a range of scale accuracy, e.g., between 75 and 75.2 kg. Uncertainty refers to imperfect or unknown information. For example, it is known that the weight of a car is within limits (1000–1500 kg), but the exact value is unknown due to missing information, such as the number of occupants and the load.

The problems of vagueness and uncertainty have received attention for long time by philosophers and logicians (e.g., [2,3]). Computational scientists have also provided new tools for dealing with uncertainty and vagueness, such as interval analysis, either classic intervals [4,5] or modal intervals [6,7], fuzzy set theory [8,9] and rough set theory [10].

Indiscernibility has also received the attention of philosophers. The identity of indiscernible [11] states is that no two distinct things are exactly alike. It is often referred to as

"Leibniz's Law" and is usually understood to mean that no two objects have exactly the same properties. The identity of indiscernibility is interesting because it raises questions about the factors that individuate qualitatively identical objects. The marks, which are presented and developed in this paper, are outlined to address indiscernibility.

For example, the temperature in a room can be measured with a thermometer at different location points within the room to obtain a spatial distribution. So the temperature is not, for example, 20 °C but an interval of values, between, for example 19 and 21 that represents the different values of the temperature in the room. It does not represent the temperature of the room, which might be necessary for the modeling of an air-conditioning system. The temperature is at this time one value of the interval [19,21], considering the points of this interval as being indistinguishable. This is a known issue in handling "lumped" or "distributed" quantities. Moreover, the thermometer used as a measurement device has a specific precision and provides a reading on a specific scale, which is likely translated to another digital scale to be used in computations. Therefore, a real number or even an interval is not able to represent the read temperature.

Until the 20th century, the preferred theory for modeling uncertainty was probability theory [12], but the introduction of fuzzy sets by Zadeh [8] had a profound impact on the notion of uncertainty. At present, sources of uncertainty remain an active challenge for the scientific community, and different research efforts are directed toward finding solutions to deal with these uncertainties, such as using Bayesian inference for predictions of turbulent flows in aerospace engineering [13], fuzzy sets for time series forecasting based on particle swarm optimization techniques [14], modal intervals for prediction modeling in grammatical evolution [15], interval analysis method based on Taylor expansion for distributed dynamic load identification [16] or rough sets to evaluate the indoor air quality [17].

In this article, marks are presented as a framework to deal with quantities represented in digital scales because this methodology can take into account many the sources of uncertainty. In any use of a mathematical model of a physical system, such as simulation, fault detection, or control, the system of marks provides values for the state variables and, simultaneously, their corresponding granularities, which represent a measure of the accumulated errors in the successive computations. This performance leads to the following:

- Define intervals of variation for these variables values.
- Decide the valid values, i.e., which have a meaning, provide by the semantic theorem.
- Warns from which simulation step the obtained values will be meaningless because the granularity is greater than the previously fixed tolerance.

In the following sections of this paper, we present and review marks theory and basic arithmetic operations. The main contributions of this paper are the following:

1. A new characterization for the difference operator;
2. A new characterization for the difference operator;
3. a new implementation of the marks library and software developments that are needed to apply these methodologies.

To demonstrate the applicability and potential of marks, a well-known benchmark in process control in which the problems of uncertainty, imprecision, and indiscernibility are present is introduced. After introducing the benchmark, three different problems built on it are presented and solved, using marks: simulation, fault detection, and control.

2. Marks

An approach to deal with the inaccuracy associated with any measurement or computation process with physical quantities is built by means of an interval tool: marks, which define intervals in which it is not possible to make any distinction between its elements, i.e., indiscernibility intervals. Marks are computational objects; however, initial marks can come from either direct or indirect readings from a measurement device. Therefore, it is necessary to represent them on a computational scale to acquire suitable computation items. When a measurement reading or a computation is obtained as a number on a numerical scale, the resulting value has to be associated to a subset of R of indiscernible numbers from this value. Each point of this subset has the same right to be considered as the value of the measurement or computation. This kind of subsets leads to the concept of a mark considered a "spot" of certain size, called the "granularity" of the mark.

Let DI_n be a digital scale in floating point notation. A mark on DI_n is a tuple of the following five real numbers:

- Center: a number c on the digital scale, which is the reading of the mark on DI_n.
- Number of digits: the number of digits n on the digital scale DI_n, $n = 15$ computing in double precision.
- Base: the *base* of the number system b, usually $b = 10$.
- Tolerance: an uncertainty that expresses the relative separation among the points of the scale that the observer considers as indiscernible from the center. It is a relative value greater or equal than b^{-n}, which is the minimum computational error on the digital scales. It is a measure to decide if the mark is to be accepted as valid.
- Granularity: an error g, coming from the inaccurate measure or computation, which will increase in every computational process. It must include the imprecision of the measure, devices calibration errors, inaccuracy of the physical system concerned, etc., and it is always expressed in relative terms, i.e., a number between 0 and 1, less than tolerance.

As the numbers n and b are specific on the digital scale DI_n, and the value of the tolerance t is assigned for the user, the mark will be denoted by $\mathfrak{m} = \langle c, g \rangle$. Tolerance t, number of digits n and base b define the *type* of the mark. The set of marks of the same type is denoted by $M(t, n, b)$.

The center of a mark is a reading in a measurement device, so it carries an error and an uncertainty associated to the problem under study. Both yield the value of its granularity. The center and granularity define the mark on the digital scale. Now, it is possible to start the computations required by the mathematical model. The errors in each step of the computations will increase the value of the granularity.

The granularity and tolerance must satisfy a minimum condition of validity of the mark as follows:

$$b^{-n} \leq g < t < 1, \quad (1)$$

These concepts are developed in [7,18], which contain the definition of a mark, its components (center, tolerance, granularity, and base number of the numerical scale), features (valid or invalid mark and associated intervals), relationships (equality and inequality), basic operators (max, min, sum, difference, product, and quotient) and general functions of marks and semantic interpretations through associated intervals to the operands and the results.

2.1. Basic Operators of Marks

Operations between marks are defined for marks of the same type and the result is also of the same type as the data. In this way, the tolerance is constant along with any computation, but the granularity increases, reflecting the step-by-step loss of information, which constitutes the deviation of the computed value from the exact value. An extract definition of elementary operators is presented after this, together with a different characterization of the operator difference.

The extension f_M of a basic operator $f \in \{\max, \min, +, -, *, /\}$ to two given marks $\mathfrak{x} = \langle c_x, g_x \rangle \in M(t, n, b)$ and $\mathfrak{y} = \langle c_y, g_y \rangle \in M(t, n, b)$ is the following:

$$f_M(\mathfrak{x}, \mathfrak{y}) = \langle di(f(c_x, c_y)), g_z \rangle \in M(t, n, b)$$

where $di(f(c_x, c_y))$ is the digital computation of the function f at (c_x, c_y) on the scale DI_n, supposing a minimum relative displacement of $di(f(c_x, c_y))$ with regard to the exact value

$f(c_x, c_y)$; and g_z is the granularity of the result, which has to be $g_z \geq \max(g_x, g_y)$ and is specified as the following:

$$g_z = \gamma_{x,y} + b^{-n}$$

where the value b^{-n} is the computation error of $di(f(c_x, c_y))$, which is the minimum error of any computation. Any computation has to carry this error by adding b^{-n}. The term $\gamma_{x,y}$ is the smallest number, verifying the following:

$$f(c_x, c_y) * [1 + \gamma_{x,y}, 1 - \gamma_{x,y}] \subseteq fR(c_x * [1 + g_x, 1 - g_x], c_y)$$
$$f(c_x, c_y) * [1 + \gamma_{x,y}, 1 - \gamma_{x,y}] \subseteq fR(c_x, c_y * [1 + g_y, 1 - g_y])$$

where fR is the *modal syntactic* extension of f [7].

For the operators min and max, no computations with the centers are necessary and the resulting granularity is $g_z = \max(g_x, g_y)$. The results are transformed to the following:

$$\max\{\mathfrak{x}, \mathfrak{y}\} = \langle \max\{c_x, c_y\}, \max\{g_x, g_y\} \rangle \tag{2}$$
$$\min\{\mathfrak{x}, \mathfrak{y}\} = \langle \min\{c_x, c_y\}, \max\{g_x, g_y\} \rangle \tag{3}$$
$$\mathfrak{x} + \mathfrak{y} = \langle c_x + c_y, \max\{g_x, g_y\} + b^{-n} \rangle \tag{4}$$
$$\mathfrak{x} - \mathfrak{y} = \begin{cases} \langle c_x - c_y, \max\left\{g_x, g_y, \left|\frac{|c_x|}{|c_x| - |c_y|}\right| g_x, \left|\frac{|c_y|}{|c_x| - |c_y|}\right| g_y\right\} + b^{-n} \rangle & \text{if } c_x \neq c_y \\ \langle 0, \max\{g_x, g_y\} \rangle & \text{if } c_x = c_y \end{cases} \tag{5}$$
$$\mathfrak{x} * \mathfrak{y} = \langle c_x * c_y, \max\{g_x, g_y\} + b^{-n} \rangle \tag{6}$$
$$\mathfrak{x} / \mathfrak{y} = \langle c_x / c_y, \max\left\{g_x, \frac{g_y}{1 - g_y})\right\} + b^{-n} \rangle \tag{7}$$

where c_x and c_y are think positive for the operators $+$ and $-$.

The difference in the particular case of two marks with equal centers is 0 without any computation. There is a similar situation with the maximum and minimum operators. In these cases, the granularity of the result must be the greatest of the operands' granularities.

Two shortcomings in the use of marks as numerical entities are evident. The first one concerns the granularity of the difference between two marks $\mathfrak{x}$ and $\mathfrak{y}$ with positive but near centers. In its formula, the difference between the two centers appears as the denominator in a fraction. Consequently, when the centers are close, the granularity can be large enough to invalidate the mark, i.e., when granularity is larger than the tolerance, thus invalidating any further result. This can be avoided by taking into account the two real numbers x and y:

$$x - y = \frac{x^m - y^m}{x^{m-1} * y^0 + x^{m-2} * y^1 + \ldots + x^1 * y^{m-2} + x^0 * y^{m-1}},$$

hence, it is possible to compute the difference $\mathfrak{x} - \mathfrak{y}$ as the following:

$$\mathfrak{x} - \mathfrak{y} = (\mathfrak{x}^m - \mathfrak{y}^m) / (\mathfrak{x}^{m-1} * \mathfrak{y}^0 + \mathfrak{x}^{m-2} * \mathfrak{y} + \ldots + \mathfrak{x} * \mathfrak{y}^{m-2} + \mathfrak{x}^0 * \mathfrak{y}^{m-1}) \tag{8}$$

where m is a natural number large enough so that $\mathfrak{x}^m$ and $\mathfrak{y}^m$ are not near. The center of the resulting mark is $c_x - c_y$, but its granularity is both different and lesser.

Firstly, to avoid overflows in the computations of $\mathfrak{x}^m$ and $\mathfrak{y}^m$ it is convenient to normalize, a priori, the two marks to be divided by the greater mark $\mathfrak{m} = \max(\mathfrak{x}, \mathfrak{y})$. For example, if, $c_y < c_x$ then, $\mathfrak{m} = \mathfrak{x}$, and the normalized marks are the following:

$$\mathfrak{u} = \mathfrak{x}/\mathfrak{m} = \langle 1, g_u \rangle, \qquad \mathfrak{v} = \mathfrak{y}/\mathfrak{m} = \langle c, g_v \rangle,$$

with $0 < c < 1$, and the difference is at this time the following:

$$\mathfrak{u} - \mathfrak{v} = (\mathfrak{u}^m - \mathfrak{v}^m) / (\mathfrak{u}^{m-1} * \mathfrak{v}^0 + \mathfrak{u}^{m-2} * \mathfrak{v} + \ldots + \mathfrak{u} * \mathfrak{v}^{m-2} + \mathfrak{u}^0 * \mathfrak{v}^{m-1}). \tag{9}$$

The power of one mark $\langle c_x, g_x\rangle^m$ with m a natural number is a particular case of a product with the same factors $\langle c_x, g_x\rangle^m = \langle c_x, g_x\rangle * \cdots * \langle c_x, g_x\rangle$ then, by induction it is the following:

$$\langle c_x, g_x\rangle^m = \langle c_x^m, g_x + (m-1)b^{-n}\rangle.$$

Then, the numerator of (9) is as follows:

$$\mathfrak{N} = \mathfrak{u}^m - \mathfrak{v}^m = \langle 1, (g_u + (m-1)b^{-n}\rangle - \langle c^m, g_v + (m-1)b^{-n}\rangle$$

and can be calculated by means of the former formula (5). If $c^m \leq 1/2$, then the following holds:

$$\mathfrak{N} = \langle 1 - c^m, \max((g_u + (m-1)b^{-n})/(1-c^m), g_v + (m-1)b^{-n})\rangle, \tag{10}$$

interchanging g_u and g_v when $c_x < c_y$.

$\mathfrak{N}$ will be a valid mark when its granularity is small, i.e., when the term c^m is less than a fixed small number $\epsilon < 1/2$, for example $\epsilon = b^{-n}$. So,

$$m = [\log(b^{-n})/\log(c)]. \tag{11}$$

The computation of the denominator of the Equation (9) is not problematic because all their terms are positive:

$$\begin{aligned}
\mathfrak{u}^{m-1} &= \langle 1, \max(g_u + (m-2)b^{-n}\rangle \\
\mathfrak{u}^{m-2} * \mathfrak{v}^1 &= \langle c, \max(g_u + (m-3)b^{-n}, g_v) + b^{-n}\rangle = \\
&= \langle c, \max(g_u + (m-2)b^{-n}, g_v + b^{-n})\rangle \\
&\cdots\cdots\cdots\cdots\cdots\cdots \\
\mathfrak{u}^{m-i} * \mathfrak{v}^{i-1} &= \langle c^{i-1}, \max(g_u + (m+i-1)b^{-n}, g_v + (i-2)*b^{-n}) + b^{-n}\rangle \\
&= \langle c^{i-1}, \max(g_u + (m-i)b^{-n}, g_v + (i-1)b^{-n})\rangle \\
&\cdots\cdots\cdots\cdots\cdots\cdots \\
\mathfrak{u}^1 * \mathfrak{v}^{m-2} &= \langle c^{m-2}, \max(g_u, g_v + (m-3)b^{-n}) + b^{-n}\rangle = \\
&= \langle c^{m-2}, \max(g_u + b^{-n}, g_v + (m-2)b^{-n})\rangle \\
\mathfrak{v}^{m-1} &= \langle c^{m-1}, g_v + (m-1)b^{-n}\rangle
\end{aligned}$$

it results to

$$\mathfrak{D} = \langle (1^m - c^m)/(1-c), \max(g_u + mb^{-n}, g_v + (2m-1)b^{-n})\rangle. \tag{12}$$

Dividing $\mathfrak{N}$ and $\mathfrak{D}$

$$\mathfrak{u} - \mathfrak{v} = \mathfrak{N}/\mathfrak{D}$$

and, eventually de-normalizing by multiplying by $\mathfrak{m}$,

$$\mathfrak{x} - \mathfrak{y} = (\mathfrak{u} - \mathfrak{v}) * \mathfrak{m}. \tag{13}$$

For example, for $t = 0.05$, $b = 10$, $n = 15$, the formula (5) gives the following:

$$\langle 3.121, 0.0001\rangle - \langle 3.1212, 0.0001\rangle = \langle -0.0002, 1.560600\rangle,$$

which is an invalid mark because the granularity is larger than 1. From (11), $m = 538995$ and (13) give the following:

$$\langle 3.121, 0.0001\rangle - \langle 3.1212, 0.0001\rangle = \langle -0.0002, 0.010101\rangle,$$

a valid mark.

The second shortcoming is the necessity to calculate the elementary functions (exp, log, power, trigonometric,...) for marks. This is possible with power series but when the convergence is slow, run time can belong. A better alternative is to use the routines ©FDLIBM developed at SunSoft, Sun Microsystems, to approximate these functions using polynomials. The computational processes for marks were developed and integrated into MATLAB in the MICELab research group (Institute of Informatics and Applications, University of Girona). The code to perform computations with marks can be found in [19].

2.2. Associated Intervals

The theory of marks is a by-product of modal intervals theory, linked by the "improper" [7] associated interval to a mark denoted by $Iv(\mathfrak{m})$:

$$Iv(\mathfrak{m}) = c * [1 + t, 1 - t], \tag{14}$$

where * is the product of a real number by an interval. Its domain, or set of its points, is referred to as Iv' and called the indiscernibility margin of $\mathfrak{m}$. Another related interval is the external shadow defined by the following:

$$Exsh(\mathfrak{m}) = Iv(\mathfrak{m}) * [1 - g, 1 + g] = c * [1 + t, 1 - t] * [1 - g, 1 + g], \tag{15}$$

necessary to obtain the semantic meaning of a computation made using marks. As $g < t$, the external shadow is an improper interval that verifies the inclusion of $Exsh(\mathfrak{m}) \subseteq Iv(\mathfrak{m})$.

2.3. Semantic Theorem for Marks

The associated intervals allow the semantic properties of Modal Intervals to be applied to the results of functions of marks. Given the marks $\mathfrak{x}_1, \ldots, \mathfrak{x}_n \in M(t, b, n)$, the continuous $\mathbb{R}^n$ to $\mathbb{R}$ function $z = f(x_1, ..., x_n)$ and the modal syntactic extension fR of f, then the following holds:

$$fR(Iv(\mathfrak{x}_1), \ldots, Iv(\mathfrak{x}_n)) \subseteq Exsh(\mathfrak{z}).$$

This inclusion confirms the important Semantic Theorem for a function of marks, which provides meaning to any valid result in the evaluation of a function. If the mark $\mathfrak{z}$ is the calculus of a function of marks, $\mathfrak{z} = f_M(\mathfrak{x}_1, \ldots, \mathfrak{x}_k)$, supposing that all the involved marks are valid, then we have the following:

$$(\forall z \in Exsh'(\mathfrak{z}))\ (\exists x_1 \in Iv'(\mathfrak{x}_1)) \ldots (\exists x_k \in Iv'(\mathfrak{x}_n))\ z = f(x_1, \ldots, x_n). \tag{16}$$

So, every point of the external shadow $Exsh'(\mathfrak{z})$ is a true value of the function f for some values of the variables in the intervals $Iv'(\mathfrak{x}_1) \ldots Iv'(\mathfrak{x}_n)$.

The external shadow interval depends on the tolerance and the granularity of the mark, which shrinks the interval width with the unavoidable increase of the granularity. As the value approaches the tolerance (the center), the interval width tends to zero. This effect causes a loss of significance, which in many cases, is possible to avoid by performing a "translation" to avoid small values of the state variables. For example, adding a constant to the values of the state variables and scaling the common tolerance, if it depends on these values.

3. Benchmark

3.1. Benchmark Description

The popular three-tank benchmark problem is used to exemplify the usefulness of marks in the context of uncertainty, vagueness, and indiscernibility [20,21]. It consists of three cylindrical tanks of liquid connected by pipes of circular section, as depicted in Figure 1. The first tank has an incoming flow, which can be controlled using a pump (actuator) and the outflow is located in the last tank.

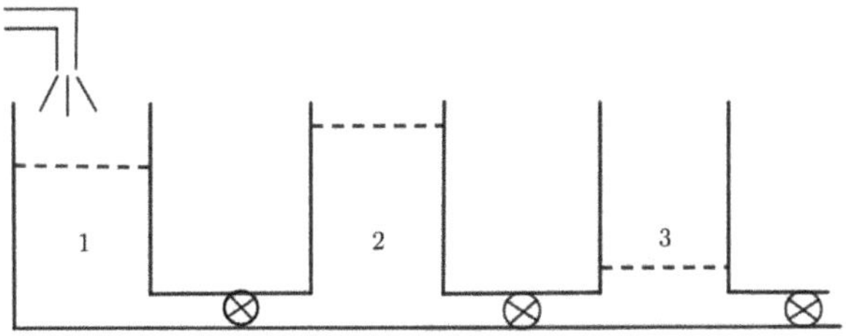

Figure 1. Schematic representation of the three-tank system.

The model in form of difference equations for this systems is

$$\begin{aligned}
&h_1(t+1) = h_1(t) + \Delta t \cdot (q_1(t) - c_{12} \cdot srh_{12} \cdot cc_1)/s_1 \\
&h_2(t+1) = h_2(t) + \Delta t \cdot (q_2(t) + c_{12} \cdot srh_{12} \cdot cc_1 - c_{23} \cdot srh_{23} \cdot cc_2)/s_1 \\
&h_3(t+1) = h_3(t) + \Delta t \cdot (q_3(t) + c_{23} \cdot srh_{23} \cdot cc_2 - c_{30} \cdot srh_{30} \cdot cc_3)/s_3 \\
&\text{with} \\
&srh_{12} = \text{sign}(h_1(t) - h_2(t)) \cdot \sqrt{|h_1(t) - h_2(t)|} \\
&srh_{23} = \text{sign}(h_2(t) - h_3(t)) \cdot \sqrt{|h_2(t) - h_3(t)|} \\
&srh_{30} = \text{sign}(h_3(t) - 0) \cdot \sqrt{|h_3(t) - 0|}
\end{aligned} \tag{17}$$

where the state variables h_1, h_2, h_3 are the level of liquid in the tanks, s_1, s_2, s_3 their respective areas, q_1, q_2, q_3 the incoming flows, c_{12}, c_{23}, c_{30} the valves constants which represent the flux between the tanks and Δt is the simulation step, in seconds.

The following values were considered: the three tanks are the same heights $h = 2$ m, areas $s_1 = s_2 = s_3 = 1$ m^2, and intermittent inputs of the maximum incoming flows (in m^3/s) are the following:

$$q_1 = 0.01 \ , \ q_2 = q_3 = 0. \tag{18}$$

For the valves' constants (in m$^{5/2}$/s) the values are

$$c_{12} = 0.009 \ , \ c_{23} = 0.008 \ , \ c_{30} = 0.007 \tag{19}$$

and the initial liquid levels (in m) are

$$h_1(0) = 0.1 \ , \ h_2(0) = 1.5 \ , \ h_3(0) = 0.6. \tag{20}$$

As an example of the application of marks for this benchmark model, we present three general problems related to many mathematical models: simulation, fault detection, and control to show the suitability of the marks for dealing with mathematical models with uncertainty and indiscernibility.

3.2. Simulation

Two different types of simulations have been performed: using real numbers and using marks, with a simulation step of $\Delta t = 5$ and 1000 steps of simulations (5000 s in all). The results using real numbers, for the three state variables are represented in Figure 2. The intermittent input flow gives the sawtooth shape for the values of h_1.

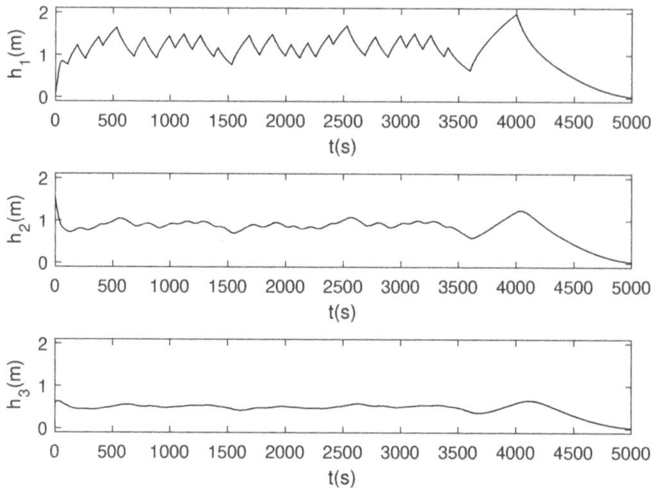

Figure 2. Three-tank system. Simulation results for the three state variables using real numbers.

For the simulations using marks, all magnitudes are considered as marks whose centers are the former real numbers and granularities have been fixed to $g = 0.00001$, for all the marks. The levels of liquid into the tanks are influenced by perturbations and the dynamics of the inputs and outputs of liquid. Calling pr this variation, the tolerance can be calculated by $t = pr/h$. Taking $pr = 0.10$, the common tolerance for all the simulations is $t = 0.05$. Unlike the granularities, the tolerances have to be equal for all the marks.

Results are shown in Figure 3 that contains the intervals associated to the marks, drawn in form of little vertical segments. Together, they are represented by the dark band of the figure. The run time for the $t = 5000$ s of the simulation is 200 s.

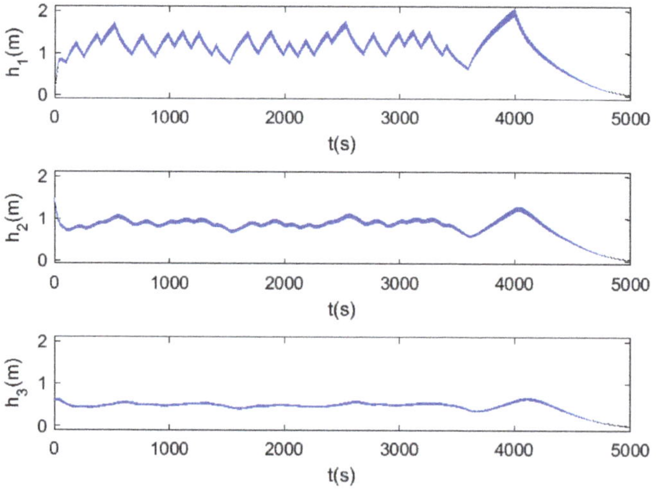

Figure 3. Three-tank system. Simulation results for the three state variables using real numbers. The dark bands are only apparent. They are the accumulation of the 1000 little vertical segments which represent the intervals associated to the resulting marks in the 1000 simulation points

In accordance with the semantic of marks (16) these results mean that, for one instant, for example $i = 500$ (time $t = 2500$ s), the model outputs the result in these marks and related associated intervals:

$$\mathfrak{h}_1(2500) = \langle 1.588474, 0.000011 \rangle \ , \ Iv'(\mathfrak{h}_1(2500)) = [1.509068, 1.667881]$$
$$\mathfrak{h}_2(2500) = \langle 1.003461, 0.000011 \rangle \ , \ Iv'(\mathfrak{h}_2(2500)) = [0.953299, 1.053623]$$
$$\mathfrak{h}_3(2500) = \langle 0.525883, 0.000011 \rangle \ , \ Iv'(\mathfrak{h}_3(2500)) = [0.499595, 0.552171].$$

An experimental state value like

$$\begin{aligned} h_{1exp}(2500) &= 1.6 \pm 0.01 \text{ m}, \\ h_{2exp}(2500) &= 1 \pm 0.01 \text{ m}, \\ h_{3exp}(2500) &= 0.52 \pm 0.01 \text{ m}. \end{aligned}$$

is contained in them and, thus, consistent with the model (17).

However, for $t = 5000$ s the results are the following:

$$\mathfrak{h}_1(5000) = \langle 0.018793, 0.000011 \rangle \ , \ Iv'(\mathfrak{h}_1(5000)) = [0.017854, 0.019733]$$
$$\mathfrak{h}_2(5000) = \langle 0.017685, 0.000011 \rangle \ , \ Iv'(\mathfrak{h}_2(5000)) = [0.016801, 0.018569]$$
$$\mathfrak{h}_3(5000) = \langle 0.012397, 0.000011 \rangle \ , \ Iv'(\mathfrak{h}_3(5000)) = [0.011778, 0.013017].$$

with the associated intervals too narrow to obtain reasonable results with the related semantic (16).

The heights values contained in the associated intervals (14) are consistent with the model, but these intervals depend on the value of the center of the mark because their widths tend to zero, as shown in the final parts of the graphics in Figure 3. To avoid this effect, in this benchmark, it is possible to change to a physical system (Figure 4), where the common height of the tanks is $h + h_{exc}$. The behaviour of the liquid levels along the simulations is the same for the two physical systems.

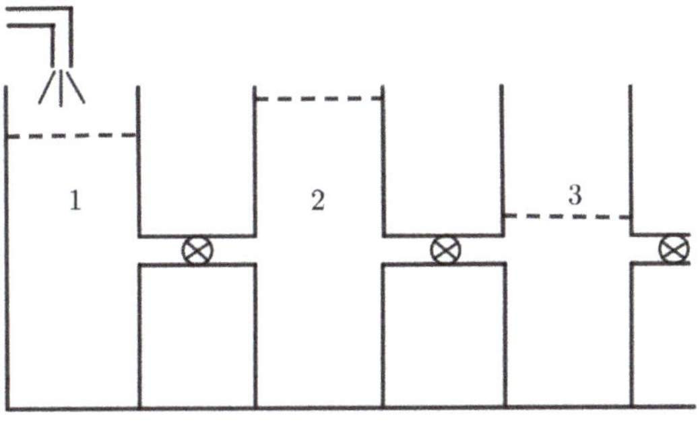

Figure 4. Schematic representation of the extended three-tank system.

To do this in the simulation algorithm, it is sufficient to add h_{exc} to the initial values of the state variables h_1, h_2 and h_3 and to scale the tolerance to $t/(h + h_{exc})$. This scaled tolerance depends on h_{exc}, which when increased for small granularities, the width of all the associated intervals moves near to $2 \cdot pr$.

For the case $h_{exc} = 2$, the tolerance is $t/(h + h_{exc}) = 0.025$. The simulation results after subtracting h_{exc} to the final values of h_1, h_2 and h_3 can be found in Figure 5.

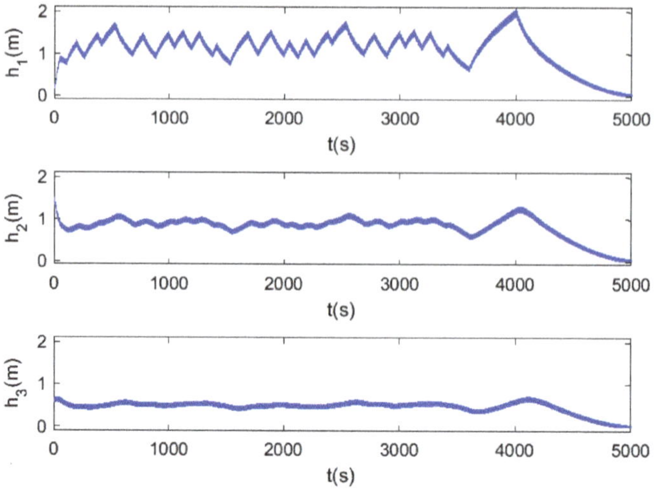

Figure 5. Extended three-tank system. Simulation results for the three state variables using real numbers.The dark bands are only apparent. They are the accumulation of the 1000 little vertical segments which represent the intervals associated to the resulting marks in the 1000 simulation points.

Now the outputs for the step number 500 ($t = 2500$ s) are the following:

$$\begin{aligned}
\mathfrak{h}_1(2500) &= \langle 1.588474, 0.000167 \rangle \ , \ Iv'(\mathfrak{h}_1(2500)) = [1.498802, 1.678147] \\
\mathfrak{h}_2(2500) &= \langle 1.003461, 0.000167 \rangle \ , \ Iv'(\mathfrak{h}_2(2500)) = [0.928407, 1.078515] \\
\mathfrak{h}_3(2500) &= \langle 0.525883, 0.000167 \rangle \ , \ Iv'(\mathfrak{h}_3(2500)) = [0.462763, 0.589003],
\end{aligned}$$

and for the step number 1000 ($t = 5000s$) are the following:

$$\begin{aligned}
\mathfrak{h}_1(5000) &= \langle 0.020353, 0.000011 \rangle \ , \ Iv'(\mathfrak{h}_1(5000)) = [0.000000, 0.070838] \\
\mathfrak{h}_2(5000) &= \langle 0.019152, 0.000011 \rangle \ , \ Iv'(\mathfrak{h}_2(5000)) = [0.000000, 0.069607] \\
\mathfrak{h}_3(5000) &= \langle 0.013426, 0.000011 \rangle \ , \ Iv'(\mathfrak{h}_3(5000)) = [0.000000, 0.063738],
\end{aligned}$$

where the effect of the small center values over the associated intervals widths has disappeared. Associated intervals are truncated to avoid negative values for the heights values.

The model results and the initial granularities are strongly dependent on one another because of the inevitable increase of the granularities throughout the simulation. So, if the initial granularity is changed to $g = 10^{-4}$, then in the simulation step $i = 903$ (time $t = 4515$ s) the resulting marks are invalid (1) (the granularity is larger than tolerance) and all the subsequent results are invalid. For $g = 10^{-3}$ in the simulation step, $i = 86$ and $g = 10^{-2}$ are only valid for the eight first simulation steps.

As a rule of thumb, starting from a granularity $g = 1e - n$, if to arrive until $g = 1e - (n+1)$ lasts p simulation steps, then to get a granularity 10 times larger, $g = 1e - (n+2)$, lasts near to $p/10$ steps more. This quasi-exponential dependence causes invalidity of non-small granularities to be reached quickly, independently from the fixed value for the tolerance.

3.3. Fault Detection

The goal is to detect the presence of a fault in a system and indicate the location of the detected fault (fault isolation). It is assumed that only the measurements of the liquid levels, which are influenced by leakage in one tank or clogging in a valve are available.

In accordance with the semantic of marks (16), a fault is detected when a measurement (within an interval of uncertainty) is outside the estimated band of associated intervals obtained by simulation using marks, indicating that it is not consistent with the model. Therefore, if the model is correct, the measurement is not. These measurements are generated simulating the behavior of the system in the following situations:

- The system is non-faulty until $t = 500$ s, and from this time on, there is a leakage in tank 1 of approximately 0.25% of the water inflow. The results are shown in Figure 6, where the bands are the results of the simulation using marks and the line is the values of the heights of the liquid in the tanks for the faulty system. The comparison shows the effect of the leakage. The line is below the band from the instant $t = 500$ s for tank 1 and over the band for tanks 2 and 3.
- The system is non-faulty until $t = 500$ s and from this time on, there is a clogging between tanks 2 and 3 of 50% of the nominal flow. Figure 7 shows the effect of the clogging. The line is below the band from the instant $t = 500$ s for tank 3 and over the band for tanks 2 and 1, some instants later.

The simulations were performed until $t = 1000$ s to underline the comparisons.

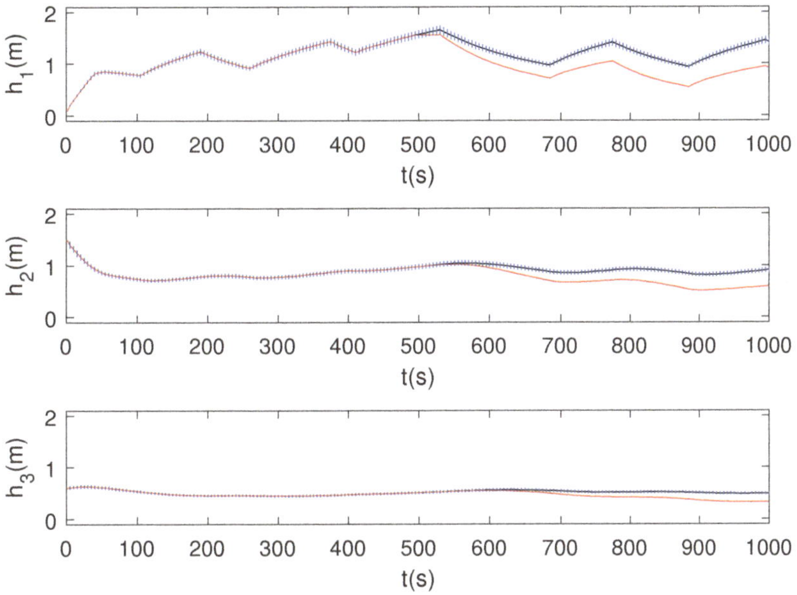

Figure 6. Faulty three-tank system: Leakage of approximately 0.25 % of the water inflow in tank 1 from $t = 500$ s. Blue bands represent the computation using marks while red lines correspond to the measured values of the three state variables.

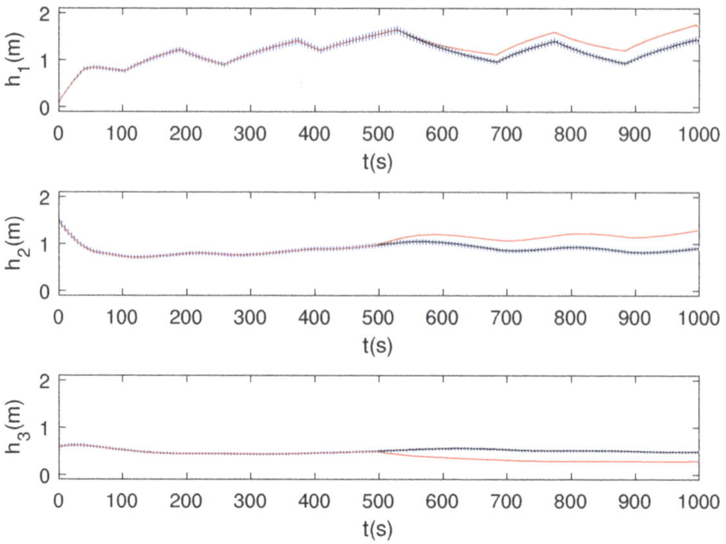

Figure 7. Faulty three-tank system: clogging between tanks 2 and 3 from $t = 500$ s. Blue bands represent the computation using marks while red lines correspond to the measured values of the three state variables.

3.4. Control

An elementary open-loop control was developed to show the usefulness of the model using marks in control systems. The process variables to be achieved are target heights of the liquid in the tanks, (or, in this test, arbitrarily chosen from the real simulation with the model, see Figure 2). For example, the ones represented by black dots in Figure 8. These outputs are controlled by the input flows to each tank in form of percentages $k_1(t), k_2(t), k_3(t)$ of the maximal flows $q_1 = q_2 = q_3 = 0.01$.

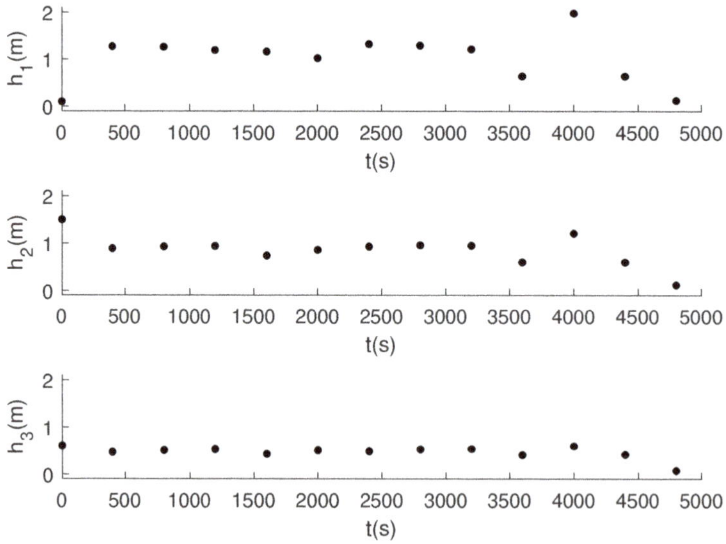

Figure 8. Three-tank system. Target heights to be achieved in open loop control.

When at instant t, the later and nearest target height is above the interval associated to the output mark $h_1(t)$ or $h_2(t)$ or $h_3(t)$, then $k_1(t) = 1$ or $k_2(t) = 1$ or $k_3(t) = 1$. When it is under, then $k_1(t) = 0$ or $k_2(t) = 0$ or $k_3(t) = 0$. Finally, when it is inside the interval, then $k_1(t) = d_1(t)$ or $k_2(t) = d_2(t)$ or $k_3(t) = d_3(t)$, where $d_1(t)$ is the relative distance between the target height and the center of the interval $h_1(t)$ (the same for $d_2(t)$ and $d_3(t)$).

The result is a set of percentages for every instant, t. The output of the model is the band of associated intervals in Figure 9. The target heights are very close to being contained in the corresponding marks. Therefore, they can be considered consistent with the model for the inputs flows defined by $k_1(t)$, $k_2(t)$ and $k_3(t)$.

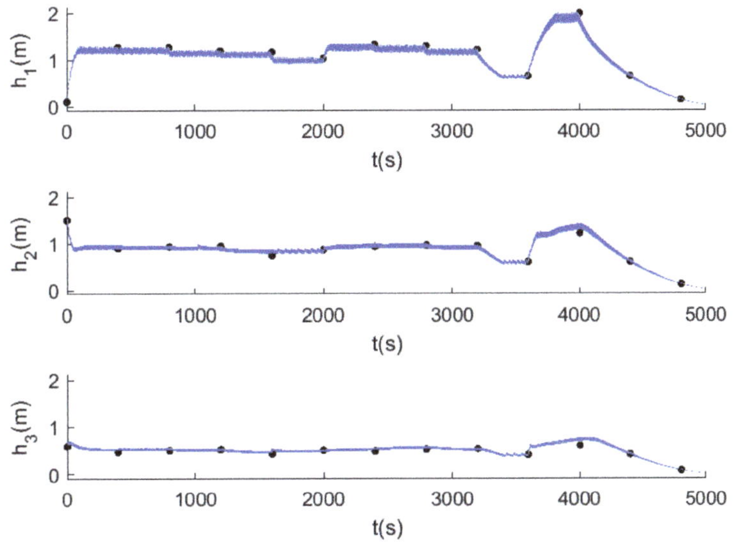

Figure 9. Three-tank system. Simulation results for the open loop control of the liquid heights.

4. Conclusions

Real numbers are the "ideal" framework for dealing with quantities associated with physical phenomena. However, real numbers are not attainable and "disappear" when the observer obtains the value of a quantity. The alternative is digital numbers but a measurement value becomes a point on a digital scale depending on the phenomenon itself, the accuracy, correctness, and errors of the devices used in the measurement, and the numbers of digits used to represent it on the digital scale.

A mark represents, in a consistent procedure, the point information provided by a digital scale. The system of marks has an internal structure that reflects not only the losses of information inherent in the readings on a digital scale and the evolution of the computations from them, but also the indiscernibility of the observed phenomena.

The computations performed using marks also reflect the gradual loss of information, due to numerical errors and truncations, and give relevant warnings for decision making, either on the acceptability of the results or on the usefulness of seeking more precision to achieve the necessary validity.

In conclusion, marks are an appropriate framework for any iterative process, within current research conditions and certain assumptions, where uncertainty is significant and can be generalized when needed. For example, if the process is a long simulation with many steps or an iterative approach with slow convergence, it is necessary to control the accumulation of experimental, scaling, and computational errors so as not to let it exceed the tolerance set by the observer.

The benchmark presented is a good example of the use of marks for an iterative process. Marks prove to be a correct and satisfactory tool for modeling physical systems with uncertainties in their variables and parameters. Marks provide a double contribution to any computational process: (1) the granularity as a good timely test for the validity of any result, and (2) they provide meaning to any valid result by means of the semantic theorem. The final semantics of a simulation using marks is just the one needed for problems like fault detection, control, or parameter identification (via optimization) of a mathematical model against a set of experimental data with uncertainties. This opens a wide field of applications for marks.

Author Contributions: Conceptualization, M.A.S., L.J., R.C., I.C and J.V.; methodology, M.A.S., L.J. and R.C.; software, M.A.S. and I.C.; validation, M.A.S. and I.C.; formal analysis, M.A.S., L.J. and R.C.; investigation, M.A.S., R.C., L.J. and J.V.; resources, M.A.S., R.C., L.J. and J.V.; data curation, M.A.S. and I.C.; writing—original draft preparation, M.A.S., R.C. and L.J.; writing—review and editing,M.A.S., L.J., R.C., I.C. and J.V.; visualization, M.A.S., L.J., R.C., I.C. and J.V.; supervision, M.A.S., L.J., R.C., I.C. and J.V.; project administration, R.C., I.C. and J.V.; funding acquisition, J.V. All authors have read and agreed to the published version of the manuscript.

Funding: This work was partially supported by the Spanish Ministry of Science and Innovation through grant PID2019-107722RB-C22/AEI/10.13039/ 501100011033 and the Government of Catalonia under 2017SGR1551.

Institutional Review Board Statement: Not applicable.

Informed Consent Statement: Not applicable.

Data Availability Statement: Data available at reference [19].

Conflicts of Interest: The authors declare no conflict of interest.

References

1. Kruse, R.; Schwecke, J.H. *Uncertainty and Vagueness in Knowledge Based Systems Numerical Methods;* Springer: Berlin/Heidelberg, Germany, 1991.
2. Russel, B. Vagueness. *Australas. J. Philos.* **1923**, *1*, 84–92. [CrossRef]
3. Black, M. Vagueness. *Philos. Sci.* **1937**, *2*, 427–455. [CrossRef]
4. Moore, R.E. *Interval Analysis;* Prentice-Hall: Englewood Cliffs, NJ, USA 1966.
5. Moore, R.E. *Methods and Applications of Interval Analysis;* SIAM: Philadelphia, PA, USA, 1979.
6. Gardenyes, E.; Mielgo, H.; Trepat, A. Modal intervals: Reasons and ground semantics. *Lect. Notes Comput. Sci.* **1986**, *212*, 27–35.
7. Sainz, M.A.; Armengol, J.; Calm, R.; Herrero, P.; Jorba, L.; Vehi, J. *Modal Interval Analysis: New Tools for Numerical Information;* Springer: Berlin/Heidelberg, Germany, 2014; Volume 2091.
8. Zadeh, L. Fuzzy sets. *Inf. Control* **1965**, *8*, 338–353. [CrossRef]
9. Moore, R.; Lodwick, W. Interval analysis and fuzzy set theory. *Fuzzy Sets Syst.* **2003**, *135*, 5–9. [CrossRef]
10. Pawlak, Z. Rough Set. *Int. J. Comput. Inf. Sci.* **1982**, *11*, 341-356 [CrossRef]
11. Leibniz, G.W. *Discourse on Metaphysics and the Monadology (Trans. George R. Montgomery);* Prometeus Books (First Published by Open Court, 1902): New York, NY, USA; 1992.
12. Valdes-Lopez, A.; Lopez-Bastida, E.; Leon-Gonzalez, J. Methodological approaches to deal with uncertainty in decision making processes. *Univ. Soc.* **2020**, *12*, 7–17.
13. Xiao, H.; Cinnella, P. Quantification of model uncertainty in RANS simulations: A review. *Prog. Aerosp. Sci.* **2019**, *108*, 1–31. [CrossRef]
14. Chen, S.M.; Zou, X.Y.; Gunawan, G.C. Fuzzy time series forecasting based on proportions of intervals and particle swarm optimization techniques. *Inf. Sci.* **2019**, *500*, 127–139. [CrossRef]
15. Contreras, I.; Calm, R.; Sainz, M.A.; Herrero, P.; Vehi, J. Combining Grammatical Evolution with Modal Interval Analysis: An Application to Solve Problems with Uncertainty. *Mathematics* **2021**, *9*, 631. [CrossRef]
16. Wang, L.; Liu, Y.; Liu, Y. An inverse method for distributed dynamic load identification of structures with interval uncertainties. *Adv. Eng. Softw.* **2019**, *131*, 77–89. [CrossRef]
17. Lei, L.; Chen, W.; Xue, Y.; Liu, W. A comprehensive evaluation method for indoor air quality of buildings based on rough sets and a wavelet neural network. *Build. Environ.* **2019**, *162*, 106296. [CrossRef]
18. Jorba, L. Intervals de Marques. Ph.D Thesis, Universitat de Barcelona, Barcelona, Spain, 2003. Available online: http://hdl.handle.net/2445/42085 (accessed on 31 August 2021).
19. Sainz, M.A.; Calm, R.; Jorba, L.; Contreras, I.; Vehi, J. Marks: A New Interval Tool. GitHub Repository. 2021. Available online: https://github.com/MiceLab/MarksLibrary (accessed on 31 August 2021).

20. Sainz, M.; Armengol, J.; Vehi, J. Fault detection and isolation of the three-tank system using the modal interval analysis. *J. Process Control* **2001**, *12*, 325–338. [CrossRef]
21. Amira. *Documentation of the Three Tank System*; Amira GmbH: Duisburg, Germany, 1994.

mathematics

MDPI

Article

A Fuzzy Inference System for Management Control Tools

Carolina Nicolas [1,*], Javiera Müller [2] and Francisco-Javier Arroyo-Cañada [3]

1 Departamento de Administración, Facultad de administración y Economía, Universidad de Santiago de Chile (USACH), Santiago 9170022, Chile
2 Escuela de Auditoría, Universidad de Valparaíso, Valparaiso 2362700, Chile; javiera.muller@uv.cl
3 Facultat d'Economia i Empresa, Universitat de Barcelona, 08034 Barcelona, Spain; fjarroyo@ub.edu
* Correspondence: carolina.nicolas@usach.cl

Abstract: Despite the importance of the role of small and medium enterprises (SMEs) in developing and growing economies, little is known regarding the use of management control tools in them. In management control in SMEs, a holistic system needs to be modeled to enable a careful study of how each lever (belief systems, boundary systems, interactive control systems, and diagnostic control systems) affects the organizational performance of SMEs. In this article, a fuzzy logic approach is proposed for the decision-making system in management control in small and medium enterprises. C. Mamdani fuzzy inference system (MFIS) was applied as a decision-making technique to explore the influence of the use of management control tools on the organizational performance of SMEs. Perceptions data analysis is obtained through empirical research.

Keywords: fuzzy logic toolbox; Mamdani method; performance; management control; small and medium enterprises (SME)

Citation: Nicolas, C.; Müller, J.; Arroyo-Cañada, F.-J. A Fuzzy Inference System for Management Control Tools. *Mathematics* **2021**, *9*, 2145. https://doi.org/10.3390/math9172145

Academic Editor: Ioan Dzitac

Received: 4 August 2021
Accepted: 29 August 2021
Published: 2 September 2021

1. Introduction

On the one hand, according to the fifth longitudinal survey of companies of the Chilean Ministry of Economy, small companies are extremely important for the national economy, as they represent 52.5% of all companies and employ 38.7% of all workers. These companies have many problems, often lacking the time, resources, or necessary information to deal with organizational performance [1].

On the other hand, management control is the process that managers use to aid the decision-making of the members within an organization. This eases the application and alignment of chosen strategies in the organization, thus achieving the pre-established objectives and benefiting the overall performance of the company [2–4].

Notwithstanding, the international literature recognizes that management control tools are fundamental for the efficient and effective management of any business [5], while information and planning systems are useful tools to obtain corporate strategic objectives [6]. When they are met, the goals and reason of existence of companies are achieved, making them an important aspect of improving organizational performance [4,5,7,8].

Besides, evidence shows that the level of use of financial management tools directly affects performance, with the most common ones (according to literature, and interpreted as levers) being budget, long-term planning, support systems for decision making, and financial and non-financial performance [7,8]. There is limited literature regarding the existence of a holistic system that aggregates all these tools, which would allow for the careful study on how each lever (pack of tools) affects the organizational performance of SMEs (Small and Medium-sized companies). In addition, there are limited empirical studies based on scientific data that examine how financial management practices affect SMEs [9]. Nonetheless, existing literature shows a positive outcome of the use of management control tools in SMEs to maximize opportunities, operational efficiency, profit, reliability of administrative information, and finances [10].

The purpose of this investigation is to explore the influence of the use of management control tools on the organizational performance of SMEs. According to the latter, the question of this research is: How do we apply an MFIS in Management Control Tools? What is the degree of use of each control lever in Chilean SMEs?

The present research explores, for the first time, the use of four levers of management control tools in SMEs (belief systems, boundary systems, interactive control systems, and diagnostic control systems) [11]. The individual influence of each of these tools on performance make up the four hypotheses of the research:

Hypothesis 1 (H1). *The beliefs lever is present as a management control tool in SMEs.*

Hypothesis 2 (H2). *The boundary lever is present as a management control tool in SMEs.*

Hypothesis 3 (H3). *The diagnostics control lever is present as a management control tool in SMEs.*

Hypothesis 4 (H4). *The interactive control lever is present as a management control tool in SMEs.*

The methodological approach taken in this study is a mixed methodology based on factorial analysis and Mamdani fuzzy inference system (MFIS). This fuzzy model is built around the MATLAB Software and its Fuzzy Logic Toolbox. The investigation is empirically applied in Chile, and the proposed fuzzy model is used to evaluate the degree of presence of these levers in SMEs, as well as their relationship with the levels of financial and non-financial performance.

Fuzzy inference systems (FIS) are methodologies that express knowledge and inaccurate data; in Web of Science, there are 1,314 records on the topic. FIS are studied mainly in computer science artificial intelligence (27%), electrical electronic engineering (20%), interdisciplinary computer science applications (10%), and others (43%). The authors with the highest number of publications are P. Melin, O. Castillo, and O. Kisi (Web of Science).

Mamdani fuzzy inference system (MFIS) shows 92 records on the topic; the main areas studied are computer science artificial intelligence (25%), electrical electronic engineering (15%), computer science theory methods (12%), environmental sciences (12%), and applied mathematics (12%). The authors with the highest number of publications in mathematics categories are B. Jayaram and M. Stepnicka (Web of Science).

Unlike other studies that apply MFIS in management, this paper proposes a diagnostic model of the four levers of management control in a holistic view; other studies are focused on customer requests or specific tools, such as the Balanced Scorecard (BSC) [12].

The importance and originality of this study are that it explores and applies the fuzzy inference system with linguistic control rules to measure the use of the management control tools in SMEs and their relationship with financial and non-financial performance. Mamdani's fuzzy inference system is advantageous for this study since it is intuitive, well-suited to human inputs, more interpretable and rule-based, and has widespread acceptance [13].

This study contributes to management control field and SMEs, which are companies that have survived the start-up stage and they are currently in the positioning and growth stages [14]. In addition, as established at the beginning, these companies are important for the national economy since they provide many jobs. From the area of management control, we can help them to continue growing by supporting them in understanding the importance of applying certain management control tools. Furthermore, this type of research helps other SMEs incorporate management tools in order to improve their organizational performance. It must also be noted that similar studies elaborated in Latin America as a whole are scarce.

The main issues addressed in this paper are: Section 1 of this paper will review the literature on management control and management control research in SMEs with fuzzy logic; Section 2 gives a brief overview of the recent methodology; Section 3 presents the results of the research, focusing on the three key themes that are: Descriptive Analysis, Measurement Scale, Fuzzy Inference Systems and Parameterization of Membership Functions, Summary of Fuzzy Indicators; Finally, the Discussion and Conclusions are presented.

2. Literature Review

Various studies identify the structural weaknesses of SMEs regarding their lack in the use of tools to create strategies that can manage projects in the medium and long term [6,7,9,10]. In addition, their methods in management control, administration, finances, accounting, and operations are done in an informal and intuitive manner, disregarding the use of decision-making tools. This is because many of them are created by the experience of staff who have worked in other companies, thus bringing with them technical but not administrative knowledge [15]. In retrospect, SMEs usually lack time, resources, or necessary information (or the skill set required to collect and evaluate this information) to measure organizational performance [1].

However, SMEs possess crucial competitive advantages: (1) Their size allows them to respond rapidly to changes in their environment, easing their integration into the productive chain, (2) They tend to be efficient providers of intermediate or final goods and services. They do, however, have the following disadvantages: (1) They are vulnerable to recessive cycles and slow economic growth, (2) They cannot surpass technical and non-technical barriers of market entry on their own, (3) They are unable to develop barriers to protect their income in specific market segments and niches [16].

2.1. Management Control

Management control is the process that managers use to aid the decision-making of the members within an organization. This eases the application and alignment of chosen strategies on the organization, thus achieving the pre-established objectives and benefiting the firm's overall performance [2–4]. Management control systems influence human behavior so that they are aligned to the goals, i.e., companies ensure that the individual's actions to achieve personal goals are aligned with the institution's corporate goals [17].

Among the various management control models, the most prominent ones are the closed cycle model from execution premium [4], which is mostly used in large and complex organizations, and the levers of control model [11], which segments management control into four levers to ensure effective management control within the organization. Diagnostics systems are used to monitor goals, which help monitor the progress of indicators, and if these are aligned with the plan, include tools such as budget, control panel, information systems, cost systems, among others. The diagnostic control by itself is not adequate to achieve effective control as tampering actions can be done to achieve objectives. This can be a risk to the organization, hence the need to couple diagnostics with other control systems [11].

Belief systems attempt to articulate organizational values with direction so that employees accomplish the objectives. They must inspire and promote behavior that aligns with the organizational values, mission, and vision [11]. In addition, boundary systems control all the behavior that workers should refrain from by defining limits and avoiding risky or negative practices (i.e., manuals, policies, contracts, documentation process, among others) [11]. Interactive control systems are used when managers obtain employee information mainly through informal means to explore the impact of certain practices, strategies, or programs to take advantage of possible opportunities. [11] Exposes the dynamic tension between the four control levers, where all must be in balance to bring forth organizational performance. Hence, there must be a holistic approach in the four levers to promote better organizational performance [11,18].

2.2. Management Control in Small and Medium Enterprises

Despite the importance of the role of SMEs in developing and growing economies, little is known regarding the use of management control tools in them [9,14], as most are used in large companies, or utilize specialized tools such as balanced scorecards and indicators to improve financial performance [18]. In addition, there is scarce research that studies management control tools as a holistic packet (which has now gained importance in the literature).

Studies regarding the influence of control tools in SMEs have mainly been elaborated in Australia, Canada, and Asia [14,19]. Research analyzing the situation in Australia, Hong Kong, Malaysia, and Singapore shows the limited use of management tools in SMEs [20]. The most utilized tools are environmental analysis tools (Strength, Weaknesses, Opportunities, Threats, Policies, Environment, Society, and Technology), forecast tools, cost-benefit tools, and budget tools.

Research reveals that the existence of control in SMEs affects different parts of the company by maximizing opportunity, operational efficiency, profit, reliability of administrative information, and finances [10]. After studying 165 industrial firms in New Zealand, [21] it was proven that applying management control improves profitability. The findings in [6] also support the idea that using management control tools has a positive impact on corporate performance, which is consistent with the literature. A study of 151 microenterprises in Malaysia discovered that performance and success in small firms are significantly affected by management initiatives [22], further adding that microenterprises should implement management training.

Medium-sized businesses use more control lever tool groups than smaller companies, which can increase their performance as indicated by the practical and theoretical hypotheses [4,11,17] and other related studies [6,10,14,20,21]. The main difference between small and medium-sized companies lies in the use of the diagnostic lever on the non-financial side (the use of strategic maps, control panel indicators, balanced scorecards, etc.). What is usually employed to plan operations from the organizational strategy [4] is coherent with the growth of companies and the literature.

There is a study elaborated by CIMA (Chartered Institute of Management Accountants) in the United Kingdom [23] that explains the practices of SMEs in the region. The results show that the use of management tools tends to be used mainly to control information rather than to make decisions. In addition, it reveals that in small companies, control tools are managed by the owners and/or managers, making it a high opportunity cost for the firm.

Literature evidence also identifies the difference in employing management control tools between family-run and non-family-run firms. In the former, management tends to use informal and subjective control systems rather than formal systems [6], largely due to the fact that family-SMEs also possess non-economic objectives. The theory also suggests that family-run businesses are based on goals that surpass financial aims, focusing on non-financial goals which influence the control methods that they employ [23]. Further evidence shows that control systems in family-run firms are mostly informal and subjective in nature [24]. The aforementioned study [6] analyzed 900 Spanish companies, of which 70.4% were family-run, and proved the hypothesis that family-run companies used inferior levels of control systems in comparison to non-family-run firms as the organizational objectives between these two differ at their core.

2.3. Management Control Research in SMEs with Fuzzy Logic

Based on fuzzy sets literature on management and performance control, it has been demonstrated that SMEs are still unsure of the value of management tools, and they still do not possess the necessary resources to be competitive in this field of knowledge. In addition, a negative relationship has been found between the number of management tools employed and the intensity (level) of their use [25]. Another fuzzy analysis in business was elaborated from the perspective of innovation and entrepreneurship, where a positive relationship between innovation and company growth was found [26]. As further evidence regarding Knowledge Management Performance Measurement shows the lack of research in this area, 49 metrics have been validated to evaluate knowledge and performance management through fuzzy methodology. This is appropriate as SMEs act in uncertain environments, and fuzzy analysis has better reach for this type of study [27]. There is literature on fuzzy analysis indicators in the Supply Chain Management of SMEs in Iran, who find themselves in an uncertain position [1]. Data suggests that SMEs in Iran take into account financial and non-financial indicators yet lack a universal consensus in the use of a balanced scorecard that is incomplete and inconsistent in its metrics and indicators. Furthermore, the correct number of metrics to effectively monitor SMEs is poorly understood. Scarce literature has been found on fuzzy analysis for the monitoring of a holistic management control system in SMEs, which happens to be the focus of this study.

2.4. Development of Hypotheses

Management control theory indicates that it has a positive influence on performance. In addition, various studies in SMEs show a positive influence of the use of management control tools on organizational performance. Henceforth, the hypothesis of the investigation is the following:

HT. *The use of management control tools has a positive influence on the organizational performance of SMEs.*

Common management tools analyzed in the literature include budget, long-term planning, decision-making support systems, and financial and non-financial performance support systems [7,8]. There is little literature that groups the tools in control levers to explore how each influences organizational performance in SMEs. Based on [11] control lever model where each of the four levers is defined (beliefs, boundary, diagnostics, interactive), we expect to find strong positive influence in financial and non-financial organizational performance through the use of control tools on each lever.

Thus, four additional hypotheses to analyze how each control lever affects organizational performance in SMEs will be employed:

Hypothesis 1 (H1). *The beliefs lever is present as a management control tool in SMEs.*

Hypothesis 2 (H2). *The boundary lever is present as a management control tool in SMEs.*

Hypothesis 3 (H3). *The diagnostics control lever is present as a management control tool in SMEs.*

Hypothesis 4 (H4). *The interactive control lever is present as a management control tool in SMEs.*

The development of each lever will be analyzed by measuring the degree of use of certain tools that are assigned to each lever. The measurement model is shown in Figure 1.

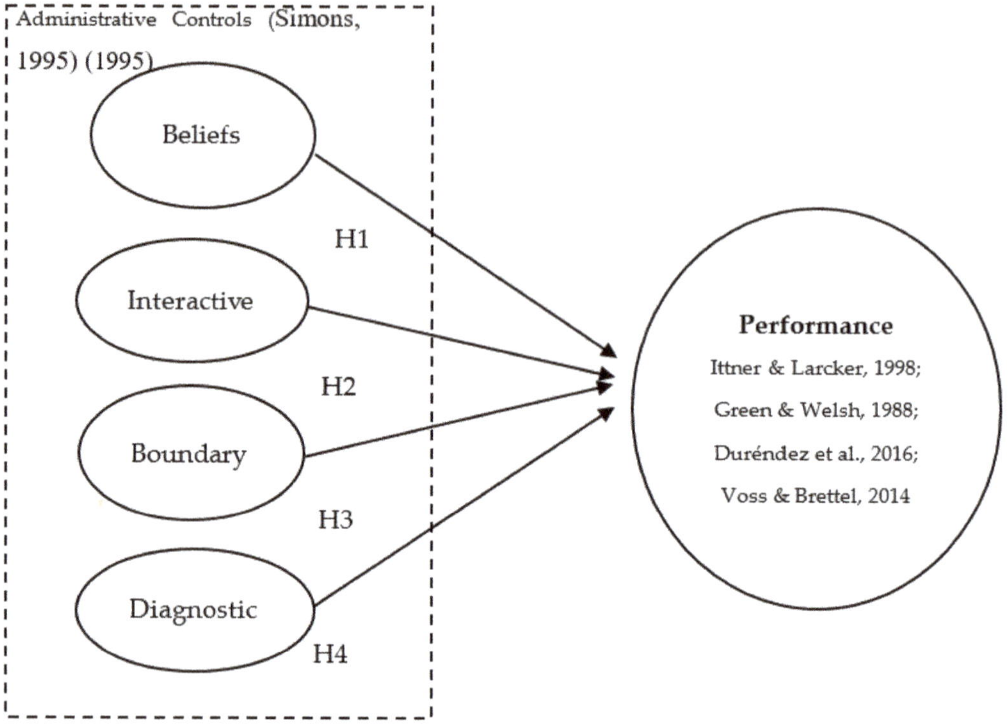

Figure 1. Theoretical model. Based on Simons, R. (1995) and Ittner and Larcker, 1998; Green and Welsh, 1988; Duréndez et al., 2016; Voss and Brettel, 2014.

3. Methodology

In order to fulfill the objective of the research and to prove the model showed in Figure 1, we implement a fuzzy inference evaluation system, thus evaluating the degree of relationship between [11] management control indicators and financial and non-financial performance. This is done by generating a fuzzy logic analysis, whose model is coupled with MATLAB's software to take advantage of the Fuzzy Logic Toolbox tool. The application of this was done on Chilean SMEs, and the fuzzy inference system (FIS) is based on [28] and the Mamdani method for fuzzification [13].

3.1. Fuzzy Inference System in Management

The origins of fuzzy logic stem from [29–31] research, Professor at the University of Carolina (USA), in his article "Fuzzy Sets". Zadeh proposed a mathematical framework for imprecise data, breaking paradigms by shifting from Boolean logic (0–1; white–black, true-false) to fuzzy logic, which implies that the elements belong to a set to a certain degree. As a consequence, a large variety of greys emerged from the traditional black and white [32,33]. In the field of social sciences, fuzzy logic delivers management techniques in an environment that has imprecision, uncertainty, incomplete information, conflictive information, truth bias, and possibility bias [34]. When discussing fuzzy logic, it must be understood that the basic underlying concepts are linguistic variables, which are variables whose values are expressed by words and not numbers. In effect, fuzzy logic must mainly be seen as a methodology to calculate words instead of numbers [28].

Fuzzy inference systems (FIS) are methodologies that express knowledge and inaccurate data, which is very representative of human thought. Therefore, this method is best employed to give answers to problems that have latent variables instead of observed variables [35]. It defines a non-linear relationship between one or more input variables and an output variable. This provides a starting point for decision-making to take place [36]. Phases:

- Fuzzification: defines the linguistic input and output variables; the linguistic and membership values.
- Fuzzy rules: specify the input and output of a fuzzy set. Fuzzy relationships show the degree of belonging or absence of association or interaction between elements from two or more sets. The fuzzy rules system uses linguistic variables as antecedents and consequent.
- Inference mechanism: Approximate Reasoning is an inference procedure used to derive conclusions from a set of fuzzy rules of the type "IF-THEN" and one (or more) input data by using the Max-Min composition, or the Max-Product composition [37,38]. "AND" or "OR" connectors are used to create the necessary rules for decision-making [13].
- Aggregation: The outputs of each rule are cross-related to obtain a unique fuzzy set.
- Defuzzification: The final stage of the process, where a precise or exact value is obtained from the fuzzy set. Defuzzification methods include centroid, bisector, mean peak, smallest of the maximum, and largest of the maximum [28].

3.2. Mamdani-Fuzzy Rule Type-Based Modelling

There are different fuzzy inference models, and their use depends on the type of problem that needs to be solved. The main difference between models lies in the consequences of the rules and in the aggregation and fuzzification methods [28]. Consequently, the research applies Mamdani's [13] model because the inputs and outputs are linguistic rules. This investigator used [29] proposal as a base regarding fuzzy algorithms for complex systems and decision-making processes [28].

Mamdani's model proposes the IF-THEN rules. This implies a series of rules. Input (regressions) matrix and as an output vector are defined as follows:

$$X = [x_1, \ldots\ldots x_2]^T \begin{bmatrix} X_{11} & X_{12} \\ X_{21} & X_{22} \\ X_{n1} & X_{n2} \end{bmatrix}$$
$$G = [g1 \ldots\ldots.gn]$$

Hence, the Mamdani fuzzy model is made of fuzzy propositions in its antecedents and consequents. The general rule is IF-THEN [39]:

Ri: if x is Ai then y is Bi; i = 1; 2; . . . ; K
Ri is the rule number.
Ai and Bi are the fuzzy sets.
x is the antecedent variable representing the input in the fuzzy system.
y is the consequent variable related to the output of the fuzzy system.

The study employs triangular fuzzy membership and trapezoidal fuzzy membership functions. This study elaborates fuzzy inference for each control lever as proposed by [11], also analyzing the distance that exists between their use and performance in SMEs.

Measuring the use of the management control tools in SMEs is a difficult and complex task to convert into quantitative values as they are partially composed of qualitative data. This implies the need to measure multiple attributes such as the existence of a mission, vision, values, holding meetings, and definition of organizational structure, among others. Hence, FIS serves as a reliable tool to tackle uncertainty in an environment rife with imperfect information.

The analysis also implies measuring the "performance" variable in a company, taking into account a variety of indicators that are impossible to measure at a unidimensional level.

Mamdani's model is built by considering a series of linguistic proposals and by elaborating different rules measured from observed data. Information is obtained by means of a survey structure that proposes a measuring tool for the levers through fuzzy inference systems.

The software employed in the research is MATLAB–Fuzzy Logic Toolbox. These are grouped in the following levers: beliefs, boundary, interactive, and diagnostics. "Performance" is also measured by considering Financial (liquidity, debt, capability to pay providers, utility) and Non-financial (organizational environment, sense of business control, improved decision making) aspects.

3.3. Sample Characteristics

The target population of this study is composed of Chilean SMEs (medium: 51 to 200 workers; small: 1 to 50 workers) that are part of the 2018 database of the Federation of Chilean Industry SOFOFA (Sociedad de Fomento Fabril) published with 4000 enterprise members. In Chile, SMEs represent 96.8% of the total companies in the country, 220,000 are SMEs, and nearly 680,000 are small businesses.

Final sample size n = 86, with a response rate of 7.1%. The sampling used is not probabilistic, which limits the conclusions of this article.

Table 1 summarizes the descriptive statistics of the surveys. This research employed SPSS to code the survey and to elaborate an exploratory factorial analysis, which is shown in the first part of this investigation.

Table 1. Profile of sample companies.

Company Profile	
Average N° of Employees	Average: 35 employees Mode: 10 employees Max: 800, Min: 2
Average Range of Annual Sales	Small 1: 2.400,01 uf a 5.000 uf.: 28% Small 2: 5.000,01 uf a 10.000 uf.: 19% Small 3: 10.000,01 uf a 25.000 uf.: 20% Medium 1: 25.000,01 uf a 50.000 uf.: 16% Medium 2: 50.000,01 uf a 100.000 uf.: 12% Other 5%
Main Sector/Industry	Commerce: 52% Agriculture: 8% Construction: 6% Hotel/Restaurant: 7%
Main Geographical Location	Metropolitan Region: 77%

Source: Prepared by the authors.

From the previous table, we can identify that the majority of companies that answered the survey are of small size, with a mode of 10 employees, and are mostly from the commercial sector of the Metropolitan Region.

Of the individuals who responded to our survey, the average work experience is 16 years, and 46% have higher education. Most of the surveys were answered by the CEO; hence, it can be foreseen that they have a high degree of knowledge of their companies.

3.4. Construction of the Variable Measurement Survey

To measure the variables in the study, we developed a structured survey to collect the empirical data. The survey design first required an exhaustive literature review on the Web of Science–Clarivate Analytics database. Following the review of published works, no empirical study was found pertaining to these variables and scales intended to measure the constructs. Hence, a proposed measuring scale for each dimension in the research is presented in Table 2. The validity of the content of the initial survey proposal was evaluated by experts in the field of management control and market research, with an additional random sample of 10 SMEs to evaluate content. An exhaustive review was elaborated to achieve a good level of acceptance of the survey in its draft, number of questions, and design. We received comments that helped us improve the design of the survey, focusing on internal validity and content, giving us an acceptable measuring scale. This allowed us to attain internal coherence for all the dimensions in the study model.

Table 2. Measurement Scales.

Input		Output	
DB Survey Code	Input Tag	DB Survey Code	Output Tag
1	Strongly disagree	1–2	No improvement
2	Disagree		
3	Neither agree nor disagree	3	Medium improvement
4	Agree	4–5	High improvement
5	Strongly agree		

Source: Prepared by the authors.

The measurement scale for the dimensions is management controls, human resources, and organizational performance measured with a Likert scale from "1" (strongly disagree) to "5" (strongly agree). The latter was selected as it is more suitable in evaluating degrees of difference rather than employing dichotomous variables to indicate the presence or absence of a particular practice. Other variables related to the questions in the study aim to define the company profiles. Each question had its own specific scale designed for its subject, producing the following variables: type of generic strategy, degree of ICT use, software used, the existence of a role in management control, types of control tools employed, questions pertaining to culture and organizational structure, number of employees, range of annual sales, sector, among others (see Appendix A.1).

The survey application method is multichannel (personal, telephone, and online). It was applied to company managers throughout the months of September and October of 2018. The process consisted of sending an email invitation to all the mentioned databases, specifically to 1200 SMEs at the national level. The objectives of the study were explained, and the participants were informed by email regarding the structured survey addressed to company managers. In the second stage of the process, an access link to the survey was sent, followed by telephone calls and corporate visits, with the aim of improving the survey response rate.

3.5. Variables Measurement

Management control variables [10] and performance [11,40–42] are measured through linguistic sentences, as proposed in the study. Given the lack of empirical research with the proposed verbal sentences to measure what the study requires, each dimension has been coupled with several authors (see Appendix A.2).

The scale employed in the survey is the Likert scale: strongly agree, agree, neither agree nor disagree, disagree, and strongly disagree. To proceed with fuzzification and defuzzification, the output tags are high improvement, medium improvement, and no improvement, respectively (see Table 2).

Appendix A.1 shows the fuzzy inference input variables. There are 25 variables that represent "Management control" grouped in: Beliefs, Interactive Control, Diagnostics, and Performance. Following [11] works and seven variables to measure "Performance" [11,40–42], all variables are qualitative. Their evaluation depends on the perception of experts. Given their nature and characteristics, fuzzy analysis is the most suitable means to measure them [43].

Thus, by evaluating the control management levers used in SMEs, the degree of truth for each linguistic sentence is measured within the [0, 1] range. This is conducted through a diagnostic instrument applied to company directors.

4. Results

In order to achieve the general objective of this research, we first present a descriptive diagnosis of the management tools used in Chilean SMEs. This is followed by a proposal to implement a fuzzy inference system of diagnosis towards management control tools.

4.1. Descriptive Analysis

The sample is diverse when compared to the generic strategy: Leader in costs: 24%, Differentiation: 22%, Focus: 14%, Mixed: 29%, Doesn't know: 11%, Management profile. 60% claim to have defined their strategic objectives.

In addition, 54% state that they are a family-run business, 55% base their decision-making process on results, and 50% have a functional structure. Regarding decision-making, 84% indicate that they have a centralized system, 68% have defined their mission and/or vision, and 73% have their organizational values well defined. The analysis demonstrates that the majority of SMEs in the beliefs lever understand and have well-defined strategies, mission, vision, values, and strategic objectives. Furthermore, the diagnostics show that 69% of decision-making is based on rational and financial quantitative information. This is linked to the large number of organizational cultures that focus on results.

In retrospect, the values and answers in the boundary lever have weaker data as the majority of managers and directors who answered explained that role hierarchy does not apply in their organization. In fact, barely more than half of the sample define role profiles to their employees and take part in formalizing roles. The interactive lever highlights that the vast majority of decision-making (84%) is centralized and that slightly more than half include staff participation in the process. Similar numbers are found in the practice of benchmarking activities. Lastly, the diagnostics system could be improved by developing navigation routes and plans based on strategic maps, increasing knowledge of the results of the company, nurturing knowledge of the state of results, and developing budget and inventory systems. According to data, the latter is the most developed.

Taking into account performance, the financial field shows improvement with 64% of answers asseverating increased utility after using management control tools within the last two years. In addition, 57% consider that their liquidity has improved, and less than half report improved levels of debt (lower debt levels). The non-financial performance indicates that 65% report improvement in the organizational environment after implementing management control tools, while 58.2% feel that their sense of control has improved due to these applications.

By segmenting SMEs into two groups (small companies and medium-sized companies), we can appreciate the difference in the use of management control tools between them. The larger a company grows, the higher the rate of use of these tools. A clear example in the beliefs lever is the important increase from 54.8% to 86% on employing strategic objectives to define strategy between small and medium-sized companies, respectively.

The interactive lever also demonstrates this difference in the section of "survey and investigate the activities of the competition", where only 58.7% of small companies do this compared to 70.3% of medium-sized companies. More interesting data pertains to the formalization of work contracts and the use of policies and procedures manuals between businesses. As observed, only 67% of small companies do this compared to 84.5% of medium-sized companies.

In the diagnostics lever the differences are very apparent in the use of costing systems (small: 56%, medium: 69%), state of results (small: 59.6%, medium: 76%), and budget (small: 68.4%, medium: 82.6%).

4.2. Measurement Scale

Before proceeding with fuzzy logic analysis, it is important to determine whether the tools employed are appropriate to measure the levers of the study. This is done by proving the reliability and validity of the chosen scale. The survey's internal consistency serves to estimate reliability using Cronbach's Alpha [44], while the scale's validity is proven through factorial analysis to evaluate whether each scale measures a single concept. The KMO statistics (Kaiser–Meyer–Olkin) is applied, and for this research, the minimum value of the limits selected for the analysis to be acceptable is 0.5 [45].

Using Bartlett's test of sphericity to evaluate the presence of correlations among variables, we obtain a level that is significantly inferior to 0.05 [45]. In the case of Cronbach's Alpha, the accepted inferior limit is 0.7, and 0.6 for new scales [46]. In retrospect, as all results are satisfactory, a factorial analysis is convenient.

Hence, the "Belief" dimension's Cronbach Alpha value is 0.935, which is a higher value than the advised minimum. For the Kaiser–Meyer–Olkin (KMO) method, the investigation obtained satisfactory results of 0.764 > 0.5. Bartlett's test returns a chi-squared range (497.57, 549.95), and $p = 0.000$. This corroborates once again the recommendation of employing factorial analysis. The results return three factors; thus, it is recommended to (1) eliminate the following items: q3_5 and q3_11. (2) Use the remaining two factors to define two new belief variables (Strategy and Belief, Planning).

The "Interactive Control" dimension returns a Cronbach Alpha value of 0.692, which is within the acceptable limits (given that this scale is new and created specifically during this research). The KMO value is 0.596, and Bartlett's test returns a chi-squared range of (35.77, 37.56). Both results are satisfactory, and the factorial analysis identified two different factors, which makes it recommendable to eliminate item Q6_4 and keep only one factor.

Regarding the "Boundary" dimension, the Cronbach Alpha value is 08.10 while KMO is 0.787. Bartlett's test returns a chi-squared range of (76.07, 84.08). The factorial analysis identified one factor, confirming that the scale measures one "Boundary" dimension.

The "Diagnostics" dimension has a Cronbach Alpha value of 0.882, KMO of 0.743, and Bartlett's test of chi-squared range (271.17, 299.73). Factorial analysis reveals the existence of three factors; thus, it is recommended to eliminate: q8_6, q8_7, and q8_9. The other two factors are defined as 'Diagnostic-financial' (q8_4, q8_5, q8_8, q8_10, q8_11) and 'Diagnostic-non-financial' (q8_1, q8_2, q8_3).

Lastly, the "Performance" variable has a Cronbach Alpha value of 0.859 and a KMO of 0.812. Bartlett's test results give a chi-squared range of (278.92, 308.28). Factorial analysis reveals the existence of one factor, confirming that the scale measures a "Performance" dimension.

4.3. Fuzzy Inference Systems and Parameterization of Membership Functions

This research has analyzed the proposed measurement tools for evaluating the presence of levers [10]) in SMEs while defining input and output variables, Table 2. The next step is to present the fuzzy inference surfaces that are present in each management control lever related to organizational performance. This is accomplished by using Fuzzy Inference Systems (FIS) with regard to their fuzzy sets and membership functions.

The proposed methodology is based on [33], where the structure stems from Mamdani's fuzzy system. The process of analysis returns a fuzzy indicator that allows us to make strategic recommendations for SMEs.

The inference mechanism output is a fuzzy output. To ensure that the output of the fuzzy system can be interpreted only by elements that process numerical information, the fuzzy output from the inference mechanism must be converted in a process called defuzzification. The output of the inference mechanism comes as a fuzzy set. A numerical value can be generated from these sets through various means, such as through Centroid (Center of Gravity), which was used in this case. The fuzzy controller employed comes from [12]. This research is elaborated through the MATLAB module and the Fuzzy Logic Toolbox, and the process was applied to all dimensions of the model in the study.

The selection of relevant fuzzy membership functions considered the extremes as trapezoidal functions as this considers them as tolerance in case an interval increases or decreases beyond the limits. For all other functions, the research used triangular functions (quad-triangular) due to their higher adaptability to the research and variables as well as increased simplicity for interpretation. Considering the different degrees of membership returned by the fuzzification process, these must be processed to generate fuzzy output.

This is achieved by the inference system, which, based on the rules, can generate the output of a fuzzy system. Fuzzy membership function of output variables, where: High improvement: [25%; 100%], Medium improvement: [50%; −50%], No improvement: [−100%; −25%]. Hence, the following analysis develops the stages demonstrated in Figure 2. As a visual example, the FIS is presented for Beliefs.

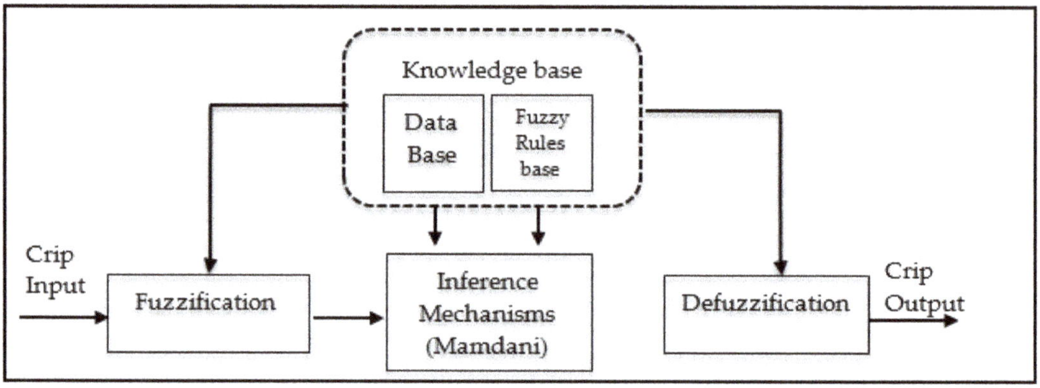

Figure 2. Stages of a Fuzzy Inference System.

4.3.1. FIS for Beliefs

The Fuzzy Inference Systems (FIS) for Beliefs and their respective fuzzy sets (with the membership parameters) are presented as follows: After factorial analysis, the dimension was divided into: "Beliefs = Corporate pillar": q3_1, q3_2, q3_3, q3_4, q3_8, and "Beliefs = Strategic proposal": q3_6, q3_7, q3_9, q3_10. The output variables are H1(1) and H1(2), respectively, as shown in Figures 3–6. The input and output variable membership functions are displayed in Figure 2: Parameterization of Fuzzy Sets for Beliefs.

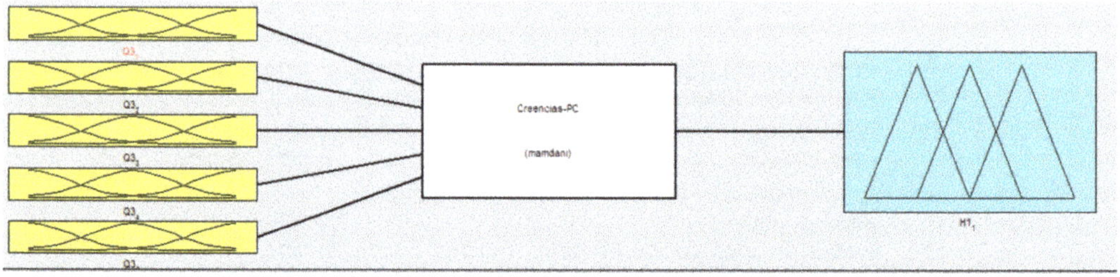

Figure 3. Inference system H1(1): Fuzzy model for Beliefs-CP dimension.

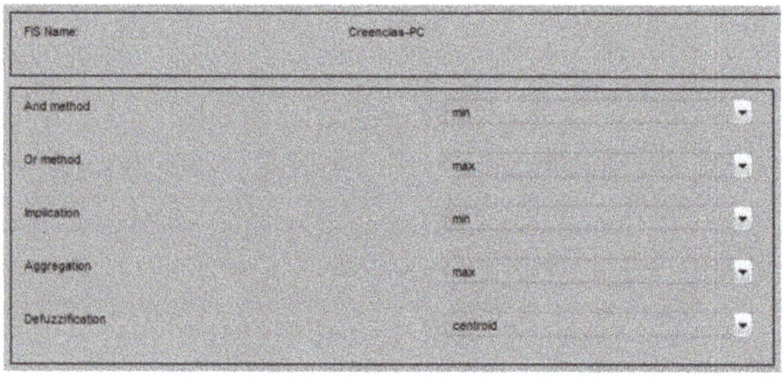

Figure 4. Employment of Centroid Defuzzification.

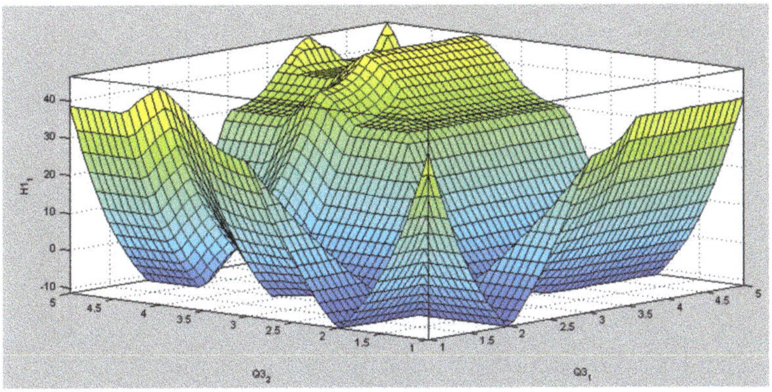

Figure 5. Surface of Fuzzy Inference for CP_Beliefs.

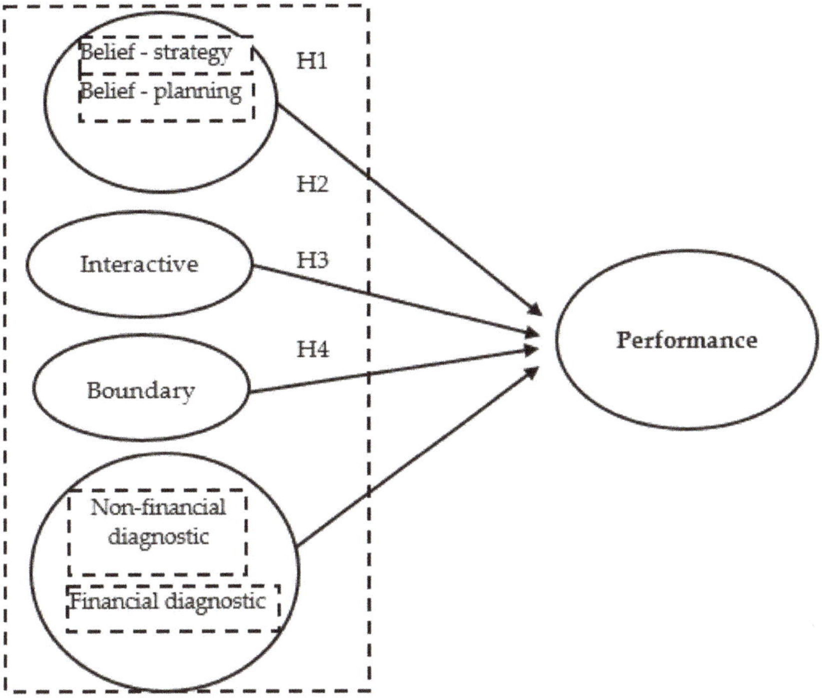

Figure 6. Proposed Final Model.

4.3.2. CP-Beliefs (Corporate Pillar) Beliefs

The CP-Beliefs lever has the following management tools: definition of mission, vision, and values. FIS analysis reveals that these are used at an average level (30.4512) by SMEs, as demonstrated in Figure 3. The presence or absence of a variable affects the stability of the Beliefs lever.

Figure 3 shows the fuzzy model structure for the CP-Beliefs dimension. It is composed of five input variables (q3_1, q3_2, q3_3, q3_4, q3_8) and an output variable H1(1).

The following, through MATLAB, considers the output variable membership function for the CP-Beliefs dimension. It is seen that the use of the lever "CP-Beliefs" has a value of "medium", highlighting the positive (30.4512). Figure 5 has two inputs and one output, which shows the output surface of the system. By default, this output was plotted with the two initial input variables.

F (x) = 30.4512. A: medium level
X(Input): Q3_1
Y(input): Q3_2
Z(output): H1_1

As shown in Figure 5, the relationship between input variables Q3_1 and Q3_2 are displayed as a large volume close to the interval 2–5, returning a result of approximately 30.45%.

For the following variables under study, Appendix A can be reviewed.

4.4. Summary of Fuzzy Indicators

Table 3 shows a summary of the fuzzy indicators obtained in this research. Observations note a considerable degree of difference between small and medium-sized companies on the presence of management control tools on their administration. Small businesses have a positive yet lower degree of presence than medium-sized companies. Both segments have weaknesses, particularly in describing role profiles, the hierarchy of roles, organizational structure, use of policies and procedure manuals, and the use of formal contracts (in general, they both hit the 'medium' level).

Table 3. Summary of Fuzzy Indicators, SMEs (Small and Medium-sized companies).

Lever	FIS—SMEs	FIS—Small	FIS—Medium
Beliefs			
Beliefs—Strategic	30.4512	17.3919	41.3900
Beliefs—Planning	24.5564	15.7527	48.7993
Interactive Control	32.4552	26.7419	39.5948
Boundary	−7.9801	−8.1384	−7.4093
Diagnostics		12.0115	34.5589
Diagnostics—F	23.4046	17.9972	36.6432
Diagnostics—N.F	0.8003	−6.5001	21.9388
Performance	12.3180	9.7492	17.3669

Source: Self-elaboration.

5. Discussion

The descriptive analysis allowed us to identify that the SMEs studied perceived that their management control tools were well defined (69%). In addition, decision-making based on rational and financial quantitative information was employed, which focuses organizational culture towards results. Regarding performance, the financial area shows improvement, with 64% of answers positively asserting that utility increased after employing management control tools for two years. Furthermore, 57% consider that their liquidity level improved, and slightly less than half report that their debt levels lowered. Regarding non-financial performance, 65% report improvements to the organizational environment after employing the tools, while 58.2% claim to sense better control in decision-making processes.

As a business grows, the use of control tools increases. A clear example can be seen in the beliefs lever values that spiked up to 54.8% and 86% in using strategic objectives to define strategy among small and medium-sized companies, respectively. It must be noted that the interactive lever clearly shows strategic surveillance strategies in the competition, with 58.7% of small companies and 70.3% of medium-sized companies practicing this activity.

Interestingly, the boundary lever is the least used among the SMEs that employ it. The values are similar to the previous results, with 67% coming from small companies and 84.5% from medium-sized companies. In retrospect, the diagnostics lever has clearer differences between small and medium-sized companies in costing system (56% vs. 69%), state of results (59.6% vs. 76%), and use of budget (68.4% vs. 86.2%).

The analysis contributes to what is highlighted in the literature regarding the structural weaknesses to create management strategies for medium and long-term projections due to a scarcity of tools. In addition, management of administration, finances, accounting, and operations is very informal and is done in an intuitive manner where few management control tools are employed in decision-making [15]. Furthermore, there is evidence of the low use of management control tools by small companies, which does improve when viewing medium-sized companies in agreement with the studies by Frost [20].

Nonetheless, we can perceive that using management tools does improve performance without substantial difference between financial and non-financial performance. This reinforces the literature that states that the use of management control tools has a positive impact on organizational performance [6,10,22]. In regard to the measuring scale employed in this study, it had been validated as it returned a high level of internal data consistency through the Cronbach Alpha [44], satisfactory statistical values with the Kaiser–Meyer–Olkin method, and acceptable numbers in Bartlett's test. Lastly, the factorial analysis identified factors that recommended that we eliminate certain variables, keeping the original theoretical model for the one that was finally used. This separated beliefs into strategy and planning and separated the diagnostics lever into financial and non-financial diagnostics. These were analyzed by means of fuzzy logic, as indicated in Figure 6.

Fuzzy Inference System (FIS) for Management Control Tool Levers

The fuzzy inference system revealed that all levers were used at 'medium' capacity by SMEs, see Table 3. The least employed lever was "Boundary", which includes the use of formal contracts, profile descriptions, manuals, organizational structure, and role hierarchy. Results show that the degree of membership was -7 without significant change between small and medium-sized companies (see Table 3). This indicator remained stable as the other variables of the lever remained in 'medium' use, but with a negative trend. In retrospect, the other levers have positive use led by interactive control, followed by diagnostics control, and lastly by beliefs. Performance is valued as 'medium' due to the median use of management control tools, grouped in control levers (Simons, 1995) by SMEs based on the literature.

We recommend training mainly the boundary lever, as it is compared to the other levers (see Table 3), as also shown in the descriptive analysis. As mentioned in the literature, it is recommended that entrepreneurs and businesspersons are trained on how to use management control tools. It is recommended that management control tools be provided to small companies. This would increase the degree of membership for medium-sized companies in all levers, including the diagnostics group, but excluding the boundary group, which remains low in all SMEs. The fuzzy inference system also revealed that the degree of membership for the beliefs lever is 'medium', which is of concern in small companies (FIS = 6.0376). Due to its positive relationship with level of performance, it is recommended to train SMEs in these control tools, as different variations (of use) may imply a highly unstable variable behavior. In addition, the fuzzy inference system explains similar results found in the descriptive analysis, where the least used lever by SMEs is "Boundary", not showing differences in degrees of membership regarding financial and non-financial performance.

Lastly, the study suggests that management control tools in SMEs have a mid-level of use or degree of membership. "Boundary" was the least used, followed by "Beliefs", "Diagnostics", and "Interactive" (as the most used). This may be because SMEs follow horizontal and organic structures, which causes information to flow and generate team collaboration for decision-making. In retrospect, it is necessary to implement training in management control for businesspersons that lead SMEs, as it could increase the company's performance and keep them sustainable over time.

6. Conclusions

The importance of the proposed system for management control in SMEs is understood by showing how the analysis with Mamdani's fuzzy inference system (MFIS) is flexible and can be adapted to the information of each organization. It considers the context of SMEs, where all the data is not always available, and several evaluations are based on the perception of the CEOs.

The second major finding was that Mamdani's model was ideal for the evaluation as the qualitative variables for analysis were ideological and linguistic.

Evaluation using fuzzy IF-THEN rules enables adjusting the analysis to the needs and available data, so the number of rules can vary, allowing for different results. These characteristics of MFIS make it widely applicable in different scenarios or countries.

The boundary lever shows the highest level of weakness in SMEs and must be given more attention as it returned a negative degree of presence in the business model. It is necessary to implement training in management control for businesspersons who run SMEs as this could increase company performance and keep them sustainable over time.

The belief, diagnostic, and interactive levers show mid-level results (of use) with a positive trend. This specifically means that in SMEs that employed the tools, in belief, tools such as the formal definition of mission, vision, and strategy. Diagnostics tools related to budget and inventory systems, whereas interactive obtains information from employees that most interact with clients to use the feedback for the future of the organization. There was no significant difference between financial and non-financial performance. Furthermore, medium-sized businesses employ more management control tools than small-sized ones, which is expected.

One of the main restrictions of the study is the sample size, which consisted of only 86 SMEs, with more belonging to small companies rather than medium-sized entities. In addition, the complexity of defining rules must be taken into account, considering the number of variables and the importance of a good team of experts to define the fuzzy rules.

This implies that future research should focus on an economic sector that would allow us to specify the degree of use of the management control tools by industry. Nevertheless, we consider the results to open a path of study and analysis to develop and apply fuzzy methodologies to the field of management accounting.

Author Contributions: Conceptualization, Formal analysis, Investigation, Methodology, Project administration, Writing—original draft, Writing—review & editing, C.N.; Formal analysis, Investigation, Writing—original draft, Writing—review & editing, J.M.; Formal analysis, Methodology, Writing—original draft, Writing—review & editing, F.-J.A.-C. All authors have read and agreed to the published version of the manuscript.

Funding: This research received no external funding.

Institutional Review Board Statement: Not applicable.

Informed Consent Statement: Not applicable.

Data Availability Statement: The data presented in this study are available on request from the corresponding author.

Conflicts of Interest: The authors declare no conflict of interest.

Appendix A

Appendix A.1 Input Variables

Beliefs	
Q3_1	The company's mission/vision is correctly defined.
Q3_2	The company's values are correctly defined.
Q3_3	I apply the institution's values in the management and decision-making of the company.
Q3_4	The generic strategy of the company is correctly defined (i.e., Cost leadership, Differentiation, or Focus).
Q3_5	The personnel is strongly committed to the company's defined strategy.
Q3_6	The strategic objectives of the company are well defined.
Q3_7	I use the strategic objectives in the management and decision-making of the company.
Q3_8	I use the strategic declarations (mission/vision) for the decision-making of the company.
Q3_9	The personnel is committed to the company's strategic declarations (mission/vision).
Q3_10	The personnel is committed to the company's strategic objectives (mission/vision).
Q3_11	Point out which generic strategy is the one that most resembles your business model. (Leadership in Costs, Differentiation, or Approach).
Interactive Control	
Q6_1	I host constant general team meetings.
Q6_2	I use various channels for internal communication.
Q6_3	I constantly stimulate personnel participation in the development of ideas to improve organizational practices (competitions, suggestion box, innovation, etc.).
Q6_4	I constantly research the competition's practices in decision-making.
Limits	
Q7_1	We use formal contracts for all personnel in this SME.
Q7_2	We use formal role profile descriptions in this SME.
Q7_3	We use policies, procedures, and/or manuals at the formal level in this SME.
Q7_4	We use a specific organizational structure in this SME.
Q7_5	The role hierarchy is well established in this SME.
Diagnostics	
Q8_1	I use the control panel or the balanced scorecard (indicators) to control management in my SME.
Q8_2	I use strategic maps to control management in my SME.
Q8_3	I use some information system to control management in my SME (i.e., Softland, SAP, Defontana, etc.).
Q8_4	I use product or service costing systems to control management in my SME.
Q8_5	I use inventory or stock surveillance to control management in my SME.
Q8_6	I use internal audits to control management in my SME.
Q8_7	I use analysis of the state of results to control management in my SME.
Q8_8	I use budgets to control management in my SME.
Q8_9	I use quality control (quality norms) to control management in my SME.
Q8_10	I know the mix of the most profitable products (volume/price) of my SME.
Q8_11	I use forecasts and demand estimations to control management in my SME.
Performance	
Q12_1	Liquidity (capability that an entity possess to obtain cash)
Q12_2	Level of debt (Total passives/Net assets, it is estimated that a healthy value is 30%).
Q12_3	Capability to pay providers.
Q12_4	Utility.
Q12_5	Organizational environment.
Q12_6	Business control.
Q12_7	Decision-making of the company.

Appendix A.2 Variables and Constructs Used in the Model

Management control	Simons, R. (1995). Kaplan, R. S., and Norton, D. P., (1996, 2004, 2008)., Anthony, R. N., and Govindarajan, V., (2008)., Frost, (2003), Widener, 2007. Krius et al., 2016.	Belief lever (defining values, strategic declarations, strategic objectives, organizational culture); Boundary lever (personal contracts, role description, policies or procedures, formal structure); Diagnostics lever (control panel, balanced scorecard, strategic maps, management information systems, costing systems, inventory, analysis of the state of results, analysis of financial situation, budget analysis, forecast); Interactive Control lever (use of communication channels with employees, regular meetings, mechanisms for individual participation, strategic surveillance).
Performance	Ittner, C. D., and Larcker, D. F. (1998). Green, S. G., and Welsh, M. A. (1988). Duréndez et al., (2016). Voss and Brettel, 2014.	Financial (liquidity, debt, capability to pay suppliers, and utility); Non-financial (organizational environment, sense of business control, improvement in decision making).

Source: Prepared by the authors.

References

1. Nia, A.S.; Saparauskas, J.; Ghorabaee, M.K. A Fuzzy ARAS method for Supply Chain Management performance measurement in SMEs under uncertainty. *Transform. Bus. Econ.* **2017**, *16*, 41.
2. Kaplan, R.S.; Norton, D.P. *The Balanced Scorecard: Translating Strategy into Action*; Harvard Business School Press: Boston, MA, USA, 1996.
3. Kaplan, R.S.; Norton, D.P. *Strategy Maps: Converting Intangible Assets into Tangible Outcomes*; Harvard Business Press: Boston, MA, USA, 2004.
4. Kaplan, R.S.; Norton, D.P. *The Execution Premium*; Harvard Business School Press: Boston, MA, USA, 2008.
5. Melnyk, S.A.; Bititci, U.; Platts, K.; Jutta, T.; Andersen, B. Is performance measurement and management fit for the future? *Manag. Account. Res.* **2014**, *255*, 173–186. [CrossRef]
6. Duréndez, A.; Ruíz-Palomo, D.; García-Pérez-de-Lema, D.; Diéguez-soto, J. Management control systems and performance in small and medium family firms. *Eur. J. Fam. Bus.* **2016**, *6*, 10–20. [CrossRef]
7. Andersen, J.; Samuelsson, J. Resource organization and firm performance: How entrepreneurial orientation and management accounting influence the profitability of growing and non-growing SMEs. *Int. J. Entrep. Behav. Res.* **2016**, *22*, 466–484. [CrossRef]
8. Chenhall, R.H.; Langfield-Smith, K. Adoption and benefits of management accounting practices: An Australian study. *Manag. Account. Res.* **1998**, *9*, 1–19. [CrossRef]
9. Ahmad, K.; Zabri, S.M. Factors explaining the use of management accounting practices in Malaysian medium-sized firms. *J. Small Bus. Enterp. Dev.* **2015**, *22*, 762–781. [CrossRef]
10. Obispo, D. Caracterización del control interno en la gestión de las empresas comerciales del Perú 2013. *Crescendo* **2015**, *6*, 64–73. [CrossRef]
11. Simons, R. *Control in an Age of Empowerment*; Harvard Business Review Press: Boston, MA, USA, 1995; pp. 80–88.
12. Nicolas, C.; Gil-Lafuente, J.; Urrutia, A.; Valenzuela, L. Fuzzy Logic Approach Applied into Balanced Scorecard. In *Advances in Intelligent Systems and Computing*; Applied Mathematics and Computational Intelligence; FIM 2015; Gil-Lafuente, A., Merigó, J., Dass, B., Verma, R., Eds.; Springer: Cham, Switzerland, 2018; Volume 730.
13. Mamdani, E.; Assilian, S. An experiment in linguistic synthesis with a fuzzy logic controller. *Int. J. Man-Mach. Stud.* **1975**, *7*, 1–13. [CrossRef]
14. Armitage, H.M.; Webb, A.; Glynn, J. The Use of Management Accounting Techniques by Small and Medium-Sized Enterprises: A Field Study of Canadian and Australian Practice. *Account. Perspect.* **2016**, *15*, 31–69. [CrossRef]
15. Castañeda, L. Los sistemas de control interno en las Mipymes y su impacto en la efectividad empresarial. *En-Contexto* **2014**, *2*, 129–146.
16. Castellanos, J.G. Pymes innovadoras. Cambio de estrategias e instrumentos. *Rev. Esc. Adm. Neg.* **2003**, *47*, 10–33.
17. Anthony, R.N.; Govindarajan, V. *Sistemas de Control Gerencial*; McGraw Hill: Madrid, Spain, 2008.
18. Baird, K.; Su, S. The association between controls, performance measures and performance. *Int. J. Product. Perform. Manag.* **2018**, *67*, 967–984. [CrossRef]
19. Malagueño, R.; Lopez-Valeiras, E.; Gomez-Conde, J. Balanced scorecard in SMEs: Effects on innovation and financial performance. *Small Bus. Econ.* **2017**, *51*, 221–244. [CrossRef]

20. Frost, F.A. The use of strategic tools by small and medium-sized enterprises: An Australasian study. *Strat. Chang.* **2003**, *12*, 49–62. [CrossRef]
21. Adler, R.; Everett, A.; Waldrom, M. Advanced management accounting techniques in manufacturing: Utilization, benefits and barriers to implementation. *Manag. Account. Res.* **2000**, *24*, 131–150. [CrossRef]
22. Munoz, J.M.; Welsh, D.H.B.; Chan, S.H.; Raven, P.V. Microenterprises in Malaysia: A preliminary study of the factors for management success. *Int. Entrep. Manag. J.* **2014**, *11*, 673–694. [CrossRef]
23. Charteres Institute of Management Accountants (CIMA). *Management Accounting Practices of UK Small-Medium-Sized Enterprises (SMEs). Improving SME Performance through Management Accounting Education*; CIMA: London, UK, 2003; Volume 9.
24. Senftlechner, D.; Martin, R.W.; Hiebl, M.R.Q. Management accounting and management control in family businesses: Past accomplishments and future opportunities. *J. Account. Organ. Chang.* **2015**, *11*, 573–606. [CrossRef]
25. Hopper, T.; Tsamenyi, M.; Uddin, S.; Wickramasinghe, D. Management accounting in less developed countries: What is known and needs knowing. *Account. Audit. Account. J.* **2009**, *22*, 469–514. [CrossRef]
26. Cerchione, R.; Esposito, E. Using knowledge management systems: A taxonomy of SME strategies. *Int. J. Inf. Manag.* **2017**, *37*, 1551–1562. [CrossRef]
27. Lee, C.S.; Wong, K.Y. A Fuzzy Logic-Based Knowledge Management Performance Measurement System for SMEs. *Cybern. Syst.* **2017**, *7*, 1–26. [CrossRef]
28. Nicolas, C. *Indicadores Clave de Gestión Sobre la Experiencia del Cliente: Un Estudio Basado en Fuzzy Text Mining*; Universidad de Barcelona: Barcelona, Spain, 2014.
29. Zadeh, L.A. The concept of a linguistic variable and its application to approximate reasoning (Part I). *Inf. Sci.* **1975**, *8*, 199–249. [CrossRef]
30. Zadeh, L.A. The concept of a linguistic variable and its application to approximate reasoning (Part II). *Inf. Sci.* **1975**, *8*, 301–357. [CrossRef]
31. Zadeh, L.A. The concept of a linguistic variable and its application to approximate reasoning (Part III). *Inf. Sci.* **1976**, *9*, 43–80. [CrossRef]
32. Gil-Aluja, J. *Elements for a Theory of Decision in Uncertainty*; Springer Science & Business Media: Berlin/Heidelberg, Germany, 1999. [CrossRef]
33. Gil Lafuente, J. *Marketing Para el Nuevo Milenio. Nuevas Técnicas Para la Gestión Comercial en la Incertidumbre*; Pirámide: Madrid, Spain, 1997.
34. Zadeh, L.A. Is there a need for fuzzy logic? *Inf. Sci.* **2008**, *178*, 2751–2779. [CrossRef]
35. Fiss, P.C. A set-theoretic approach to organizational configurations. *Acad. Manag. Rev.* **2007**, *32*, 1180–1198. [CrossRef]
36. Jang, J.-S.; Sun, C.-T. Neuro-fuzzy modeling and control. *Proc. IEEE* **1995**, *83*, 378–406. [CrossRef]
37. Jang, J.S.R.; Sun, C.T.; Mizutani, E. *Neuro Fuzzy and Soft Computing*; Prentice-Hall: Upper Saddle River, NJ, USA, 1997.
38. Medina, S. Estado de la cuestión acerca del uso de la lógica difusa en problemas financieros. *Cuad. Adm.* **2006**, *19*, 195–223.
39. Tutmez, B.; Tercan, A.E. Assessment of Uncertainty in Geological Sites Based on Data Clustering and Conditional Prob-abilities. *J. Uncertain Syst.* **2007**, *1*, 207–221.
40. Ittner, C.D.; Larcker, D.F. Are Nonfinancial Measures Leading Indicators of Financial Performance? An Analysis of Customer Satisfaction. *J. Account. Res.* **1998**, *36*, 1–35. [CrossRef]
41. Green, S.G.; Welsh, M.A. Cybernetics and dependence: Reframing the control concept. *Acad. Manag. Rev.* **1988**, *13*, 287–301. [CrossRef]
42. Voss, U.; Brettel, M. The effectiveness of Management Control in Small Firms: Perspectives from Resource Dependence Theory. *J. Small Bus. Manag.* **2014**, *52*, 569–587. [CrossRef]
43. Serrano-García, J.; Acevedo-Alvarez, C.A.; Castelblanco-Gomez, J.M.; Arbelaez-Toro, J.J. Measuring organizational capabilities for technological innovation through a fuzzy inference system. *Technol. Soc.* **2017**, *50*, 93–109. [CrossRef]
44. Cronbach, L.J. Coefficient alpha and the internal structure of tests. *Psychometrika* **1951**, *16*, 297–334. [CrossRef]
45. Field, A. *Discovering Statistics Using SPSS for Windows*; Sage Publications: London, UK, 2000.
46. Nunnally, J.C. *Psychometric Theory*, 2nd ed.; McGraw-Hill: New York, NY, USA, 1978.

mathematics

MDPI

Article

Fuzzy Differential Subordinations Obtained Using a Hypergeometric Integral Operator

Georgia Irina Oros

Department of Mathematics and Computer Science, University of Oradea, 1 Universitatii Street, 410087 Oradea, Romania; georgia_oros_ro@yahoo.co.uk or goros@uoradea.ro

Abstract: This paper is related to notions adapted from fuzzy set theory to the field of complex analysis, namely fuzzy differential subordinations. Using the ideas specific to geometric function theory from the field of complex analysis, fuzzy differential subordination results are obtained using a new integral operator introduced in this paper using the well-known confluent hypergeometric function, also known as the Kummer hypergeometric function. The new hypergeometric integral operator is defined by choosing particular parameters, having as inspiration the operator studied by Miller, Mocanu and Reade in 1978. Theorems are stated and proved, which give corollary conditions such that the newly-defined integral operator is starlike, convex and close-to-convex, respectively. The example given at the end of the paper proves the applicability of the obtained results.

Keywords: analytic function; univalent function; fuzzy differential subordination; fuzzy best dominant; confluent hypergeometric function; integral operator

Citation: Oros, G.I. Fuzzy Differential Subordinations Obtained Using a Hypergeometric Integral Operator. *Mathematics* **2021**, *9*, 2539. https://doi.org/10.3390/math9202539

Academic Editors: Ioan Dzitac and Sorin Nadaban

Received: 14 September 2021
Accepted: 6 October 2021
Published: 10 October 2021

1. Introduction

The introduction of the fuzzy set concept by Lotfi A. Zadeh, in the paper "Fuzzy Sets" [1] in 1965, did not suggest the extraordinary evolution of the concept which followed. Received with distrust at first, the concept is very popular nowadays, being adapted to many research topics. Mathematicians were also interested in embedding the concept of fuzzy set in their research and it was indeed included in many mathematical approaches. The review paper included in the present special issue, devoted to the celebration of the 100th anniversary of Zadeh's birth [2], shows how fuzzy set theory has evolved related to certain branches of science, and points out the contribution of one of Zadeh's disciples, Professor I. Dzitac, to the development of soft computing methods connected with fuzzy set theory. Professor I. Dzitac has celebrated his friendship with the multidisciplinary scientist, Lotfi A. Zadeh, by writing the introductory paper of a special issue on fuzzy logic dedicated to the centenary of Zadeh's birth [3].

As far as complex analysis is concerned, fuzzy set theory has been included in studies related to geometric function theory in 2011, when the first paper appeared introducing the notion of subordination in fuzzy set theory [4] which has had its inspiration in the classical aspects of subordination introduced by Miller and Mocanu [5,6]. The next papers published followed the line of research set by Miller and Mocanu and referred to fuzzy differential subordination, adapting notions from the already well-established theory of differential subordination [7–9]. The idea was soon picked up by researchers in geometric function theory and all the classical lines of research in this topic were adapted to the new fuzzy aspects. A review paper published in 2017 [10] included in its references the first published papers related to this topic, validating its development. The dual notion of fuzzy differential superordination was also introduced in 2017 [11].

An important topic in geometric function theory is conducting studies which involve operators. Such studies for obtaining new fuzzy subordination results were published soon after the notion was introduced, in 2013 [12], continued during the next years [13–16] and later added the superordination results [17–19]. During the last years, many papers were published which show that the research on this topic is in continuous development process and we mention only a few here [20–24].

Following this line of research, a new hypergeometric integral operator is introduced in this paper using a confluent (or Kummer) hypergeometric function and having, as inspiration, the operator studied by Miller, Mocanu and Reade in 1978, by taking specific values for parameters involved in its definition. Fuzzy differential subordinations are obtained and the fuzzy best dominants are given, which facilitate obtaining sufficient conditions for univalence of this operator.

2. Preliminaries

The research presented in this paper is done in the general environment known in the theory of differential subordination given in the monograph [25] combined with fuzzy set notions introduced in [4,7].

The unit disc of the complex plane is denoted by U. $\mathcal{H}(U)$ stands for the class of holomorphic functions in U. Consider the subclass, $\mathcal{A}_n = \{f \in \mathcal{H}(U) : f(z) = z + a_{n+1}z^{n+1} + \cdots, z \in U\}$, with $\mathcal{A}_1 = \mathcal{A}$.

For $a \in \mathbb{C}$, $n \in \mathbb{N}^*$ the following subclass of holomorphic functions is obtained: $\mathcal{H}[a,n] = \{f \in \mathcal{H}(U) : f(z) = a + a_n z^n + a_{n+1}z^{n+1} + \cdots, z \in U\}$, with $\mathcal{H}_0 = \mathcal{H}[0,1]$.

For $\alpha < 1$, let $\mathcal{S}^*(\alpha) = \{f \in \mathcal{A} : \mathrm{Re}\left(\frac{zf'(z)}{f(z)}\right) > \alpha\}$ denote the class of starlike functions of order α. For $\alpha = 0$, the class of starlike functions is denoted by $\mathcal{S}^*$.

For $\alpha < 1$, let $\mathcal{K}(\alpha) = \{f \in \mathcal{A} : \mathrm{Re}\left(\frac{zf''(z)}{f'(z)} + 1\right) > \alpha\}$ denote the class of convex functions of order α. For $\alpha = 0$, the class of convex functions is denoted by $\mathcal{K}$.

The subclass of close-to-convex functions is defined as: $\mathcal{C} = \{f \in \mathcal{H}(U) : \exists\, \varphi \in \mathcal{K}, \mathrm{Re}\left(\frac{f'(z)}{\varphi'(z)}\right) > 0, z \in U\}$.

It is also said that function f is close-to-convex with respect to function φ.

Definition 1 ([4]). *Let $D \subset \mathbb{C}$ and $z_0 \in D$ be a fixed point. We take the functions $f, g \in \mathcal{H}(D)$. The function f is said to be fuzzy subordinate to g and write $f \prec_{\mathcal{F}} g$ or $f(z) \prec_{\mathcal{F}} g(z)$, if there exists a function $F : \mathbb{C} \to [0,1]$, such that*

(i) $f(z_0) = g(z_0)$,
(ii) $F(f(z)) \le F(g(z))$, *for all* $z \in D$.

Remark 1. *(a) Such a function $F : \mathbb{C} \to [0,1]$ can be considered $F(z) = \frac{|z|}{1+|z|}$, $F(z) = \frac{1}{1+|z|}$, $F(z) = \frac{\sqrt{|z|}}{1+\sqrt{|z|}}$.*

(b) Relation (ii) is equivalent to $f(D) \subset g(D)$.

Definition 2 ([7], Definition 2.2). *Let $\psi : \mathbb{C}^3 \times D \to \mathbb{C}$, $a \in \mathbb{C}$, and let h be univalent in U, with $h(z_0) = a$, g be univalent in D, with $g(z_0) = a$, and p be analytic in D, with $p(z_0) = a$. Likewise, $\psi(p(z), zp'(z), z^2p''(z); z)$ is analytic in D and $F : \mathbb{C} \to [0,1]$, $F(z) = \frac{|z|}{1+|z|}$. If p is analytic in D and satisfies the (second-order) fuzzy differential subordination*

$$F\Big(\psi(p(z), zp'(z), z^2p''(z); z)\Big) \le F(h(z)), \quad z \in U, \tag{1}$$

i.e., $\psi(p(z), zp'(z), z^2p''(z); z) \prec_{\mathcal{F}} h(z)$, or

$$\frac{|\psi(p(z), zp'(z), z^2p''(z); z)|}{1 + |\psi(p(z), zp'(z), z^2p''(z); z)|} \le \frac{|h(z)|}{1 + |h(z)|}, \quad z \in D, \tag{2}$$

then p is called a fuzzy solution of the fuzzy differential subordination. The univalent function q is called a fuzzy dominant of fuzzy solutions of the differential subordination, or more simply, a fuzzy dominant, if $\frac{|p(z)|}{1+|p(z)|} \le \frac{|q(z)|}{1+|q(z)|}$, or $p(z) \prec_{\mathcal{F}} q(z)$, $z \in D$, for all p satisfying (1) or (2). A fuzzy *dominant* $\widetilde{q}$ *that satisfies* $\frac{|\widetilde{q}(z)|}{1+|\widetilde{q}(z)|} \leq \frac{|q(z)|}{1+|q(z)|}$, *or* $\widetilde{q}(z) \prec_{\mathcal{F}} q(z), z \in D$, *for all fuzzy dominants q of* (1) *or* (2) *is said to be the fuzzy best dominant of* (1) *or* (2). *Note that the fuzzy best dominant is unique up to a rotation in D.*

Lemma 1 ([25], Theorem 2.2). *Let* $\delta, \omega \in \mathbb{C}, \omega \neq 0$, *and h be a convex function in D, and* $F : \mathbb{C} \to [0,1]$, $F(z) = \frac{|z|}{1+|z|}$, $z \in D$. *We suppose that the Briot–Bouquet differential equation*

$$q(z) + \frac{zq'(z)}{\delta + \omega q(z)} = h(z),\ z \in D,\ q(z_0) = h(z_0) = a$$

has a solution $q \in \mathcal{H}(D)$, *which verifies* $q(z) \prec_{\mathcal{F}} h(z), z \in D$, *or* $\frac{|q(z)|}{1+|q(z)|} \leq \frac{|h(z)|}{1+|h(z)|}, z \in D$.

If the function $p \in \mathcal{H}[a,n]$, $\psi : \mathbb{C}^2 \times D \to \mathbb{C}$, $\psi(p(z), zp'(z)) = p(z) + \frac{zp'(z)}{\delta+\omega p(z)}$ *is analytic in D, with* $\psi(p(z_0), z_0 p'(z_0)) = h(z_0)$, $z_0 \in D$, *then*

$$\psi(p(z), zp'(z)) \prec_{\mathcal{F}} h(z),\ z \in D, \tag{3}$$

or

$$\frac{|\psi(p(z), zp'(z))|}{1+|\psi(p(z), zp'(z))|} \leq \frac{|h(z)|}{1+|h(z)|} \tag{4}$$

implies

$$p(z) \prec_{\mathcal{F}} q(z),\ \text{or}\ \frac{|p(z)|}{1+|p(z)|} \leq \frac{|q(z)|}{1+|q(z)|},\ z \in D,$$

and q is the fuzzy best dominant of the fuzzy differential subordination (3) *or* (4).

The confluent (or Kummer) hypergeometric function has been investigated connected to univalent functions more intensely starting from 1985 when it was used by L. de Branges in the proof of Bieberbach's conjecture [26]. The applications of hypergeometric functions in univalent function theory is very well pointed out in the review paper, recently published by H.M. Srivastava [27].

Definition 3 ([25]). *Let u and v be complex numbers with* $v \neq 0, -1, -2, \ldots$, *and consider the function defined by*

$$\phi(u,v;z) = \frac{\Gamma(v)}{\Gamma(u)} \sum_{k=0}^{\infty} \frac{\Gamma(u+k)}{\Gamma(v+k)} \frac{z^k}{k!} = \tag{5}$$

$$1 + \frac{u}{v}\frac{z}{1!} + \frac{u(u+1)}{v(v+1)}\frac{z^2}{2!} + \cdots + \frac{u(u+1)\ldots(u+n-1)}{v(v+1)\ldots(v+n-1)}\frac{z^n}{n!} + \ldots,$$

where $(e)_k = \frac{\Gamma(e+k)}{\Gamma(e)} = e(e+1)(e+2)\ldots(e+k-1)$, *and* $(e)_0 = 1$, *called the confluent (or Kummer) hypergeometric function is analytic in* $\mathbb{C}$.

Remark 2. *(a) For* $z = 0$, $\phi(u,v;0) = 1$ *and* $\phi(u,v;z) \neq 0, z \in U$,
(b) For $u \neq 0$, $\phi'(u,v;0) = \frac{u}{v} \neq 0$.

The operator used for obtaining the original results presented in this paper was obtained using a confluent (or Kummer) hypergeometric function and a general operator studied in 1978 by S.S. Miller, P.T. Mocanu and M.O. Reade [28] by taking specific values for parameters $\beta, \gamma, \alpha, \delta$:

$$J(f)(z) = \left[\frac{\beta+\gamma}{z^{\gamma}\phi(z)} \int_0^z f^{\alpha}(t)\varphi(t)t^{\delta-1}dt\right]^{\frac{1}{\beta}}. \tag{6}$$

A confluent (or Kummer) hypergeometric function was recently used in many papers for defining new interesting operators as it can be seen in [29–32].

Two more lemmas from differential subordination theory that are necessary in the proofs of the original results are listed next:

Lemma 2 ([33], Theorem 4.6.3, p. 84)**.** *A necessary and sufficient condition for a function* $f \in \mathcal{H}(U)$ *to be close-to-convex is given by:*

$$\int_{\theta_1}^{\theta_2} Re\left[1+\frac{zf''(z)}{f'(z)}\right]d\theta > -\pi,\ z_0 = r_0 e^{i\theta_0},$$

for all θ_1, θ_2 *with* $0 \le \theta_1 < \theta_2 < 2\pi$, $r \in (0,1)$.

Lemma 3 ([25], Theorem Marx–Strohhäcker, p. 9)**.** *If* $f \in \mathcal{K}$ *then* $Re\frac{zf'(z)}{f(z)} > \frac{1}{2}$, *i.e.,* $f \in \mathcal{S}^*\left(\frac{1}{2}\right)$, *for* $z \in U$.

3. Main Results

The new hypergeoemtric integral operator is defined using Definition 3 and the integral operator given by relation (6).

Definition 4. *Let* $\beta > 1$, $\gamma > 0$ *and the confluent (Kummer) hypergeometric function* ϕ *given by (5).*

Let $M : \mathcal{H}(U) \to \mathcal{H}(U)$ *be given by:*

$$M(z) = \frac{\beta}{z^{\beta-1}}\int_0^z \left[\frac{\Gamma(v)}{\Gamma(u)}\sum_{k=0}^{\infty}\frac{\Gamma(u+k)}{\Gamma(v+k)}\frac{t^k}{k!}\right]t^{\beta-1}dt,\ z \in U. \tag{7}$$

Remark 3. *(a) For* $\beta > 1$, $\gamma > 0$ *and* $\phi(u,v;z) = \frac{\Gamma(v)}{\Gamma(u)}\sum\limits_{k=0}^{\infty}\frac{\Gamma(u+k)}{\Gamma(v+k)}\frac{z^k}{k!} = 1 + \frac{u}{v}\frac{z}{1!} + \frac{u(u+1)}{v(v+1)}\frac{z^2}{2!} + \ldots$, $u, v \in \mathbb{C}$, $v \neq 0, -1, -2, \ldots$, *we have*

$$M(z) = \frac{\beta}{z^{\beta-1}}\int_0^z \left[1+\frac{u}{v}\frac{t}{1!}+\frac{u(u+1)}{v(v+1)}\frac{t^2}{2!}+\ldots\right]^{\gamma} t^{\beta-1}dt = \tag{8}$$

$$\frac{\beta}{z^{\beta-1}}\int_0^z \left(1+\gamma\frac{u}{v}t+p_2t^2+\ldots\right)t^{\beta-1}dt =$$

$$\frac{\beta}{z^{\beta-1}}\int_0^z \left[t^{\beta-1}+\gamma\frac{u}{v}t^{\beta}+\ldots\right]dt = z+\gamma\frac{u}{v}\frac{\beta}{\beta+1}z^2+p_2\frac{\beta}{\beta+2}z^3+\ldots,$$

which is the analytic expression of the operator M.

(b) For $z \in U$, $M'(z) = 1 + 2\gamma\frac{u}{v}\frac{\beta}{\beta+1}z + \ldots$, *with* $M'(0) = 1$.

Using the method of differential subordination, next, a theorem is proved, giving the best dominant of a certain fuzzy differential subordination. Using specific functions as the fuzzy best dominant, conditions for starlikeness and convexity of the operator M are obtained as corollaries.

Theorem 1. *For* $\beta, \gamma \in \mathbb{C}$, $\beta > 1$, $\gamma > 0$, *let the fuzzy function* $F : \mathbb{C} \to [0,1]$ *be given by*

$$F(z) = \frac{|z|}{1+|z|},\ z \in U, \tag{9}$$

and consider a holomorphic function in U *given by the equation*

$$h(z) = q(z) + \frac{zq'(z)}{\beta-1+\gamma q(z)},\ h(0) = q(0), \tag{10}$$

when q is a univalent solution in U which satisfies the fuzzy differential subordination:

$$\frac{|q(z)|}{1+|q(z)|} \leq \frac{|h(z)|}{1+|h(z)|} \text{ i.e., } q(z) \prec_{\mathcal{F}} h(z),\ z \in U. \tag{11}$$

Consider $\phi(u,v;z)$, a confluent (or Kummer) hypergeometric function given by (5) and the operator $M(z)$ given by (7).

If $1+\gamma\frac{z\phi'(u,v;z)}{\phi(u,v;z)}$ is analytic in U, and

$$\frac{\left|1+\gamma\frac{z\phi'(u,v;z)}{\phi(u,v;z)}\right|}{1+\left|1+\gamma\frac{z\phi'(u,v;z)}{\phi(u,v;z)}\right|} \leq \frac{|h(z)|}{1+|h(z)|} \text{ i.e., } 1+\gamma\frac{z\phi'(u,v;z)}{\phi(u,v;z)} \prec_{\mathcal{F}} h(z), \tag{12}$$

then

$$\frac{\left|\frac{zM'(z)}{M(z)}\right|}{1+\left|\frac{zM'(z)}{M(z)}\right|} \leq \frac{|q(z)|}{1+|q(z)|} \text{ i.e., } \frac{zM'(z)}{M(z)} \prec_{\mathcal{F}} q(z),\ z \in U,$$

and q is the fuzzy best dominant.

Proof. Relation (7) is equivalent to

$$z^{\beta-1}M(z) = \beta\int_0^z \phi^{\gamma}(u,v;t)t^{\beta-1}dt. \tag{13}$$

Differentiating (13) and after short calculation we obtain

$$(\beta-1)M(z)+zM'(z) = \beta\phi^{\gamma}(u,v;z)\cdot z, \tag{14}$$

which is equivalent to

$$M(z)\left[\beta-1+\frac{zM'(z)}{M(z)}\right] = \beta\phi^{\gamma}(u,v;z)\cdot z. \tag{15}$$

We let

$$p(z) = \frac{zM'(z)}{M(z)} = \frac{z\left(1+2\gamma\frac{u}{v}\frac{\beta}{\beta+1}z+\dots\right)}{z\left(1+\gamma\frac{u}{v}\frac{\beta}{\beta+1}z+\dots\right)} = \frac{1+2\gamma\frac{u}{v}\frac{\beta}{\beta+1}z+\dots}{1+\gamma\frac{u}{v}\frac{\beta}{\beta+1}z+\dots}, \tag{16}$$

$p(0)=1$.

Using (16) in (15), we get

$$M(z)[\beta-1+p(z)] = \beta\phi^{\gamma}(u,v;z)\cdot z. \tag{17}$$

Differentiating (17), we obtain

$$\frac{zM'(z)}{M(z)} + \frac{zp'(z)}{\beta-1+p(z)} = \gamma\frac{z\phi'(u,v;z)}{\phi(u,v;z)}+1. \tag{18}$$

Using (16) in (18), we have

$$p(z) + \frac{zp'(z)}{\beta-1+p(z)} = \gamma\frac{z\phi'(u,v;z)}{\phi(u,v;z)}+1. \tag{19}$$

Using (19) in (12), we get

$$\frac{\left|p(z)+\frac{zp'(z)}{\beta-1+p(z)}\right|}{1+\left|p(z)+\frac{zp'(z)}{\beta-1+p(z)}\right|} \leq \frac{|h(z)|}{1+|h(z)|}, z \in U. \tag{20}$$

In order to obtain the desired relation, Lemma 1 will be applied. To apply the lemma, let the function $\psi : \mathbb{C}^2 \times U \to \mathbb{C}$,

$$\psi(r,s;z) = r + \frac{s}{\beta-1+r}, \quad r,s \in \mathbb{C}. \tag{21}$$

For $r = p(z)$, $s = zp'(z)$, relation (21) becomes

$$\psi(p(z), zp'(z)) = p(z) + \frac{zp'(z)}{\beta-1+p(z)}, \quad z \in U. \tag{22}$$

Using (22) in (20), we have

$$\frac{|\psi(p(z), zp'(z))|}{1+|\psi(p(z), zp'(z))|} \leq \frac{|h(z)|}{1+|h(z)|}, z \in U. \tag{23}$$

Applying Lemma 1, for $\delta = \beta - 1$, $\omega = \gamma \neq 0$, we obtain

$$\psi(p(z), zp'(z)) \prec_{\mathcal{F}} h(z), \; z \in U. \tag{24}$$

Using (22) in (24), we get

$$p(z) + \frac{zp'(z)}{\beta-1+p(z)} \prec_{\mathcal{F}} h(z), \; z \in U. \tag{25}$$

According to Lemma 1, relation (25) implies

$$p(z) \prec_{\mathcal{F}} q(z), \; z \in U. \tag{26}$$

Using (16) in (26) we have

$$\frac{zM'(z)}{M(z)} \prec_{\mathcal{F}} q(z), \; z \in U. \tag{27}$$

Since q satisfies the differential Equation (10), q is the fuzzy best dominant. □

Remark 4. *Using particular expressions for the fuzzy best dominant q, sufficient conditions for starlikeness of the operator $M(z)$ given by (7) can be obtained.*

If in Theorem 1, function $q(z) = \frac{1-z}{1+z}$ is considered the following corollary is obtained.

Corollary 1. *For $\beta, \gamma \in \mathbb{C}$, $\beta > 1$, $\gamma > 0$, let the fuzzy function $F : \mathbb{C} \to [0,1]$ given by (9) and consider a holomorphic function in U given by the equation*

$$h(z) = \frac{(\beta-1)(1-z^2)+1-4z+z^2}{(\beta-1)(1+z)^2+1-z^2},$$

$h(0) = q(0) = 1$, where function $q(z) = \frac{1-z}{1+z}$ is a univalent solution in U, which satisfies the fuzzy differential subordination

$$\frac{|1-z|}{|1+z|+|1-z|} \leq \frac{|(\beta-1)(1-z^2)+1-4z+z^2|}{\left|(\beta-1)(1+z)^2+1-z^2\right|+|(\beta-1)(1-z^2)+1-4z+z^2|},$$

i.e.,

$$\frac{1-z}{1+z} \prec_{\mathcal{F}} \frac{(\beta-1)(1-z^2)+1-4z+z^2}{(\beta-1)(1+z)^2+1-z^2}, \ z \in U.$$

Consider $\phi(u,v;z)$ to be a confluent (or Kummer) hypergeometric operator given by (5), and the operator $M(z)$ given by (7).

If $1+\gamma\frac{z\phi'(u,v;z)}{\phi(u,v;z)}$ is analytic in U, and

$$\frac{\left|1+\gamma\frac{z\phi'(u,v;z)}{\phi(u,v;z)}\right|}{1+\left|1+\gamma\frac{z\phi'(u,v;z)}{\phi(u,v;z)}\right|} \leq \frac{|(\beta-1)(1-z^2)+1-4z+z^2|}{\left|(\beta-1)(1+z)^2+1-z^2\right|+|(\beta-1)(1-z^2)+1-4z+z^2|}$$

$$\text{i.e., } 1+\gamma\frac{z\phi'(u,v;z)}{\phi(u,v;z)} \prec_{\mathcal{F}} \frac{(\beta-1)(1-z^2)+1-4z+z^2}{(\beta-1)(1+z)^2+1-z^2},$$

then

$$\frac{\left|\frac{zM'(z)}{M(z)}\right|}{1+\left|\frac{zM'(z)}{M(z)}\right|} \leq \frac{\left|\frac{1-z}{1+z}\right|}{1+\left|\frac{1-z}{1+z}\right|} \ \text{i.e., } \frac{zM'(z)}{M(z)} \prec_{\mathcal{F}} \frac{1-z}{1+z}, \ z \in U, \text{ or } M \in \mathcal{S}^*$$

and $q(z) = \frac{1-z}{1+z}$ is the fuzzy best dominant.

Proof. By using the function $q(z) = \frac{1-z}{1+z}$ in relation (27) from the proof of Theorem 1, the following fuzzy subordination is obtained:

$$\frac{zM'(z)}{M(z)} \prec_{\mathcal{F}} q(z) = \frac{1-z}{1+z}, \ z \in U. \tag{28}$$

Since $\text{Re}\left(\frac{zq''(z)}{q'(z)}+1\right) = \frac{1-\rho^2}{1+2\rho\cos\alpha+\rho^2} > 0$, $0<\rho<1$, the q is convex, and $\text{Re}\frac{1-z}{1+z} > 0$, $z \in U$, differential subordination (28) is equivalent to

$$\text{Re}\frac{zM'(z)}{M(z)} > \text{Re}\frac{1-z}{1+z} > 0, \ z \in U, \ \text{ hence } M \in \mathcal{S}^*. \tag{29}$$

□

Remark 5. *Using the convex function $q(z) = \frac{1+z}{1-z}$ as the fuzzy best dominant in Theorem 1, sufficient conditions for the convexity of the operator $M(z)$ given by (7) can be obtained as a corollary.*

Corollary 2. *For $\beta, \gamma \in \mathbb{C}$, $\beta > 1$, $\gamma > 0$, let the fuzzy function $F: \mathbb{C} \to [0,1]$ given by (9) and consider a holomorphic function in U given by the equation*

$$h(z) = \frac{(\beta-1)(1-z^2)+\gamma(1+z)^2}{(\beta-1)(1-z)^2+\gamma(1-z^2)},$$

$h(0) = q(0) = 1$, where function $q(z) = \frac{1+z}{1-z}$ is a univalent solution in U which satisfies the fuzzy differential subordination

$$\frac{|1+z|}{|1-z|+|1+z|} \leq \frac{\left|(\beta-1)(1-z^2)+\gamma(1+z)^2\right|}{\left|(\beta-1)(1-z)^2+\gamma(1-z^2)\right|+\left|(\beta-1)(1-z^2)+\gamma(1+z)^2\right|},$$

i.e.,

$$\frac{1+z}{1-z} \prec_{\mathcal{F}} \frac{(\beta-1)(1-z^2)+\gamma(1+z)^2}{(\beta-1)(1-z)^2+\gamma(1-z^2)}, \ z \in U.$$

Consider $\phi(u,v;z)$ the confluent (or Kummer) hypergeometric operator given by (5), and the operator $M(z)$ given by (7).

If

$$1+(\gamma-1)\frac{z\phi'(u,v;z)}{\phi(u,v;z)}+\frac{(\gamma+1)z\phi'(u,v;z)+z^2\gamma\phi''(u,v;z)}{\phi(u,v;z)+\gamma z\phi'(u,v;z)}=E(u,v;z)$$

is analytic in U, and

$$\frac{|E(u,v;z)|}{1+|E(u,v;z)|}\leq\frac{\left|(\beta-1)(1-z^2)+\gamma(1+z)^2\right|}{\left|(\beta-1)(1-z)^2+\gamma(1-z^2)\right|+\left|(\beta-1)(1-z^2)+\gamma(1+z)^2\right|} \quad (30)$$

$$\text{i.e., } E(u,v;z)\prec_{\mathcal{F}}\frac{(\beta-1)(1-z^2)+\gamma(1+z)^2}{(\beta-1)(1-z)^2+\gamma(1-z^2)},$$

then

$$\frac{\left|1+\frac{zM''(z)}{M'(z)}\right|}{1+\left|1+\frac{zM''(z)}{M'(z)}\right|}\leq\frac{\left|\frac{1+z}{1-z}\right|}{1+\left|\frac{1+z}{1-z}\right|}\ \text{ i.e., } 1+\frac{zM''(z)}{M'(z)}\prec_{\mathcal{F}}\frac{1+z}{1-z},\ z\in U, \text{ or } M\in\mathcal{K}$$

and $q(z)=\frac{1+z}{1-z}$ is the fuzzy best dominant.

Proof. Differentiating relation (14), from the proof of Theorem 1, we have

$$\beta M'(z)+zM''(z)=\beta\phi^{\gamma-1}(u,v;z)\left[\phi(u,v;z)+\gamma z\phi'(u,v;z)\right],$$

which is equivalent to

$$M'(z)\left[(\beta-1)+1+\frac{zM''(z)}{M'(z)}\right]=\beta\phi^{\gamma-1}(u,v;z)\left[\phi(u,v;z)+\gamma z\phi'(u,v;z)\right] \quad (31)$$

Let

$$1+\frac{zM''(z)}{M'(z)}=p(z), z\in U, p(0)=1. \quad (32)$$

By replacing (32) in (31) we obtain

$$M'(z)[(\beta-1)+p(z)]=\beta\phi^{\gamma-1}(u,v;z)\left[\phi(u,v;z)+\gamma z\phi'(u,v;z)\right]. \quad (33)$$

Differentiating relation (33) we get

$$\frac{zM''(z)}{M'(z)}+\frac{zp'(z)}{\beta-1+p(z)}=(\gamma-1)\frac{z\phi'(u,v;z)}{\phi(u,v;z)}+\frac{z(\gamma+1)\phi'(u,v;z)+z^2\gamma\phi''(u,v;z)}{\phi(u,v;z)+\gamma z\phi'(u,v;z)}.$$

After some computations, we have

$$1+\frac{zM''(z)}{M'(z)}+\frac{zp'(z)}{\beta-1+p(z)}=1+(\gamma-1)\frac{z\phi'(u,v;z)}{\phi(u,v;z)}+\frac{z(\gamma+1)\phi'(u,v;z)+z^2\gamma\phi''(u,v;z)}{\phi(u,v;z)+\gamma z\phi'(u,v;z)}. \quad (34)$$

Using relation (32) in (34) we can write

$$p(z)+\frac{zp'(z)}{\beta-1+p(z)}=E(u,v;z),\ z\in U. \quad (35)$$

By considering (35) in (30), the following inequality emerges

$$\frac{\left|p(z)+\frac{zp'(z)}{\beta-1+p(z)}\right|}{1+\left|p(z)+\frac{zp'(z)}{\beta-1+p(z)}\right|} \leq \frac{\left|(\beta-1)(1-z^2)+\gamma(1+z)^2\right|}{\left|(\beta-1)(1-z)^2+\gamma(1-z^2)\right|+\left|(\beta-1)(1-z^2)+\gamma(1+z)^2\right|}. \quad (36)$$

In order to obtain the expected result, Lemma 1 will be used. For that, let $\psi : \mathbb{C}^2 \times U \to \mathbb{C}$, given by (21) and for $r = p(z)$ and $s = zp'(z)$ from relation (35), we have

$$\psi(p(z), zp'(z)) = E(u, v; z), \ \ z \in U. \quad (37)$$

Using (37) in (30), we get

$$\frac{|\psi(p(z), zp'(z))|}{1+|\psi(p(z), zp'(z))|} \leq \frac{\left|(\beta-1)(1-z^2)+\gamma(1+z)^2\right|}{\left|(\beta-1)(1-z)^2+\gamma(1-z^2)\right|+\left|(\beta-1)(1-z^2)+\gamma(1+z)^2\right|}. \quad (38)$$

Using Lemma 1, for $\delta = \beta - 1, \omega = \gamma \neq 0$, we have

$$\psi(p(z), zp'(z)) \prec_{\mathcal{F}} h(z) = \frac{(\beta-1)(1-z^2)+\gamma(1+z)^2}{(\beta-1)(1-z)^2+\gamma(1-z^2)}, \quad (39)$$

which, according to Lemma 1, implies

$$p(z) \prec_{\mathcal{F}} q(z) = \frac{1+z}{1-z}, \ z \in U. \quad (40)$$

Using in (40) relation (32) we have

$$1 + \frac{zM''(z)}{M'(z)} \prec_{\mathcal{F}} q(z) = \frac{1+z}{1-z}, \ z \in U. \quad (41)$$

Since q is convex, relation (41) is equivalent to

$$\mathrm{Re}\left(1 + \frac{zM''(z)}{M'(z)}\right) > \mathrm{Re}\frac{1+z}{1-z} > 0, \ z \in U, \text{hence } M \in \mathcal{K}. \quad (42)$$

□

Remark 6. *Using Lemma 3 and the convexity property proved for the operator $M(z)$, the following corollary can be stated giving the property of the integral operator $M(z)$ given by (7) to be starlike of order $\frac{1}{2}$.*

Corollary 3. *Let the operator $M(z)$ be given by (7). Then $M(z) \in \mathcal{K}$ implies $M(z) \in \mathcal{S}^*\left(\frac{1}{2}\right)$.*

Proof. Since $\mathrm{Re}\left[1 + \frac{zM''(z)}{M'(z)}\right] > 0$ using Lemma 3 we obtain $\mathrm{Re}\frac{zM'(z)}{M(z)} > \frac{1}{2}$, hence $M \in \mathcal{S}^*\left(\frac{1}{2}\right)$. □

Remark 7. *Using function $q(z) = \frac{1-2z}{1+z}$ as fuzzy best dominant in Theorem 1, we get the following corollary, which gives a sufficient condition for the operator $M(z)$ given by (7) to be convex of order $\left(-\frac{1}{2}\right)$.*

Corollary 4. *For $\beta, \gamma \in \mathbb{C}$, $\beta > 1$, $\gamma > 0$, let the fuzzy function $F : \mathbb{C} \to [0,1]$ given by (9) and consider the holomorphic function in U given by the equation*

$$h(z) = \frac{(\beta-1)(1+z)(1-2z) + (1-2z)^2 - 3z}{(\beta-1)(1+z)^2 + (1-2z)(1+z)},$$

$h(0) = q(0) = 1$, where function $q(z) = \frac{1-2z}{1+z}$ is a univalent solution in U which satisfies the fuzzy differential subordination

$$\frac{|1-2z|}{|1+z| + |1-2z|} \leq$$

$$\frac{\left|(\beta-1)(1+z)(1-2z) + (1-2z)^2 - 3z\right|}{\left|(\beta-1)(1+z)^2 + (1-2z)(1+z)\right| + \left|(\beta-1)(1+z)(1-2z) + (1-2z)^2 - 3z\right|},$$

i.e.,

$$\frac{1-2z}{1+z} \prec_{\mathcal{F}} \frac{(\beta-1)(1+z)(1-2z) + (1-2z)^2 - 3z}{(\beta-1)(1+z)^2 + (1-2z)(1+z)}, \quad z \in U.$$

Consider the confluent (or Kummer) hypergeometric function $\phi(u,v;z)$ given by (5), and the operator $M(z)$ given by (7).

If

$$E(u,v;z) = 1 + (\gamma-1)\frac{z\phi'(u,v;z)}{\phi(u,v;z)} + \frac{(\gamma+1)z\phi'(u,v;z) + z^2\gamma\phi''(u,v;z)}{\phi(u,v;z) + \gamma z\phi'(u,v;z)}$$

is analytic in U, and

$$\frac{|E(u,v;z)|}{1 + |E(u,v;z)|} \leq$$

$$\frac{\left|(\beta-1)(1+z)(1-2z) + (1-2z)^2 - 3z\right|}{\left|(\beta-1)(1+z)^2 + (1-2z)(1+z)\right| + \left|(\beta-1)(1+z)(1-2z) + (1-2z)^2 - 3z\right|}$$

$$\text{i.e., } E(u,v;z) \prec_{\mathcal{F}} \frac{(\beta-1)(1+z)(1-2z) + (1-2z)^2 - 3z}{(\beta-1)(1+z)^2 + (1-2z)(1+z)},$$

then

$$\frac{\left|1 + \frac{zM''(z)}{M'(z)}\right|}{1 + \left|1 + \frac{zM''(z)}{M'(z)}\right|} \leq \frac{\left|\frac{1-2z}{1+z}\right|}{1 + \left|\frac{1-2z}{1+z}\right|} \text{ i.e., } 1 + \frac{zM''(z)}{M'(z)} \prec_{\mathcal{F}} \frac{1-2z}{1+z}, \ z \in U, \text{ or } M \in \mathcal{K}\left(-\frac{1}{2}\right),$$

and $q(z) = \frac{1-2z}{1+z}$ is the fuzzy best dominant.

Proof. Using in (41) $q(z) = \frac{1-2z}{1+z}$, the relation becomes

$$1 + \frac{zM''(z)}{M'(z)} \prec_{\mathcal{F}} \frac{1-2z}{1+z}, \ z \in U. \tag{43}$$

Since $\mathrm{Re}\left(\frac{zq''(z)}{q'(z)} + 1\right) = Re\frac{1-z}{1+z} > 0$, we have that q is convex and $\mathrm{Re}\frac{1-2z}{1+z} = -\frac{1}{2}$. Then relation (43) is equivalent to

$$\mathrm{Re}\left(1 + \frac{zM''(z)}{M'(z)}\right) > \mathrm{Re}\frac{1-2z}{1+z} > -\frac{1}{2}, \ z \in U, \ \text{hence } M \in \mathcal{K}\left(-\frac{1}{2}\right). \tag{44}$$

□

Remark 8. *Using Corollary 4 we next prove that the operator $M(z)$ given by (7) is close-to-convex in U.*

Theorem 2. *Let $M(z)$ be given by (7) satisfying the condition $Re\left(1+\frac{zM''(z)}{M'(z)}\right) > -\frac{1}{2}, z \in U$. Then $M(z)$ is close-to-convex.*

Proof. For obtaining the desired result, Lemma 2 is applied. We calculate

$$\int_{\theta_1}^{\theta_2} \text{Re}\left(1+\frac{zM''(z)}{M'(z)}\right)d\theta = \int_{\theta_1}^{\theta_2} -\frac{1}{2}d\theta = -\frac{1}{2}\theta|_{\theta_1}^{\theta_2} = -\frac{1}{2}(\theta_2-\theta_1) > -\frac{1}{2}2\pi = -\pi.$$

From Lemma 2, this means that $M(z) \in \mathcal{C}$. □

Example 1. *Let $u=-1$, $v=\frac{1}{3}$, $\phi\left(-1,\frac{1}{3};z\right) = 1+3iz$, and*
$\frac{z\phi'\left(-1,\frac{1}{3};z\right)}{\phi\left(-1,\frac{1}{3};z\right)} = \frac{3iz}{1+3iz} = 1 - \frac{1-3\sin\alpha - 3i\cos\alpha}{10-6\sin\alpha}$.
For $\beta=2$, $\gamma=1$, we calculate

$$M(z) = \frac{2}{z}\int_0^z (1+3it)tdt = \frac{2}{z}\left(\frac{z^2}{2}+iz^3\right) = z+2iz^2,$$

and $\frac{zM'(z)}{M(z)} = 1+\frac{2iz}{1+2iz}$.

Using Corollary 1, we have:

Let $\beta=2$, $\gamma=1$, fuzzy function $F(z)=\frac{|z|}{1+|z|}$, $z\in U$, and $h(z)=\frac{1-2z}{1+z}$, $h(0)=1$, with univalent solution $q(z)=\frac{1-z}{1+z}$, $z\in U$, which satisfy the fuzzy differential subordination

$$\frac{|1-z|}{|1+z|+|1-z|} \le \frac{|1-2z|}{|1+z|+|1-2z|}, \text{ i.e., } q(z)=\frac{1-z}{1+z} \prec_{\mathcal{F}} \frac{1-2z}{1+z} = h(z),$$

and $\phi\left(-1,\frac{1}{3};z\right) = 1+3iz$, Kummer hypergeometric function.

If $\frac{1+6iz}{1+3iz}$ is holomorphic in U, and

$$\frac{|1+6iz|}{|1+3iz|+|1+6iz|} \le \frac{|1-2z|}{|1+z|+|1-2z|}, \text{ i.e., } \frac{1+6iz}{1+3iz} \prec_{\mathcal{F}} \frac{1-2z}{1+z},$$

then

$$\frac{|3iz|}{|1+3iz|+|3iz|} \le \frac{|1-z|}{|1+z|+|1-z|}, \text{ i.e., } \frac{3iz}{1+3iz} \prec_{\mathcal{F}} \frac{1-z}{1+z}, \ z\in U.$$

In

$$Re\frac{z\phi'\left(-1,\frac{1}{3};z\right)}{\phi\left(-1,\frac{1}{3};z\right)} = Re\left(1-\frac{1-3\sin\alpha-3i\cos\alpha}{10-6\sin\alpha}\right) = 1-\frac{1-3\sin\alpha}{10-6\sin\alpha}$$

$$= \frac{10-6\sin\alpha-1+3\sin\alpha}{4+6(1-\sin\alpha)} = \frac{9-3\sin\alpha}{4+6(1-\sin\alpha)} = \frac{6+3(1-\sin\alpha)}{4+6(1-\sin\alpha)} > 0.$$

4. Conclusions

Since the study of operators using the fuzzy differential subordination theory presents interest at this time, and many new and interesting results have been obtained recently, the research regarding this topic is further conducted in this paper. A new hypergeometric integral operator $M(z)$ is introduced in this paper in relation (7) by using a confluent (or Kummer) hypergeometric function and, having as inspiration, the operator studied by Miller, Mocanu and Reade in 1978 [28] and taking specific values for parameters $\beta, \gamma, \alpha, \delta$

involved in its definition. Using the notion of fuzzy differential subordination and results related to it, in the first theorem proved, the fuzzy best dominant of a certain fuzzy differential subordination is given. Using particular functions as fuzzy best dominants, several corollaries are stated, giving sufficient conditions for the operator $M(z)$ to be starlike, convex, starlike of order $\frac{1}{2}$ and convex of order $(-\frac{1}{2})$, respectively. The second theorem proved shows the property of the operator $M(z)$ to be close-to-convex. For further study, the properties already proved, related to starlikeness and convexity of the operator $M(z)$, could inspire applications in introducing special classes of analytic functions. The operator could also be studied using the dual theory of fuzzy differential superordination, possibly obtaining sandwhich-type theorems, connecting with the present results a usual outcome in geometric function theory. Since particular values for parameters have been used for defining this operator, it might be interesting to try using other values for obtaining certain potentially interesting operators. It being well-known how hypergeometric functions have numerous applications in physics, engineering and statistics, applications of the operators involving those functions could prove useful in other disciplines. The theory of fuzzy differential subordination is still very new and one cannot predict what applications in real life or other scientific domains it might have. Those are subjects for investigation in long-term future studies.

Funding: This research received no external funding.

Institutional Review Board Statement: Not applicable.

Informed Consent Statement: Not applicable.

Data Availability Statement: Not applicable.

Conflicts of Interest: The author declares no conflict of interest.

References

1. Zadeh, L.A. Fuzzy Sets. *Inf. Control* **1965**, *8*, 338–353. [CrossRef]
2. Dzitac, S.; Nădăban, S. Soft Computing for Decision-Making in Fuzzy Environments: A Tribute to Professor Ioan Dzitac. *Mathematics* **2021**, *9*, 1701. [CrossRef]
3. Dzitac, I. Zadeh's Centenary. *Int. J. Comput. Commun. Control* **2021**, *16*, 4102. [CrossRef]
4. Oros, G.I.; Oros, G. The notion of subordination in fuzzy sets theory. *Gen. Math.* **2011**, *19*, 97–103.
5. Miller, S.S.; Mocanu, P.T. Second order-differential inequalities in the complex plane. *J. Math. Anal. Appl.* **1978**, *65*, 298–305. [CrossRef]
6. Miller, S.S.; Mocanu, P.T. Differential subordinations and univalent functions. *Michig. Math. J.* **1981**, *28*, 157–171. [CrossRef]
7. Oros, G.I.; Oros, G. Fuzzy differential subordination. *Acta Univ. Apulensis* **2012**, *3*, 55–64.
8. Oros, G.I.; Oros, G. Dominants and best dominants in fuzzy differential subordinations. *Stud. Univ. Babeş-Bolyai Math.* **2012**, *57*, 239–248.
9. Oros, G.I.; Oros, G. Briot-Bouquet fuzzy differential subordination. *An. Univ. Oradea Fasc. Mat.* **2012**, *19*, 83–87.
10. Dzitac, I.; Filip, F.G.; Manolescu, M.J. Fuzzy Logic Is Not Fuzzy: World-renowned Computer Scientist Lotfi A. Zadeh. *Int. J. Comput. Commun. Control* **2017**, *12*, 748–789. [CrossRef]
11. Atshan, W.G.; Hussain, K.O. Fuzzy Differential Superordination. *Theory Appl. Math. Comput. Sci.* **2017**, *7*, 27–38.
12. Alb Lupaş, A. A note on special fuzzy differential subordinations using generalized Sălăgean operator and Ruscheweyh derivative. *J. Comput. Anal. Appl.* **2013**, *15*, 1476–1483.
13. Alb Lupaş, A.; Oros, G. On special fuzzy differential subordinations using Sălăgean and Ruscheweyh operators. *Appl. Math. Comput.* **2015**, *261*, 119–127.
14. Venter, A.O. On special fuzzy differential subordination using Ruscheweyh operator. *Analele Univ. Oradea Fasc. Mat.* **2015**, *22*, 167–176.
15. Eş, A.H. On fuzzy differential subordination. *Math. Moravica* **2015**, *19*, 123–129.
16. Majeed, A.H. Fuzzy differential subordinations properties of analytic functions involving generalized differential operator. *Sci. Int. Lahore* **2018**, *30*, 297–302.
17. Ibrahim, R.W. On the subordination and superordination concepts with applications. *J. Comput. Theor. Nanosci.* **2017**, *14*, 2248–2254. [CrossRef]
18. Altai, N.H.; Abdulkadhim, M.M.; Imran, Q.H. On first order fuzzy differential superordination. *J. Sci. Arts* **2017**, *173*, 407–412.
19. Thilagavathi, K. Fuzzy subordination and superordination results for certain subclasses of analytic functions associated with Srivastava-Attitya operator. *Int. J. Pure Appl. Math.* **2018**, *118*, 921–929.

20. El-Deeb, S.M.; Alb Lupaş, A. Fuzzy differential subordinations associated with an integral operator. *An. Univ. Oradea Fasc. Mat.* **2020**, *27*, 133–140.
21. Alb Lupaş, A.; Oros, G.I. New Applications of Sălăgean and Ruscheweyh Operators for Obtaining Fuzzy Differential Subordinations. *Mathematics* **2021**, *9*, 2000. [CrossRef]
22. El-Deeb, S.M.; Oros, G.I. Fuzzy differential subordinations connected with the linear operator. *Math. Bohem.* **2021**, 1–10. [CrossRef]
23. Oros, G.I. New fuzzy differential subordinations. *Commun. Fac. Sci. Univ. Ank. Ser. A1 Math. Stat.* **2021**, *70*, 229–240.
24. Srivastava, H.M.; El-Deeb, S.M. Fuzzy Differential Subordinations Based upon the Mittag-Leffler Type Borel Distribution. *Symmetry* **2021**, *13*, 1023. [CrossRef]
25. Miller, S.S.; Mocanu, P.T. *Differential Subordinations. Theory and Applications*; Marcel Dekker, Inc.: New York, NY, USA; Basel, Switzerland, 2000.
26. de Branges, L. A proof of the Bieberbach conjecture. *Acta Math.* **1985**, *154*, 137–152. [CrossRef]
27. Srivastava, H.M. Operators of basic (or q-) calculus and fractional q-calculus and their applications in geometric function theory of complex analysis. *Iran. J. Sci. Technol. Trans. A Sci.* **2020**, *44*, 327–344. [CrossRef]
28. Miller, S.S.; Mocanu, P.T.; Reade, M.O. Starlike integral operators. *Pac. J. Math.* **1978**, *79*, 157–168. [CrossRef]
29. Al-Janaby, H.F.; Ghanim, F. A subclass of Noor-type harmonic p-valent functions based on hypergeometric functions. *Kragujev. J. Math.* **2021**, *45*, 499–519. [CrossRef]
30. Alb Lupaş, A.; Oros, G.I. Differential Subordination and Superordination Results Using Fractional Integral of Confluent Hypergeometric Function. *Symmetry* **2021**, *13*, 327. [CrossRef]
31. Ghanim, F.; Al-Shaqsi, K.; Darus, M.; Al-Janaby, H.F. Subordination Properties of Meromorphic Kummer Function Correlated with Hurwitz–Lerch Zeta-Function. *Mathematics* **2021**, *9*, 192. [CrossRef]
32. Oros, G.I. Study on new integral operators defined using confluent hypergeometric function. *Adv. Differ. Equ.* **2021**, *2021*, 342. [CrossRef]
33. Mocanu, P.T.; Bulboacă, T.; Sălăgean, Ş.G. *Geometric Theory of Analytic Functions*; Casa Cărţii de Ştiinâă: Cluj-Napoca, Romania, 1999.

Σ mathematics

MDPI

Article

Multigranulation Roughness of Intuitionistic Fuzzy Sets by Soft Relations and Their Applications in Decision Making

Muhammad Zishan Anwar [1], Shahida Bashir [1], Muhammad Shabir [2] and Majed G. Alharbi [3,*]

1 Department of Mathematics, University of Gujrat, Gujrat 50700, Pakistan; 18026109-002@uog.edu.pk (M.Z.A.); shahida.bashir@uog.edu.pk (S.B.)
2 Department of Mathematics, Quaid-i-Azam University, Islamabad 45320, Pakistan; mshabirbhatti@qau.edu.pk
3 Department of Mathematics, College of Science and Arts, Almithnab, Qassim University, Buridah 51931, Saudi Arabia
* Correspondence: 3661@qu.edu.sa

Abstract: Multigranulation rough set (MGRS) based on soft relations is a very useful technique to describe the objectives of problem solving. This MGRS over two universes provides the combination of multiple granulation knowledge in a multigranulation space. This paper extends the concept of fuzzy set Shabir and Jamal in terms of an intuitionistic fuzzy set (IFS) based on multi-soft binary relations. This paper presents the multigranulation roughness of an IFS based on two soft relations over two universes with respect to the aftersets and foresets. As a result, two sets of IF soft sets with respect to the aftersets and foresets are obtained. These resulting sets are called lower approximations and upper approximations with respect to the aftersets and with respect to the foresets. Some properties of this model are studied. In a similar way, we approximate an IFS based on multi-soft relations and discuss their some algebraic properties. Finally, a decision-making algorithm has been presented with a suitable example.

Keywords: intuitionistic fuzzy set; soft relation; multigranulation roughness; decision making

MSC: 03E72; 20F10

Citation: Anwar, M.Z.; Bashir, S.; Shabir, M.; Alharbi, M.G. Multigranulation Roughness of Intuitionistic Fuzzy Sets by Soft Relations and Their Applications in Decision Making. *Mathematics* **2021**, *9*, 2587. https://doi.org/10.3390/math9202587

Academic Editor: Sorin Nadaban

Received: 14 September 2021
Accepted: 8 October 2021
Published: 15 October 2021

1. Introduction

In our real world, many problems naturally involve uncertainty. This uncertainty can be observed in several fields, such as environmental science, medical science, economics and engineering. Researchers are active and interested to address uncertainty. In this respect, many theories have been presented, such as the probability theory, fuzzy set (FS) theory, rough set (RS) theory, intuitionistic fuzzy set (IFS) theory and soft set (SS) theory etc.

Fuzzy set (FS) proposed by Zadeh in [1] is a framework to address partial truth, uncertainty and impreciseness. Zadeh's FS is a very crucial, innovative and ingenious set because of its importance in multiple research dimensions. Often, we are encountered by ill-defined situations which are addressed through quantitative expressions. To evaluate better results from these critical situations, the FS is much useful by using qualitative expressions due to its degree of membership. The FS represents degree of membership for each element of the universe of a discourse to a subset of it, and later on, Attanassov presented intuitionistic fuzzy set (IFS) [2] which avails the opportunity to model the problem precisely based on the observations and treat more accurately to uncertainty quantification. Attanassov discussed the literature based on theory and fundamentals of IFSs in [3]. An IFS is a very useful concept with its applications in many different fields, such as electoral system, market prediction, machine learning, pattern recognition, career determination and medical diagnosis [4]. The description in terms of membership degree only in many cases is insufficient because the presence of non-membership degree is helpful to deal with uncertainty in good manner.

Molodtsov [5] presented an untraditional approach known as soft set (SS) theory for handling the vagueness and uncertainty. A collection of approximate descriptions of an element in terms of parameters by a set-valued map is known as a soft set. This theory has become a successful approach to different problems in different fields due to its rich operations. In decision-making problems, this is an applicable tool using the RSs [6]. Many researchers hybridized the models of SSs with different applicable theories [7–9]. Maji et al. defined Fuzzy SS (FSS) and Intuitionistic fuzzy soft set (IFSS) [10,11]. After that, several extensions of SSs have been presented, such as the vague SS [12], the soft RS (SRS) [13,14], the generalized FSS [15], the trapezoidal FSS [16], interval-valued FSS [17]. Agarwal built a framework of the generalized IFSS [18]. Feng et al. [19] pointed out some errors in generalized IFSS [18] and rebuilt the generalized IFSS. Many authors combined the concepts of IFSs and fuzzy RSs (FRSs). Samanta and Mondal [20] presented the IF rough set (IFRS) model. In [21], the combination of RS and FS has been studied. To overcome the unnaturalness of FRSs, Sang et al. [22] proposed a newly defined IFRS model.

Pawlak presented rough sets (RSs) to deal with incomplete data, vagueness and uncertainty [6,23]. To solve different problems based on incomplete data, many researchers showed interest in RSs. The RS theory is an untraditional technique to discuss data investigation, representation of vague or inexact data and reasoning based reduction of vague data [24]. Recently, researchers have investigated RSs in the light of dataset features, varieties of remedy and procedure control. Many extensions of RSs have been presented for many requirements, such as the RS model based on reflexive relations, equivalence relations and tolerance relations, FRS model, SRS model and rough FS (RFS) model [25–27]. As it is commonly known, many problems have different universes of discourse [28], such as the objections of customers and their solutions in enterprise management, the characteristics of customers and the features of products in personalized marketing, the mechanical defects and their solutions in machine diagnosis, the symptoms of diseases and drugs in diseased diagnosis. To formalize these problems, the RS models have been generalized over two universes [29–31]. Pie et al. [32] built a framework of the RS model on algebraic characteristics over two universes. According to the inter-relationship between two universes, Liu et al. [33] connected the graded RS with appropriate parameters. Ma and Sun [34] proposed a framework of probability RS to deal with impreciseness [19,35].

Granular computing is very useful to describe the objectives of a problem solving through multiple binary relations [36]. Under a single granulation, a set is characterized by lower and upper approximations in the light of granular computing. By using multiple equivalence relations, multigranulation rough set (MGRS) approximations have been investigated. Rauszer [37] presented a framework of a multi-agent system based on an equivalence relation where each agent has its own knowledge base. Khan and Banerjee [38,39] considered the agent as a "source" in a more general setting but many other scholars considered "agent" as granulation [40–42]. Khan et al. [38,39] presented two approximation operators in terms of multiple source approximation system [35]. In the same sense, Qian et al. [43] presented the MGRS theory. There are two types of MGRS, named optimistic MGRS (OMGRS) and pessimistic MGRS (PMGRS) [41]. Later, many extensions of MGRS have been presented by different authors [44,45]. The real dataset involving multiple and overlapping knowledge has been dealt with by presenting MGRS and covering RSs. These two special models have been generalized as many hybrid models such as covering MGRS [43] and MGRS based on multiple equivalence relations [41] etc. Liu et al. [46] used a covering approximation space and presented four types of covering MGRS models. Later on, Xu et al. presented a covering MGRS based on order relations [47], fuzzy compatible relations [48] and generalized relations [49] by relaxing the conditions of equivalence relations. After that, many researchers proposed different MGRS models according to different dataset needs. Dou et al. [50] presented a useful MGRS model for variable-precision and discussed its properties. Ju et al. [51] proposed a heuristic algorithm using their newly defined variable-precision MGRS model for computing reduction. Feng et al. [52] presented a three-way decision-based type-1 variable-precision multigranulation decision-theoretic fuzzy rough set [53]. In management science and various professional fields, the MGRS showed its importance and extensions of MGRS have made their role with respect to the nature of problems [54–58]. Qian et al. discussed the risk attitudes by presenting OMGRS and PMGRS models using multiple binary relations. Huang et al. [59] combined MGRS and IFS and presented an IFMGRS model. Pang et al. [60] combined MGRS with three-way decision making and proposed a multi-criteria decision-making (MCGDM) model. Different experts have different experiences and expertises to solve different decision-making problems. Better decisions can be made by taking opinions of multiple experts compared with taking one expert's opinion only. In view of this logic, Zadeh [36] introduced granular computing knowledge through multiple relations. Sun and Ma [61] proposed the MGRS model over two universes. For selective dataset approximation, Tan et al. [35] presented the MGRS model with granularity selection algorithm [56]. Xu et al. [62] combined Pawlak RS model, FRS model and MGRS model in terms of granular computing and proposed multi-granulation fuzzy rough set (MGFRS) model. Recently, Shabir et al. [63] proposed a useful model of MGRS with multi-soft binary relations. After that, Shabir et al. [64] extended that optimistic MGRS in terms of FS which is called optimistic multigranulation fuzzy rough set (OMGFRS). The existing MGRS models have obvious disadvantages regarding FSs.

1. The existing MGRS models with FSs are unable to manage the real life situations where only degree of membership is discussed;
2. Decision experts have hesitation to make better decision due to no consideration of their own subjective consciousness.

To manage these above critical situations, we extended the model of MGFRS based on soft binary relations in [64] in terms of IFS. We used IFSs instead of FSs to present an optimistic multigranulation intuitionistic fuzzy rough set (OMGIFRS) model.

The organization of the remaining paper is as follows. In Section 2, some basic definitions and fundamental concepts of FS, IFS, RS, SS, FRS, IFSS, MGRS, and soft relation are given. Section 3 presents the optimistic granulation roughness of an IFS based on two soft relations over two universes with their basic properties and examples. OMGIFRS over two universes and their properties are discussed in Section 4. Section 5 presents the decision making algorithm with a practical example about decision making problems. In Section 6, we made a comparison of our proposed model with other existing theories. Finally, we conclude our research work in Section 7.

2. Preliminaries and Basic Concepts

In this section, some fundamental notions about IFS, RS, MGRS, SS, soft binary relation, and IFSS are given. Throughout this paper, U_1 and U_2 represent two non-empty finite sets unless stated otherwise.

Definition 1 ([2]). *Let U be a non-empty universe. An IFS B in the universe U is an object having the form $B = \{\langle x, \mu_B(x), \gamma_B(x)\rangle : x \in U\}$, where $\mu_B : U \to [0,1]$ and $\gamma_B : U \to [0,1]$ satisfying $0 \leq \mu_B(x) + \gamma_B(x) \leq 1$ for all $x \in U$. The values $\mu_B(x)$ and $\gamma_B(x)$ are called degree of membership and degree of non-membership of $x \in U$ to B, respectively. The number $\pi_B(x) = 1 - \mu_B(x) - \gamma_B(x)$ is called the degree of hesitancy of $x \in U$ to B. The collection of all IFSs in U is denoted by $IF(U)$. In the remaining paper, we shall write an IFS by $B = \langle \mu_B, \gamma_B\rangle$ instead of $B = \{\langle x, \mu_B(x), \gamma_B(x)\rangle : x \in U\}$. Let $B = \langle \mu_B, \gamma_B\rangle$ and $B_1 = \langle \mu_{B_1}, \gamma_{B_1}\rangle$ be two IFSs in U. Then, $B \subseteq B_1$ if and only if $\mu_B(x) \leq \mu_{B_1}(x)$ and $\gamma_{B_1}(x) \leq \gamma_B(x)$ for all $x \in U$. Two IFSs B and B_1 are said to be equal if and only if $B \subseteq B_1$ and $B_1 \subseteq B$.*

Definition 2 ([2]). *The union and intersection of two IFSs B and B_1 in U are denoted and defined by $B \cup B_1 = \langle \mu_B \cup \mu_{B_1}, \gamma_B \cap \gamma_{B_1}\rangle$ and $B \cap B_1 = \langle \mu_B \cap \mu_{B_1}, \gamma_B \cup \gamma_{B_1}\rangle$ where $(\mu_B \cup \mu_{B_1})(x) = \sup\{\mu_B(x), \mu_{B_1}(x)\}$, $(\gamma_B \cap \gamma_{B_1})(x) = \inf\{\gamma_B(x), \gamma_{B_1}(x)\}$, $(\mu_B \cap \mu_{B_1})(x) = \inf\{\mu_B(x), \mu_{B_1}(x)\}$, $(\gamma_B \cup \gamma_{B_1})(x) = \sup\{\gamma_B(x), \gamma_{B_1}(x)\}$, for all $x \in U$.*

Next, we define two special types of IFSs as:

The IF universe set $U = 1_U =< 1, 0 >$ and IF empty set $\Phi = 0_U =< 0, 1 >$, where $1(x) = 1$ and $0(x) = 0$ for all $x \in U$. The complement of an IFS $B =< \mu, \gamma >$ is denoted and defined as $B^c =< \gamma, \mu >$. See Table 1 for acronyms.

Table 1. List of acronyms.

Acronyms	Representations
FSs	Fuzzy sets
IFSs	Intuitionistic fuzzy sets
RSs	Rough sets
SSs	Soft sets
FSSs	Fuzzy soft sets
IFSSs	Intuitionistic fuzzy soft sets
SRSs	Soft rough sets
IFRSs	Intuitionistic fuzzy rough sets
FRSs	Fuzzy rough sets
RFSs	Rough fuzzy sets
MGRS	Multigranulation rough set
OMGRS	Optimistic multigranulation rough set
PMGRS	Pessimistic multigranulation rough set
IFMGRS	Intuitionistic fuzzy multigranulation rough set
MCGDM	Multi-critria group decision making
OMGFRS	Optimistic multigranulation fuzzy rough set
OMGIFRS	Optimistic multigranulation intuitionistic fuzzy rough set
IFN	Intuitionistic fuzzy number
IFV	Intuitionistic fuzzy value

For a fixed $x \in U$, the pair $(\mu_B(x), \gamma_B(x))$ is called intuitionistic fuzzy value (IFV) or intuitionistic fuzzy number (IFN). In order to define the order between two IFNs, Chen and Tan [65] introduced the score function as $S(x) = \mu_B(x) - \gamma_B(x)$ and Hong and Choi [66] defined the accuracy function as $H(x) = \mu_B(x) + \gamma_B(x)$, where $x \in U$. Xu [62,67] used both the score and accuracy functions to form the order relation between any pair of IFVs (x, y) as given below:

(a) if $S(x) > S(y)$, then $x > y$;
(b) if $S(x) = S(y)$, then
 (1) if $H(x) = H(y)$, then $x = y$;
 (2) if $H(x) < H(y)$, then $x < y$.

Definition 3 ([6])**.** *Let σ be an equivalence relation on a universe U. For any $M \subseteq U$, the Pawlak lower and upper approximations of M with respect to σ are defined by*

$$\begin{aligned} \underline{\sigma}(M) &= \{u \in U : [u]_\sigma \subseteq M\}, \\ \overline{\sigma}(M) &= \{u \in U : [u]_\sigma \cap M \neq \emptyset\}, \end{aligned}$$

where $[u]_\sigma$ is the equivalence class of u with respect to σ. The set $BM_\sigma = \overline{\sigma}(M) - \underline{\sigma}(M)$ is the boundary region of $M \subseteq U$. If $BM_\sigma(M) = \emptyset$, then M is defineable (exact), otherwise, M is rough with respect to σ.

Qian et al. [40] extended the Pawlak rough set model to the MGRS model, where the set approximations are defined by using multi-equivalence relations on a universe.

Definition 4 ([40]). *Let $\sigma_1, \sigma_2, \sigma_3, \ldots, \sigma_n$ be n equivalence relations on a universe U. For any $M \subseteq U$, the Pawlak lower and upper approximations of M are defined by*

$$\begin{aligned} \underline{M}_{\Sigma_{i=1}^{n}\sigma_i} &= \{u \in U : [u]_{\sigma_i} \subseteq M \text{ for some } i,\ 1 \leq i \leq n\}, \\ \overline{M}^{\Sigma_{i=1}^{n}\sigma_i} &= (\underline{M^c}_{\Sigma_{i=1}^{n}\sigma_i})^c, \end{aligned}$$

where $[u]_{\sigma_i}$ is the equivalence class of u with respect to σ_i.

Definition 5 ([5]). *A pair (F, A) is called an SS over U if F is a mapping given by $F : A \to P(U)$, where A is a subset of E (the set of parameters) and $P(U)$ is the power set of U. Thus, $F(e)$ is a subset of U for all $e \in A$. Hence, a SS over U is a parameterized collection of subsets of U.*

Definition 6 ([68]). *Let (σ, A) be an SS over $U_1 \times U_1$. Then, (σ, A) is called a soft binary relation on U_1. In fact, (σ, A) is a parameterized collection of binary relations on U_1, that is, we have a binary relation $\sigma(e)$ on U_1 for each parameter $e \in A$.*

Li et al. [69] presented the generalization of the soft binary relation over U_1 to U_2, as follows.

Definition 7 ([69]). *A soft binary relation (σ, A) from U_1 to U_2 is an SS over $U_1 \times U_2$, that is, $\sigma : A \to P(U_1 \times U_2)$, where A is a subset of the set of parameters E.*

Of course, (σ, A) is a parameterized collection of binary relations from U_1 to U_2. That is, for each $e \in A$, we have a binary relation $\sigma(e)$ from U_1 to U_2.

Definition 8 ([11]). *A pair (F, A) is called an IFSS over U if F is a mapping given by $F : A \to IF(U)$ and A is a subset of E (the set of parameters). Thus, $F(e)$ is an IFS in U for all $e \in A$. Hence, an IFSS over U is a parameterized collection of IF sets in U.*

Definition 9 ([11]). *For two IFSSs (F, A) and (G, B) over a common universe U, we say that (F, A) is an IF soft subset of (G, B) if (1) $A \subseteq B$ and (2) $F(e)$ is an IF subset of $G(e)$ for all $e \in A$. Two IFSSs (F, A) and (G, B) over a common universe U are said to be IF soft equal if (F, A) is an IF soft subset of (G, B) and (G, B) is an IF soft subset of (F, A). The union of two IFSSs (F, A) and (G, A) over the common universe U is the IFSS (H, A), where $H(e) = F(e) \cup G(e)$ for all $e \in A$. The intersection of two IFSSs (F, A) and (G, A) over the common universe U is the IFSS (K, A), where $K(e) = F(e) \cap G(e)$ for all $e \in A$.*

Definition 10 ([70]). *Let (σ, A) be a soft binary relation from U_1 to U_2 and $B = \langle \mu_B, \gamma_B \rangle$ be an IFS in U_2. Then, lower approximation $\underline{\sigma}^B = (\underline{\sigma}^{\mu_B}, \underline{\sigma}^{\gamma_B})$ and upper approximation $\overline{\sigma}^B = (\overline{\sigma}^{\mu_B}, \overline{\sigma}^{\gamma_B})$ of $B = \langle \mu_B, \gamma_B \rangle$ with respect to aftersets are defined as follows:*

$$\underline{\sigma}^{\mu_B}(e)(u_1) = \begin{cases} \wedge_{a \in u_1\sigma(e)} \mu_B(a) & \text{if } u_1\sigma(e) \neq \emptyset; \\ 1 & \text{if } u_1\sigma(e) = \emptyset; \end{cases}$$

$$\underline{\sigma}^{\gamma_B}(e)(u_1) = \begin{cases} \vee_{a \in u_1\sigma(e)} \gamma_B(a) & \text{if } u_1\sigma(e) \neq \emptyset; \\ 0 & \text{if } u_1\sigma(e) = \emptyset; \end{cases}$$

and

$$\overline{\sigma}^{\mu_B}(e)(u_1) = \begin{cases} \vee_{a \in u_1\sigma(e)} \mu_B(a) & \text{if } u_1\sigma(e) \neq \emptyset; \\ 0 & \text{if } u_1\sigma(e) = \emptyset; \end{cases}$$

$$\overline{\sigma}^{\gamma_B}(e)(u_1) = \begin{cases} \wedge_{a \in u_1\sigma(e)} \gamma_B(a) & \text{if } u_1\sigma(e) \neq \emptyset; \\ 1 & \text{if } u_1\sigma(e) = \emptyset, \end{cases}$$

where $u_1\sigma(e) = \{a \in U_2 : (u_1, a) \in \sigma(e)\}$ *and is called the afterset of* u_1 *for* $u_1 \in U_1$ *and* $e \in A$.

- $\underline{\sigma}^{\mu_B}(e)(u_1)$ indicates the degree to which u_1 definitely has the property e;
- $\underline{\sigma}^{\gamma_B}(e)(u_1)$ indicates the degree to which u_1 probably does not have the property e;
- $\overline{\sigma}^{\mu_B}(e)(u_1)$ indicates the degree to which u_1 probably has the property e;
- $\overline{\sigma}^{\gamma_B}(e)(u_1)$ indicates the degree to which u_1 definitely does not have the property e.

Definition 11 ([70]). *Let* (σ, A) *be a soft binary relation from* U_1 *to* U_2 *and* $B = \langle \mu_B, \gamma_B \rangle$ *be an IFS in* U_1. *Then, lower approximation* ${}^B\underline{\sigma} = ({}^{\mu_B}\underline{\sigma}, {}^{\gamma_B}\underline{\sigma})$ *and upper approximation* ${}^B\overline{\sigma} = ({}^{\mu_B}\overline{\sigma}, {}^{\gamma_B}\overline{\sigma})$ *of* $B = \langle \mu_B, \gamma_B \rangle$ *with respect to foresets are defined as follows:*

$$ {}^{\mu_B}\underline{\sigma}(e)(u_2) = \begin{cases} \wedge_{a \in \sigma(e)u_2} \mu_B(a) & \text{if } \sigma(e)u_2 \neq \emptyset; \\ 1 & \text{if } \sigma(e)u_2 = \emptyset; \end{cases} $$

$$ {}^{\gamma_B}\underline{\sigma}(e)(u_2) = \begin{cases} \vee_{a \in \sigma(e)u_2} \gamma_B(a) & \text{if } \sigma(e)u_2 \neq \emptyset; \\ 0 & \text{if } \sigma(e)u_2 = \emptyset; \end{cases} $$

and

$$ {}^{\mu_B}\overline{\sigma}(e)(u_2) = \begin{cases} \vee_{a \in \sigma(e)u_2} \mu_B(a) & \text{if } \sigma(e)u_2 \neq \emptyset; \\ 0 & \text{if } \sigma(e)u_2 = \emptyset; \end{cases} $$

$$ {}^{\gamma_B}\overline{\sigma}(e)(u_2) = \begin{cases} \wedge_{a \in \sigma(e)u_2} \gamma_B(a) & \text{if } \sigma(e)u_2 \neq \emptyset; \\ 1 & \text{if } \sigma(e)u_2 = \emptyset, \end{cases} $$

where $\sigma(e)u_2 = \{a \in U_1 : (a, u_2) \in \sigma(e)\}$ *and is called the foreset of* u_2 *for* $u_2 \in U_2$ *and* $e \in A$.

Of course, $\underline{\sigma}^B : A \to IF(U_1)$, $\overline{\sigma}^B : A \to IF(U_1)$ and ${}^B\underline{\sigma} : A \to IF(U_2)$, ${}^B\overline{\sigma} : A \to IF(U_2)$.

Theorem 1 ([70]). *Let* (σ, A) *be a soft binary relation from* U_1 *to* U_2, *that is* $\sigma : A \to P(U_1 \times U_2)$. *For any IFSs* $B = \langle \mu_B, \gamma_B \rangle$, $B_1 = \langle \mu_{B_1}, \gamma_{B_1} \rangle$ *and* $B_2 = \langle \mu_{B_2}, \gamma_{B_2} \rangle$ *of* U_2, *the following are true:*

(1) *If* $B_1 \subseteq B_2$ *then* $\underline{\sigma}^{B_1} \subseteq \underline{\sigma}^{B_2}$;
(2) *If* $B_1 \subseteq B_2$ *then* $\overline{\sigma}^{B_1} \subseteq \overline{\sigma}^{B_2}$;
(3) $\underline{\sigma}^{B_1} \cap \underline{\sigma}^{B_2} = \underline{\sigma}^{B_1 \cap B_2}$;
(4) $\overline{\sigma}^{B_1} \cap \overline{\sigma}^{B_2} \supseteq \overline{\sigma}^{B_1 \cap B_2}$;
(5) $\underline{\sigma}^{B_1} \cup \underline{\sigma}^{B_2} \subseteq \underline{\sigma}^{B_1 \cup B_2}$;
(6) $\overline{\sigma}^{B_1} \cup \overline{\sigma}^{B_2} = \overline{\sigma}^{B_1 \cup B_2}$;
(7) $\underline{\sigma}^{1_{U_2}} = 1U_1$ *if* $u_1\sigma(e) \neq \emptyset$;
(8) $\overline{\sigma}^{1_{U_2}} = 1U_1$ *if* $u_1\sigma(e) \neq \emptyset$;
(9) $\underline{v}^B = \left(\overline{\sigma}^{B^c}\right)^c$ *if* $u_1\sigma(e) \neq \emptyset$;
(10) $\overline{\sigma}^B = \left(\underline{\sigma}^{B^c}\right)^c$ *if* $u_1\sigma(e) \neq \emptyset$;
(11) $\underline{\sigma}^{0_{U_2}} = 0_{U_1} = \overline{\sigma}^{0_{U_2}}$ *if* $u_1\sigma(e) \neq \emptyset$.

Theorem 2 ([70]). *Let* (σ, A) *be a soft binary relation from* U_1 *to* U_2, *that is* $\sigma : A \to P(U_1 \times U_2)$. *For any IFSs* $B = \langle \mu_B, \gamma_B \rangle$, $B_1 = \langle \mu_{B_1}, \gamma_{B_1} \rangle$ *and* $B_2 = \langle \mu_{B_2}, \gamma_{B_2} \rangle$ *of* U_1, *the following are true:*

(1) *If* $B_1 \subseteq B_2$, *then* ${}^{B_1}\underline{\sigma} \subseteq^{B_2} \underline{\sigma}$;
(2) *If* $B_1 \subseteq B_2$, *then* ${}^{B_1}\overline{\sigma} \subseteq^{B_2} \overline{\sigma}$;
(3) ${}^{B_1}\underline{\sigma} \cap^{B_2} \underline{\sigma} =^{B_1 \cap B_2} \underline{\sigma}$;
(4) ${}^{B_1}\overline{\sigma} \cap^{B_2} \overline{\sigma} \supseteq^{B_1 \cap B_2} \overline{\sigma}$;
(5) ${}^{B_1}\underline{\sigma} \cup^{B2} \underline{\sigma} \subseteq^{B_1 \cup B_2} \underline{\sigma}$;
(6) ${}^{B_1}\overline{\sigma} \cup^{B_2} \overline{\sigma} =^{B_1 \cup B_2} \overline{\sigma}$;
(7) $\underline{\sigma}^{1_{U_1}} = 1_{U_2}$ *if* $u_2\sigma(e) \neq \emptyset$;
(8) $\overline{\sigma}^{1_{U_1}} = 1_{U_2}$ *if* $u_2\sigma(e) \neq \emptyset$;

(9) $^{B}\underline{\sigma} = \left(^{B^c}\overline{\sigma}\right)^c$ *if* $u_2\sigma(e) \neq \emptyset$;

(10) $^{B}\overline{\sigma} = \left(^{B^c}\underline{\sigma}\right)^c$ *if* $u_2\sigma(e) \neq \emptyset$;

(11) $^{0_{U_1}}\underline{\sigma} = 0_{U_2} =^{0_{U_1}} \overline{\sigma}$.

3. Roughness of an Intuitionistic Fuzzy Set by Two Soft Relations

In this section, we discuss the optimistic roughness of an IFS by two soft binary relations from U_1 to U_2. We approximate an IFS of universe U_2 in universe U_1 and an IFS of U_1 in U_2 by using aftersets and foresets of soft binary relations, respectively. In this way, we obtain two IFSSs corresponding to IFSs in $U_2(U_1)$. We also study some properties of these approximations.

Definition 12. *Let U_1 and U_2 be two non-empty sets, (σ_1, A) and (σ_2, A) be two soft binary relations from U_1 to U_2 and $B = \langle\mu_B, \gamma_B\rangle$ be an IFS in U_2. Then, the optimistic lower approximation $\underline{\sigma_1 + \sigma_2}_o^B = \left(\underline{\sigma_1 + \sigma_2}_o^{\mu_B}, \underline{\sigma_1 + \sigma_2}_o^{\gamma_B}\right)$ and the upper approximation $^o\overline{\sigma_1 + \sigma_2}^B = (^o\overline{\sigma_1 + \sigma_2}^{\mu_B}, ^o\overline{\sigma_1 + \sigma_2}^{\gamma_B})$ of $B = \langle\mu_B, \gamma_B\rangle$ are IF soft sets over U_1 and are defined as:*

$$\underline{\sigma_1 + \sigma_2}_o^{\mu_B}(e)(u_1) = \begin{cases} \wedge\{\mu_B(u_2) : u_2 \in (u_1\sigma_1(e) \cup u_1\sigma_2(e))\}, & \text{if } u_1\sigma_1(e) \cup u_1\sigma_2(e) \neq \emptyset; \\ 1 & \text{otherwise;} \end{cases}$$

$$\underline{\sigma_1 + \sigma_2}_o^{\gamma_B}(e)(u_1) = \begin{cases} \vee\{\gamma_B(u_2) : u_2 \in (u_1\sigma_1(e) \cap u_1\sigma_2(e))\}, & \text{if } u_1\sigma_1(e) \cap u_1\sigma_2(e) \neq \emptyset; \\ 0 & \text{otherwise;} \end{cases}$$

and

$$^o\overline{\sigma_1 + \sigma_2}^{\mu_B}(e)(u_1) = \begin{cases} \vee\{\mu_B(u_2) : u_2 \in (u_1\sigma_1(e) \cap u_1\sigma_2(e))\}, & \text{if } u_1\sigma_1(e) \cap u_1\sigma_2(e) \neq \emptyset; \\ 0 & \text{otherwise;} \end{cases}$$

$$^o\overline{\sigma_1 + \sigma_2}^{\gamma_B}(e)(u_1) = \begin{cases} \wedge\{\gamma_B(u_2) : u_2 \in (u_1\sigma_1(e) \cup u_1\sigma_2(e))\}, & \text{if } u_1\sigma_1(e) \cup u_1\sigma_2(e) \neq \emptyset; \\ 1 & \text{otherwise;} \end{cases}$$

for all $u_1 \in U_1$, where $u_1\sigma_1(e) = \{u_2 \in U_2 : (u_1, u_2) \in \sigma_1(e)\}$ and $u_1\sigma_2(e) = \{u_2 \in U_2 : (u_1, u_2) \in \sigma_2(e)\}$ are called the aftersets of u_1 for $u_1 \in U_1$ and $e \in A$. Obviously, $(\underline{\sigma_1 + \sigma_2}_o^B(e)), A)$ and $(^o\overline{\sigma_1 + \sigma_2}^B(e))$ are two IFS soft sets over U_1.

Definition 13. *Let U_1 and U_2 be two non-empty sets, (σ_1, A) and (σ_2, A) be two soft binary relations from U_1 to U_2 and $B = \langle\mu_B, \gamma_B\rangle$ be an IFS in U_1. Then, the optimistic lower approximation $^B\underline{\sigma_1 + \sigma_2}_o = \left(^{\mu_B}\underline{\sigma_1 + \sigma_2}_o, ^{\gamma_B}\underline{\sigma_1 + \sigma_2}_o\right)$ and the optimistic upper approximation $^B\overline{\sigma_1 + \sigma_2} = (^{\mu_B}\overline{\sigma_1 + \sigma_2}^o, ^{\gamma_B}\overline{\sigma_1 + \sigma_2}^o)$ of $B = \langle\mu_B, \gamma_B\rangle$ are IF soft sets over U_2 and are defined as:*

$$^{\mu_B}\underline{\sigma_1 + \sigma_2}_o(e)(u_2) = \begin{cases} \wedge\{\mu_B(u_1) : u_1 \in (\sigma_1u_2(e) \cup \sigma_2u_2(e))\}, & \text{if } \sigma_1u_2(e) \cup \sigma_2u_2(e) \neq \emptyset; \\ 1 & \text{otherwise;} \end{cases}$$

$$^{\gamma_B}\underline{\sigma_1 + \sigma_2}_o(e)(u_2) = \begin{cases} \vee\{\gamma_B(u_1) : u_1 \in (\sigma_1u_2(e) \cap \sigma_2u_2(e))\}, & \text{if } \sigma_1u_2(e) \cap \sigma_2u_2(e) \neq \emptyset; \\ 0 & \text{otherwise;} \end{cases}$$

and

$$^{\mu_B}\overline{\sigma_1 + \sigma_2}^o(e)(u_2) = \begin{cases} \vee\{\mu_B(u_1) : u_1 \in (\sigma_1u_2(e) \cap \sigma_2u_2(e))\}, & \text{if } \sigma_1u_2(e) \cap \sigma_2u_2(e) \neq \emptyset; \\ 0 & \text{otherwise;} \end{cases}$$

$$^{\gamma_B}\overline{\sigma_1 + \sigma_2}^o(e)(u_2) = \begin{cases} \wedge\{\gamma_B(u_1) : u_1 \in (\sigma_1u_2(e) \cup \sigma_2u_2(e))\}, & \text{if } \sigma_1u_2(e) \cup \sigma_2u_2(e) \neq \emptyset; \\ 1 & \text{otherwise;} \end{cases}$$

for all $u_2 \in U_2$ *where* $\sigma_1(e)u_2 = \{u_1 \in U_1 : (u_1, u_2) \in \sigma_1(e)\}$ *and* $\sigma_2(e)u_2 = \{u_1 \in U_1 : (a, u_2) \in \sigma_2(e)\}$ *are called the foresets of* u_2 *for* $u_2 \in U_2$ *and* $e \in A$. *Obviously,* $({}^{B}\underline{\sigma_1 + \sigma_2}_o(e)), A)$ *and* $({}^{B}\overline{\sigma_1 + \sigma_2}^o(e))$ *are two IFS soft sets over* U_2.

Of course, $\underline{\sigma_1 + \sigma_2}_o^B(e) : A \to IF(U_1)$, ${}^o\overline{\sigma_1 + \sigma_2}^B(e) : A \to IF(U_1)$ and ${}^B\underline{\sigma_1 + \sigma_2}_o(e) : A \to IF(U_2)$, ${}^B\overline{\sigma_1 + \sigma_2}^o(e) : A \to IF(U_2)$.

The following example explains the above definitions.

Example 1. *Let* $U_1 = \{1, 2, 3\}, U_2 = \{a, b, c\}$ *and* $A = \{e_1, e_2\}$, *and* (σ_1, A) *and* (σ_2, A) *be soft binary relations from* U_1 *to* U_2 *defined by*

$$\begin{aligned} \sigma_1(e_1) &= \{(1,a),(1,b),(2,a)\}, \sigma_1(e_2) = \{(2,b),(3,a)\}, \\ \sigma_2(e_1) &= \{(2,b),(2,c),(3,a)\} \text{ and } \sigma_2(e_2) = \{(1,c),(3,b),(3,c)\}. \end{aligned}$$

Then, their aftersets and foresets are

$$\begin{aligned} 1\sigma_1(e_1) &= \{a,b\},\ 2\sigma_1(e_1) = \{a\},\ 3\sigma_1(e_1) = \varnothing, \\ 1\sigma_1(e_2) &= \varnothing,\ 2\sigma_1(e_2) = \{b\},\ 3\sigma_1(e_2) = \{a\} \text{ and} \\ 1\sigma_2(e_1) &= \varnothing,\ 2\sigma_2(e_1) = \{b,c\},\ 3\sigma_2(e_1) = \{a\}, \\ 1\sigma_2(e_2) &= \{c\},\ 2\sigma_2(e_2) = \varnothing,\ 3\sigma_2(e_2) = \{b,c\}, \end{aligned}$$

$$\begin{aligned} \sigma_1(e_1)a &= \{1,2\},\ \sigma_1(e_1)b = \{1\},\ \sigma_1(e_1)c = \varnothing, \\ \sigma_1(e_2)a &= \{3\},\ \sigma_1(e_2)b = \{2\},\ \sigma_1(e_2)c = \varnothing \text{and} \\ \sigma_2(e_1)a &= \{3\},\ \sigma_2(e_1)b = \{2\},\ \sigma_2(e_1)c = \{2\}, \\ \sigma_2(e_2)a &= \varnothing,\ \sigma_2(e_2)b = \{3\},\ \sigma_2(e_2)c = \{1,3\}. \end{aligned}$$

(1) Define $B_1 = \langle \mu_{B_1}, \gamma_{B_1} \rangle : U_2 \to [0,1]$ (given in Table 2).

Table 2. Intuitionistic fuzzy set B_1.

B_1	a	b	c
μ_{B_1}	0.5	0.4	0.3
γ_{B_1}	0.4	0.5	0.7

The optimistic multigranulation lower and upper approximations of B_1 with respect to the aftersets are given in Tables 3 and 4.

Table 3. Optimistic multigranulation lower approximation of B_1.

	1	2	3
$\underline{\sigma_1 + \sigma_2}_o^{\mu_{B_1}}(e_1)$	0.4	0.3	0.5
$\underline{\sigma_1 + \sigma_2}_o^{\gamma_{B_1}}(e_1)$	0.5	0.7	0.4
$\underline{\sigma_1 + \sigma_2}_o^{\mu_{B_1}}(e_2)$	0.3	0.4	0.3
$\underline{\sigma_1 + \sigma_2}_o^{\gamma_{B_1}}(e_2)$	0.7	0.5	0.7

Table 4. Optimistic multigranulation upper approximation of B_1.

	1	**2**	**3**
${}^{o}\overline{\sigma_1+\sigma_2}^{\mu_{B_1}}(e_1)$	0	0	0
${}^{o}\overline{\sigma_1+\sigma_2}^{\gamma_{B_1}}(e_1)$	1	1	1
${}^{o}\overline{\sigma_1+\sigma_2}^{\mu_{B_1}}(e_2)$	0	0	0
${}^{o}\overline{\sigma_1+\sigma_2}^{\gamma_{B_1}}(e_2)$	1	1	1

(2) Define $B = \langle \mu_B, \gamma_B \rangle : U_1 \to [0,1]$ as given in Table 5.

Table 5. Intuitionistic fuzzy set B.

B	**1**	**2**	**3**
μ_B	0.3	0.7	0.6
γ_B	0.6	0.3	0.2

The optimistic multigranulation lower and upper approximations of B with respect to the foresets are given in Tables 6 and 7.

Table 6. Optimistic multigranulation lower approximation of B.

	a	*b*	*c*
${}^{\mu_B}\underline{\sigma_1+\sigma_2}_o(e_1)$	0.3	0.3	0.7
${}^{\gamma_B}\underline{\sigma_1+\sigma_2}_o(e_1)$	0.6	0.6	0.3
${}^{\mu_B}\underline{\sigma_1+\sigma_2}_o(e_2)$	0.6	0.6	0.3
${}^{\gamma_B}\underline{\sigma_1+\sigma_2}_o(e_2)$	0.2	0.2	0.6

Table 7. Optimistic multigranulation upper approximation of B.

	a	*b*	*c*
${}^{\mu_B}\overline{\sigma_1+\sigma_2}^{o}(e_1)$	0	0	0
${}^{\gamma_B}\overline{\sigma_1+\sigma_2}^{o}(e_1)$	1	1	1
${}^{\mu_B}\overline{\sigma_1+\sigma_2}^{o}(e_2)$	0	0	0
${}^{\gamma_B}\overline{\sigma_1+\sigma_2}^{o}(e_2)$	1	1	1

Proposition 1. *Let $(\sigma_1, A), (\sigma_2, A)$ be two soft relations from U_1 to U_2, that is $\sigma_1 : A \to P(U_1 \times U_2)$ and $\sigma_2 : A \to P(U_1 \times U_2)$ and $B \in IF(U_2)$. Then, the following hold with respect to the aftersets:*

1. $\underline{\sigma_1+\sigma_2}_o^B \leqslant \underline{\sigma_1}^B \vee \underline{\sigma_2}^B$;
2. ${}^{o}\overline{\sigma_1+\sigma_2}^B \leqslant \overline{\sigma_1}^B \wedge \overline{\sigma_2}^B$.

Proof. (1) Let $u_1 \in U_1$. Then, $\underline{\sigma_1+\sigma_2}_o^{\mu_B}(e)(u_1) = \wedge\{\mu_B(u_2) : u_2 \in (u_1\sigma_1(e) \cup u_1\sigma_2(e))\} \leqslant (\wedge\{\mu_B(u_2) : u_2 \in (u_1\sigma_1(e)\}) \vee (\wedge\{\mu_B(u_2) : u_2 \in (u_1\sigma_2(e)\}) = \underline{\sigma_1}^{\mu_B}(e)(u_1) \vee \underline{\sigma_2}^{\mu_B}(e)(u_1)$.

Similarly, let $u_1 \in U_1$. Then $\underline{\sigma_1+\sigma_2}_o^{\gamma_B}(e)(u_1) = \vee\{\gamma_B(u_2) : u_2 \in (u_1\sigma_1(e) \cap u_1\sigma_2(e))\} \geqslant (\vee\{\gamma_B(u_2) : u_2 \in (u_1\sigma_1(e)\}) \wedge (\vee\{\gamma_B(u_2) : u_2 \in (u_1\sigma_2(e)\}) = \underline{\sigma_1}^{\gamma_B}(e)(u_1) \wedge \underline{\sigma_2}^{\gamma_B}(e)(u_1)$.

Hence, $\underline{\sigma_1+\sigma_2}_o^B \leqslant \underline{\sigma_1}^B \vee \underline{\sigma_2}^B$.

(2) The properties can be proved similarly to (1). □

Proposition 2. *Let $(\sigma_1, A), (\sigma_2, A)$ be two soft relations from U_1 to U_2, that is $\sigma_1 : A \to P(U_1 \times U_2)$ and $\sigma_2 : A \to P(U_1 \times U_2)$ and $B \in IF(U_1)$. Then, the following hold with respect to the foresets:*

1. ${}^{B}\underline{\sigma_1+\sigma_2}_o \leqslant^{B} \underline{\sigma_1} \vee^{B} \underline{\sigma_2}$;
2. ${}^{B}\overline{\sigma_1+\sigma_2}^{o} \leqslant^{B} \overline{\sigma_1} \wedge^{B} \overline{\sigma_2}$.

Proof. The proof is similar to the proof of Proposition 1. □

For the converse, we have the following example.

Example 2 (Continued from Example 1). *According to Example 1, we have the following:*

$$\begin{aligned}
\underline{\sigma_1}^{\mu_{B_1}}(e_1)(2) &= 0.5 \textit{ and } \underline{\sigma_1}^{\gamma_{B_1}}(e_1)(2) = 0.4; \\
\underline{\sigma_2}^{\mu_{B_1}}(e_1)(2) &= 0.3 \textit{ and } \underline{\sigma_2}^{\gamma_{B_1}}(e_1)(2) = 0.7; \\
\overline{\sigma_1}^{\mu_{B_1}}(e_1)(2) &= 0.5 \textit{ and } \overline{\sigma_1}^{\mu_{B_1}}(e_1)(2) = 0.4; \\
\overline{\sigma_2}^{\mu_{B_1}}(e_1)(2) &= 0.4 \textit{ and } \overline{\sigma_2}^{\gamma_{B_1}}(e_1)(2) = 0.5.
\end{aligned}$$

Hence,

$$\begin{aligned}
\underline{\sigma_1 + \sigma_2}_o^{\mu_{B_1}}(e_1)(2) &= 0.3 \not\geq 0.5 = \underline{\sigma_1}^{\mu_{B_1}}(e_1)(2) \vee \underline{\sigma_2}^{\mu_{B_1}}(e_1)(2); \\
\underline{\sigma_1 + \sigma_2}_o^{\gamma_{B_1}}(e_1)(2) &= 0.7 \not\leq 0.4 = \underline{\sigma_1}^{\gamma_{B_1}}(e_1)(2) \wedge \underline{\sigma_2}^{\gamma_{B_1}}(e_1)(2); \\
{}^{o}\overline{\sigma_1 + \sigma_2}^{\mu_{B_1}}(e_1)(2) &= 0 \not\geq 0.4 = \overline{\sigma_1}^{\mu_{B_1}}(e_1)(2) \wedge \overline{\sigma_2}^{\mu_{B_1}}(e_1)(2); \\
{}^{o}\overline{\sigma_1 + \sigma_2}^{\gamma_{B_1}}(e_1)(2) &= 1 \not\leq 0.5 = \overline{\sigma_1}^{\gamma_{B_1}}(e_1)(2) \vee \overline{\sigma_1}^{\gamma_{B_1}}(e_1)(2).
\end{aligned}$$

In addition,

$$\begin{aligned}
{}^{\mu_B}\underline{\sigma_1}(e_1)(a) &= 0.3 \textit{ and } {}^{\gamma_B}\underline{\sigma_1}(e_1)(a) = 0.6 \\
{}^{\mu_B}\underline{\sigma_2}(e_1)(a) &= 0.6 \textit{ and } {}^{\gamma_B}\underline{\sigma_2}(e_1)(a) = 0.2 \\
{}^{\mu_B}\overline{\sigma_1}(e_1)(a) &= 0.7 \textit{ and } {}^{\gamma_B}\overline{\sigma_1}(e_1)(a) = 0.3 \\
{}^{\mu_B}\overline{\sigma_2}(e_1)(a) &= 0.6 \textit{ and } {}^{\gamma_B}\overline{\sigma_2}(e_1)(a) = 0.2
\end{aligned}$$

Hence,

$$\begin{aligned}
{}^{\mu_B}\underline{\sigma_1 + \sigma_2}_o(e_1)(a) &= 0.3 \not\geq 0.7 =^{\mu_B} \underline{\sigma_1}(e_1)(a) \vee^{\mu_B} \underline{\sigma_2}(e_1)(a); \\
{}^{\gamma_B}\underline{\sigma_1 + \sigma_2}_o(e_1)(a) &= 0.3 \not\leq 0.2 =^{\gamma_B} \underline{\sigma_1}(e_1)(a) \wedge^{\gamma_B} \underline{\sigma_2}(e_1)(a); \\
{}^{\mu_B}\overline{\sigma_1 + \sigma_2}^{o}(e_1)(a) &= 0 \not\geq 0.6 =^{\mu_B} \overline{\sigma_1}(e_1)(a) \wedge^{\mu_B} \overline{\sigma_2}(e_1)(a); \\
{}^{\gamma_B}\overline{\sigma_1 + \sigma_2}^{o}(e_1)(a) &= 1 \not\leq 0.3 =^{\gamma_B} \overline{\sigma_1}(e_1)(a) \vee^{\gamma_B} \overline{\sigma_2}(e_1)(a).
\end{aligned}$$

Proposition 3. *Let $(\sigma_1, A), (\sigma_2, A)$ be two soft relations from U_1 to U_2, that is $\sigma_1 : A \to P(U_1 \times U_2)$ and $\sigma_2 : A \to P(U_1 \times U_2)$ and $B \in IF(U_2)$. Then, the following hold:*

(1) $\underline{\sigma_1 + \sigma_2}_o^{1_{U_2}} = 1_{U_1}$ for all $e \in A$;
(2) ${}^{o}\overline{\sigma_1 + \sigma_2}^{1_{U_2}} = 1_{U_1}$ if $u_1\sigma_1(e) \cap u_1\sigma_2(e) \neq \emptyset$ and $u_1\sigma_1(e) \cup u_1\sigma_2(e) \neq \emptyset$;
(3) $\underline{\sigma_1 + \sigma_2}_o^{0_{U_2}} = 0_{U_1}$ if $u_1\sigma_1(e) \cup u_1\sigma_2(e) \neq \emptyset$ and $u_1\sigma_1(e) \cap u_1\sigma_2(e) \neq \emptyset$;
(4) ${}^{o}\overline{\sigma_1 + \sigma_2}^{0_{U_2}} = 0_{U_1}$ for all $e \in A$.

Proof. (1) Let $u_1 \in U_1$ and $1_{U_2} = \langle 1, 0 \rangle$ be the universal set of U_2. If $u_1\sigma_1(e) \cup u_1\sigma_2(e) = \emptyset$, then $\underline{\sigma_1 + \sigma_2}_o^{1}(e)(u_1) = 1$ and $\underline{\sigma_1 + \sigma_2}_o^{0}(e)(u_1) = 0$.

If $u_1\sigma_1(e) \cup u_1\sigma_2(e) \neq \emptyset$, then $\underline{\sigma_1 + \sigma_2}_o^{1}(e)(u_1) = \wedge\{1(u_2) : u_2 \in (u_1\sigma_1(e) \cup u_1\sigma_2(e))\} = \wedge\{1 : u_2 \in (u_1\sigma_1(e) \cup u_1\sigma_2(e))\} = 1$,

and $\underline{\sigma_1 + \sigma_2}_o^{0}(e)(u_1) = \vee\{0(u_2) : u_2 \in (u_1\sigma_1(e) \cap u_1\sigma_2(e))\} = \vee\{0 : u_2 \in (u_1\sigma_1(e) \cap u_1\sigma_2(e))\} = 0$.

(2) The properties can be proved similarly to (1).

(3) Let $u_1 \in U_1$ and $0_{U_2} = \langle 0, 1 \rangle$ be the universal set of U_2. If $u_1\sigma_1(e) \cup u_1\sigma_2(e) \neq \emptyset$, then $\underline{\sigma_1 + \sigma_2}_o^{0}(e)(u_1) = \wedge\{0(u_2) : u_2 \in (u_1\sigma_1(e) \cup u_1\sigma_2(e))\} = \wedge\{0 : u_2 \in (u_1\sigma_1(e) \cup u_1\sigma_2(e))\} = 0$,

and $\underline{\sigma_1 + \sigma_2}_o^{1}(e)(u_1) = \vee\{1(u_2) : u_2 \in (u_1\sigma_1(e) \cap u_1\sigma_2(e))\} = \vee\{1 : u_2 \in (u_1\sigma_1(e) \cap u_1\sigma_2(e))\} = 1$.

(4) The properties can be proved similarly to (3). □

Proposition 4. *Let $(\sigma_1, A), (\sigma_2, A)$ be two soft relations from U_1 to U_2, that is $\sigma_1 : A \to P(U_1 \times U_2)$ and $\sigma_2 : A \to P(U_1 \times U_2)$ and and $B \in IF(U_1)$. Then, the following hold:*

(1) ${}^{1_{U_1}}\underline{\sigma_1 + \sigma_2}_o = 1_{U_2}$ for all $e \in A$;

(2) ${}^{1_{U_1}}\overline{\sigma_1 + \sigma_2}^o = 1_{U_2}$ for all $e \in A$,for all $e \in A$, if $\sigma_1(e)u_2 \cap \sigma_2(e)u_2 \neq \emptyset$ and $\sigma_1(e)u_2 \cup \sigma_2(e)u_2 \neq \emptyset$;

(3) ${}^{0_{U_1}}\underline{\sigma_1 + \sigma_2}_o = 0_{U_2}$ for all $e \in A$,for all $e \in A$, if $\sigma_1(e)u_2 \cup \sigma_2(e)u_2 \neq \emptyset$ and $\sigma_1(e)u_2 \cap \sigma_2(e)u_2 \neq \emptyset$;

(4) ${}^{0_{U_1}}\overline{\sigma_1 + \sigma_2}^o$ for all $e \in A$.

Proof. The proof is similar to the proof of Proposition 3. □

Proposition 5. *Let $(\sigma_1, A), (\sigma_2, A)$ be two soft relations from U_1 to U_2, that is $\sigma_1 : A \to P(U_1 \times U_2)$ and $\sigma_2 : A \to P(U_1 \times U_2)$ and $B, B_1, B_2 \in IF(U_2)$. Then, the following properties hold:*

(1) If $B_1 \subseteq B_2$ then $\underline{\sigma_1 + \sigma_2}_o^{B_1} \subseteq \underline{\sigma_1 + \sigma_2}_o^{B_2}$;

(2) If $B_1 \subseteq B_2$ then ${}^o\overline{\sigma_1 + \sigma_2}^{B_1} \subseteq^o \overline{\sigma_1 + \sigma_2}^{B_2}$;

(3) $\underline{\sigma_1 + \sigma_2}_o^{B_1 \cap B_2} = \underline{\sigma_1 + \sigma_2}_o^{B_1} \cap \underline{\sigma_1 + \sigma_2}_o^{B_2}$;

(4) $\underline{\sigma_1 + \sigma_2}_o^{B_1 \cup B_2} \supseteq \underline{\sigma_1 + \sigma_2}_o^{B_1} \cup \underline{\sigma_1 + \sigma_2}_o^{B_2}$;

(5) ${}^o\overline{\sigma_1 + \sigma_2}^{B_1 \cup B_2} =^o \overline{\sigma_1 + \sigma_2}^{B_1} \cup^o \overline{\sigma_1 + \sigma_2}^{B_2}$;

(6) ${}^o\overline{\sigma_1 + \sigma_2}^{B_1 \cap B_2} \subseteq^o \overline{\sigma_1 + \sigma_2}^{B_1} \cap^o \overline{\sigma_1 + \sigma_2}^{B_2}$.

Proof. (1) Since $B_1 \subseteq B_2$ so $\mu_{B_1} \leqslant \mu_{B_2}$ and $\gamma_{B_1} \geqslant \gamma_{B_2}$. Thus $\underline{\sigma_1 + \sigma_2}_o^{\mu_{B_1}}(e)(u_1) = \wedge\{\mu_{B_1}(u_2) : u_2 \in (u_1\sigma_1(e) \cup u_1\sigma_2(e))\} \leqslant \wedge\{\mu_{B_2}(u_2) : u_2 \in (u_1\sigma_1(e) \cup u_1\sigma_2(e))\} = \underline{\sigma_1 + \sigma_2}_o^{\mu_{B_2}}(e)(u_1)$,

and $\underline{\sigma_1 + \sigma_2}_o^{\gamma_{B_1}}(e)(u_1) = \vee\{\gamma_{B_1}(u_2) : u_2 \in (u_1\sigma_1(e) \cap u_1\sigma_2(e))\} \geqslant \vee\{\gamma_{B_2}(u_2) : u_2 \in (u_1\sigma_1(e) \cap u_1\sigma_2(e))\} = \underline{\sigma_1 + \sigma_2}_o^{\gamma_{B_2}}(e)(u_1)$.

(2) The properties can be proved similarly to (1).

(3) Let $u_1 \in U_1$. If $u_1\sigma_1(e) \cup u_1\sigma_2(e) = \emptyset$, then $\underline{\sigma_1 + \sigma_2}_o^{\mu_{B_1 \cap B_2}}(e)(u_1) = 1 = \underline{\sigma_1 + \sigma_2}_o^{\mu_{B_1}}(e)(u_1) \cap \underline{\sigma_1 + \sigma_2}_o^{\mu_{B_2}}(e)(u_1)$

and $\underline{\sigma_1 + \sigma_2}_o^{\gamma_{B_1 \cap B_2}}(e)(u_1) = 0 = \underline{\sigma_1 + \sigma_2}_o^{\gamma_{B_1}}(e)(u_1) \cup \underline{\sigma_1 + \sigma_2}_o^{\gamma_{B_2}}(e)(u_1)$.

If $u_1\sigma_1(e) \cup u_1\sigma_2(e) \neq \emptyset$,

then $\underline{\sigma_1 + \sigma_2}_o^{\mu_{B_1 \cap B_2}}(e)(u_1) = \wedge\{(\mu_{B_1} \wedge \mu_{B_2})(u_2) : u_2 \in (u_1\sigma_1(e) \cup u_1\sigma_2(e))\} = \wedge\{\mu_{B_1}(u_2) \wedge \mu_{B_2}(u_2) : u_2 \in (u_1\sigma_1(e) \cup u_1\sigma_2(e))\}$

$= (\wedge\{\mu_{B_1}(u_2) : u_2 \in (u_1\sigma_1(e) \cup u_1\sigma_2(e))\}) \wedge (\wedge\{\mu_{B_2}(u_2) : u_2 \in (u_1\sigma_1(e) \cup u_1\sigma_2(e))\})$

$= \underline{\sigma_1 + \sigma_2}_o^{\mu_{B_1}}(e)(u_1) \cap \underline{\sigma_1 + \sigma_2}_o^{\mu_{B_2}}(e)(u_1)$.

In addition, $\underline{\sigma_1 + \sigma_2}_o^{\gamma_{B_1 \cap B_2}}(e)(u_1) = \vee\{(\mu_{B_1} \vee \mu_{B_2})(u_2) : u_2 \in (u_1\sigma_1(e) \cup u_1\sigma_2(e))\} = \vee\{\mu_{B_1}(u_2) \vee \mu_{B_2}(u_2) : u_2 \in (u_1\sigma_1(e) \cup u_1\sigma_2(e))\}$

$= (\vee\{\mu_{B_1}(u_2) : u_2 \in (u_1\sigma_1(e) \cup u_1\sigma_2(e))\}) \vee (\vee\{\mu_{B_2}(u_2) : u_2 \in (u_1\sigma_1(e) \cup u_1\sigma_2(e))\})$

$= \underline{\sigma_1 + \sigma_2}_o^{\mu_{B_1}}(e)(u_1) \cup \underline{\sigma_1 + \sigma_2}_o^{\mu_{B_2}}(e)(u_1)$.

This shows that $\underline{\sigma_1 + \sigma_2}_o^{B_1 \cap B_2} = \underline{\sigma_1 + \sigma_2}_o^{B_1} \cap \underline{\sigma_1 + \sigma_2}_o^{B_2}$.

(4) The properties can be proved similarly to (3).

(5) Let $u_1 \in U_1$. If $u_1\sigma_1(e) \cap u_1\sigma_2(e) = \emptyset$, then ${}^o\overline{\sigma_1 + \sigma_2}^{\mu_{B_1 \cup B_2}}(e)(u_1) = 0 =^o \overline{\sigma_1 + \sigma_2}^{\mu_{B_1}}(e)(u_1) \cup^o \overline{\sigma_1 + \sigma_2}^{\mu_{B_2}}(e)(u_1)$

and ${}^o\overline{\sigma_1 + \sigma_2}^{\gamma_{B_1 \cup B_2}}(e)(u_1) = 1 =^o \overline{\sigma_1 + \sigma_2}^{\mu_{B_1}}(e)(u_1) \cap^o \overline{\sigma_1 + \sigma_2}^{\mu_{B_2}}(e)(u_1)$.

If $u_1\sigma_1(e) \cap u_1\sigma_2(e) \neq \emptyset$,

then ${}^o\overline{\sigma_1 + \sigma_2}^{\mu_{B_1 \cup B_2}}(e)(u_1) = \vee\{(\mu_{B_1} \vee \mu_{B_2})(u_2) : u_2 \in (u_1\sigma_1(e) \cap u_1\sigma_2(e))\} = \vee\{\mu_{B_1}(u_2) \vee \mu_{B_2}(u_2) : u_2 \in (u_1\sigma_1(e) \cap u_1\sigma_2(e))\}$

$= (\vee\{\mu_{B_1}(u_2) : u_2 \in (u_1\sigma_1(e) \cap u_1\sigma_2(e))\}) \vee (\vee\{\mu_{B_2}(u_2) : u_2 \in (u_1\sigma_1(e) \cap u_1\sigma_2(e))\})$

$=^o \overline{\sigma_1 + \sigma_2}^{\mu_{B_1}}(e)(u_1) \cup^o \overline{\sigma_1 + \sigma_2}^{\mu_{B_2}}(e)(u_1)$.

In addition, ${}^o\overline{\sigma_1 + \sigma_2}^{\gamma_{B_1 \cup B_2}}(e)(u_1) = \wedge\{(\gamma_{B_1} \wedge \gamma_{B_2})(u_2) : u_2 \in (u_1\sigma_1(e) \cup u_1\sigma_2(e))\} = \wedge\{\gamma_{B_1}(u_2) \wedge \gamma_{B_2}(u_2) : u_2 \in (u_1\sigma_1(e) \cup u_1\sigma_2(e))\}$

$= (\wedge\{\gamma_{B_1}(u_2) : u_2 \in (u_1\sigma_1(e) \cup u_1\sigma_2(e))\}) \wedge (\wedge\{\gamma_{B_2}(u_2) : u_2 \in (u_1\sigma_1(e) \cup u_1\sigma_2(e))\})$
$=^o \overline{\sigma_1 + \sigma_2}^{\gamma_{B_1}}(e)(u_1) \cap^o \overline{\sigma_1 + \sigma_2}^{\gamma_{B_2}}(e)(u_1)$.
This shows that ${}^o\overline{\sigma_1 + \sigma_2}^{B_1 \cup B_2} =^o \overline{\sigma_1 + \sigma_2}^{B_1} \cup^o \overline{\sigma_1 + \sigma_2}^{B_2}$.
(6) The properties can be proved similarly to (5). □

Proposition 6. *Let $(\sigma_1, A), (\sigma_2, A)$ be two soft relations from U_1 to U_2, that is $\sigma_1 : A \to P(U_1 \times U_2)$ and $\sigma_2 : A \to P(U_1 \times U_2)$ and $B, B_1, B_2 \in IF(U_2)$. Then, the following properties hold:*
(1) If $B_1 \subseteq B_2$ then ${}^{B_1}\underline{\sigma_1 + \sigma_2}_o \subseteq^{B_2} \underline{\sigma_1 + \sigma_2}_o$;
(2) If $B_1 \subseteq B_2$ then ${}^{B_1}\overline{\sigma_1 + \sigma_2}^o \subseteq^{B_2} \overline{\sigma_1 + \sigma_2}^o$;
(3) ${}^{B_1 \cap B_2}\underline{\sigma_1 + \sigma_2}_o =^{B_1} \underline{\sigma_1 + \sigma_2}_o \cap^{B_2} \underline{\sigma_1 + \sigma_2}_o$;
(4) ${}^{B_1 \cup B_2}\underline{\sigma_1 + \sigma_2}_o \supseteq^{B_1} \underline{\sigma_1 + \sigma_2}_o \cup^{B_2} \underline{\sigma_1 + \sigma_2}_o$;
(5) ${}^{B_1 \cup B_2}\overline{\sigma_1 + \sigma_2}^o =^{B_1} \overline{\sigma_1 + \sigma_2}^o \cup^{B_2} \overline{\sigma_1 + \sigma_2}^o$;
(6) ${}^{B_1 \cap B_2}\overline{\sigma_1 + \sigma_2}^o \subseteq^{B_1} \overline{\sigma_1 + \sigma_2}^o \cap^{B_2} \overline{\sigma_1 + \sigma_2}^o$.

Proof. The proof is similar to the proof of Proposition 5. □

4. Roughness of an Intuitionistic Fuzzy Set over Two Universes by Multi-Soft Relations

In this section, we discuss the optimistic roughness of an IFS by multi-soft binary relations from U_1 to U_2 and approximate an IFS of universe U_2 in universe U_1 and an IFS U_1 in U_2 by using aftersets and foresets of soft binary relations, respectively. In this way, we obtain two intuitionistic fuzzy soft sets corresponding to IFSs in $U_2(U_1)$. We also study some properties of these approximations.

Definition 14. *Let U_1 and U_2 be two non-empty finite universes, π be a family of soft binary relations from U_1 to U_2. Then, we say (U_1, U_2, π) the multigranulation generalized soft approximation space over two universes.*

Definition 15. *Let (U_1, U_2, π) be the multigranulation generalized soft approximation space over two universes U_1 and U_2, where $\pi = \sigma_1, \sigma_2, \sigma_3, ...\sigma_m$ and $B = \langle \mu_B, \gamma_B \rangle$ be an IFS in U_2. Then, the optimistic lower approximation $\underline{\Sigma_{i=1}^m \sigma_i}_o^B = \left(\underline{\Sigma_{i=1}^m \sigma_i}_o^{\mu_B}, \underline{\Sigma_{i=1}^m \sigma_i}_o^{\gamma_B}\right)$ and the optimistic upper approximation ${}^o\overline{\Sigma_{i=1}^m \sigma_i}^B = \left({}^o\overline{\Sigma_{i=1}^m \sigma_i}^{\mu_B}, {}^o\overline{\Sigma_{i=1}^m \sigma_i}^{\gamma_B}\right)$ of $B = \langle \mu_B, \gamma_B \rangle$ are IF soft sets over U_1 with respect to the aftersets of soft relations $(\sigma_i, A) \in \pi$ and are defined as:*

$$\underline{\Sigma_{i=1}^m \sigma_i}_o^{\mu_B}(e)(u_1) = \begin{cases} \wedge\{\mu_B(u_2) : u_2 \in \cup_{i=1}^m u_1\sigma_i(e)\}, & \text{if } \cup_{i=1}^m u_1\sigma_i(e) \neq \emptyset; \\ 1 & \text{otherwise;} \end{cases}$$

$$\underline{\Sigma_{i=1}^m \sigma_i}_o^{\gamma_B}(e)(u_1) = \begin{cases} \vee\{\gamma_B(u_2) : u_2 \in \cap_{i=1}^m u_1\sigma_i(e)\}, & \text{if } \cap_{i=1}^m u_1\sigma_i(e) \neq \emptyset; \\ 0 & \text{otherwise;} \end{cases}$$

and

$${}^o\overline{\Sigma_{i=1}^m \sigma_i}^{\mu_B}(e)(u_1) = \begin{cases} \vee\{\mu_B(u_2) : u_2 \in \cap_{i=1}^m u_1\sigma_i(e)\}, & \text{if } \cap_{i=1}^m u_1\sigma_i(e) \neq \emptyset; \\ 0 & \text{otherwise;} \end{cases}$$

$${}^o\overline{\Sigma_{i=1}^m \sigma_i}^{\gamma_B}(e)(u_1) = \begin{cases} \wedge\{\gamma_B(u_2) : u_2 \in \cup_{i=1}^m u_1\sigma_i(e)\}, & \text{if } \cup_{i=1}^m u_1\sigma_i(e) \neq \emptyset; \\ 1 & \text{otherwise,} \end{cases}$$

where $u_1\sigma_i(e) = \{u_2 \in U_2 : (u_1, u_2) \in \sigma_i(e)\}$ are called the aftersets of u_1 for $u_1 \in U_1$ and $e \in A$. Obviously, $(\underline{\Sigma_{i=1}^m \sigma_i}_o^B, A)$ and $({}^o\overline{\Sigma_{i=1}^m \sigma_i}^B, A)$ are two IFS soft sets over U_1.

Definition 16. *Let (U_1, U_2, π) be the multigranulation generalized soft approximation space over two universes U_1 and U_2, where $\pi = \sigma_1, \sigma_2, \sigma_3, ...\sigma_m$ and $B = \langle \mu_B, \gamma_B \rangle$ be an IFS in U_1. Then, the optimistic lower approximation ${}^B\underline{\Sigma_{i=1}^m \sigma_i}_o = \left({}^{\mu_B}\underline{\Sigma_{i=1}^m \sigma_i}_o, {}^{\gamma_B}\underline{\Sigma_{i=1}^m \sigma_i}_o\right)$ and the optimistic upper* *approximation* ${}^{B}\overline{\Sigma_{i=1}^{m}\sigma_i}^{o} = \left({}^{\mu_B}\overline{\Sigma_{i=1}^{m}\sigma_i}^{o}, {}^{\gamma_B}\overline{\Sigma_{i=1}^{m}\sigma_i}^{o}\right)$ *of* $B = \langle \mu_B, \gamma_B \rangle$ *are IF soft sets over* U_2 *with respect to the foresets of soft relations* $(\sigma_i, A) \in \pi$ *and are defined as:*

$$\mu_B \underline{\Sigma_{i=1}^{m}\sigma_i}_o(e)(u_2) = \begin{cases} \wedge\{\mu_B(u_1) : u_1 \in \cup_{i=1}^{m}\sigma_i(e)u_2\}, & \text{if } \cup_{i=1}^{m}\sigma_i(e)u_2 \neq \emptyset; \\ 1 & \text{otherwise;} \end{cases}$$

$$\gamma_B \underline{\Sigma_{i=1}^{m}\sigma_i}_o(e)(u_2) = \begin{cases} \vee\{\gamma_B(u_1) : u_1 \in \cap_{i=1}^{m}\sigma_i(e)u_2\}, & \text{if } \cap_{i=1}^{m}\sigma_i(e)u_2 \neq \emptyset; \\ 0 & \text{otherwise;} \end{cases}$$

and

$$\mu_B \overline{\Sigma_{i=1}^{m}\sigma_i}^{o}(e)(u_2) = \begin{cases} \vee\{\mu_B(u_1) : u_1 \in \cap_{i=1}^{m}\sigma_i(e)u_2\}, & \text{if } \cap_{i=1}^{m}\sigma_i(e)u_2 \neq \emptyset; \\ 0 & \text{otherwise;} \end{cases}$$

$$\gamma_B \overline{\Sigma_{i=1}^{m}\sigma_i}^{o}(e)(u_2) = \begin{cases} \wedge\{\gamma_B(u_1) : u_1 \in \cup_{i=1}^{m}\sigma_i(e)u_2\}, & \text{if } \cup_{i=1}^{m}\sigma_i(e)u_2 \neq \emptyset; \\ 1 & \text{otherwise,} \end{cases}$$

where $\sigma_1(e)u_2 = \{u_1 \in U_1 : (u_1, u_2) \in \sigma_i(e)\}$ *are called the foresets of* u_2 *for* $u_2 \in U_2$ *and* $e \in A$. *Obviously,* $({}^{B}\underline{\Sigma_{i=1}^{m}\sigma_i}_o, A)$ *and* $({}^{B}\overline{\Sigma_{i=1}^{m}\sigma_i}^{o}, A)$ *are two IFS soft sets over* U_2.

Moreover, $\underline{\Sigma_{i=1}^{m}\sigma_i}_o^{B} : A \to IF(U_1), {}^{o}\overline{\Sigma_{i=1}^{m}\sigma_i}^{B} : A \to IF(U_1)$ and ${}^{B}\underline{\Sigma_{i=1}^{m}\sigma_i}_o : A \to IF(U_2), {}^{B}\overline{\Sigma_{i=1}^{m}\sigma_i}^{o} : A \to IF(U_2)$.

Proposition 7. *Let* (U_1, U_2, π) *be the multigranulation generalized soft approximation space over two universes* U_1 *and* U_2 *and* $B = \langle \mu_B, \gamma_B \rangle$ *be an IFS in* U_2. *Then, the following properties for* $\underline{\Sigma_{i=1}^{m}\sigma_i}_o^{B}, {}^{o}\overline{\Sigma_{i=1}^{m}\sigma_i}^{B}$ *hold:*

(1) $\underline{\Sigma_{i=1}^{m}\sigma_i}_o^{B} \subseteq \vee_{i=1}^{m}\underline{\sigma_i}^{B}$;

(2) ${}^{o}\overline{\Sigma_{i=1}^{m}\sigma_i}^{B} \subseteq \wedge_{i=1}^{m}\overline{\sigma_i}^{B}$.

Proof. The proof is similar to the proof of Proposition 1. □

Proposition 8. *Let* (U_1, U_2, π) *be the multigranulation generalized soft approximation space over two universes* U_1 *and* U_2 *and* $B = \langle \mu_B, \gamma_B \rangle$ *be an IFS in* U_1. *Then, the following properties for* ${}^{B}\underline{\Sigma_{i=1}^{m}\sigma_i}_o, {}^{B}\overline{\Sigma_{i=1}^{m}\sigma_i}^{o}$ *hold:*

(1) ${}^{B}\underline{\Sigma_{i=1}^{m}\sigma_i}_o \subseteq \vee_{i=1}^{m}\, {}^{B}\underline{\sigma_i}$;

(2) ${}^{B}\overline{\Sigma_{i=1}^{m}\sigma_i}^{o} \subseteq \wedge_{i=1}^{m}\, {}^{B}\overline{\sigma_i}$.

Proof. The proof is similar to the proof of Proposition 2. □

Proposition 9. *Let* (U_1, U_2, π) *be the multigranulation generalized soft approximation space over two universes* U_1 *and* U_2. *Then, the following properties with respect to the aftersets hold:*

(1) $\underline{\Sigma_{i=1}^{m}\sigma_i}_o^{1_{U_2}} = 1_{U_1}$ *for all* $e \in A$;

(2) ${}^{o}\overline{\Sigma_{i=1}^{m}\sigma_i}^{1_{U_2}} = 1_{U_1}$ *if* $\cap_{i=1}^{m}u_1\sigma_i(e) \neq \emptyset$ *and* $\cup_{i=1}^{m}u_1\sigma_i(e) \neq \emptyset$, *for some* $i \leq m$;

(3) $\underline{\Sigma_{i=1}^{m}\sigma_i}_o^{0_{U_2}} = 0_{U_1}$ *if* $\cup_{i=1}^{m}u_1\sigma_i(e) \neq \emptyset$ *and* $\cap_{i=1}^{m}u_1\sigma_i(e) \neq \emptyset$ *, for some* $i \leq m$;

${}^{o}\overline{\Sigma_{i=1}^{m}\sigma_i}^{0_{U_2}} = 0_{U_1}$ *for all* $e \in A$.

Proof. The proof is similar to the proof of Proposition 3. □

Proposition 10. *Let* (U_1, U_2, π) *be the multigranulation generalized soft approximation space over two universes* U_1 *and* U_2. *Then, the following properties with respect to the foresets hold:*

(1) ${}^{1_{U_1}}\underline{\Sigma_{i=1}^{m}\sigma_i}_o = 1_{U_2}$ *for all* $e \in A$;

(2) ${}^{1_{U_1}}\overline{\Sigma_{i=1}^{m}\sigma_i}^{o} = 1_{U_2}$ *if* $\cap_{i=1}^{m}\sigma_i(e)u_2 \neq \emptyset$ *and* $\cup_{i=1}^{m}\sigma_i(e)u_2 \neq \emptyset$, *for some* $i \leq m$;

(3) ${}^{0_{U_1}}\underline{\Sigma_{i=1}^m \sigma_i}_o = 0_{U_2}$ *if* $\cup_{i=1}^m \sigma_i(e)u_2 \neq \emptyset$ *and* $\cap_{i=1}^m \sigma_i(e)u_2 \neq \emptyset$, *for some* $i \leq m$;
(4) ${}^{0_{U_1}}\overline{\Sigma_{i=1}^m \sigma_i}^o = 0_{U_2}$ *for all* $e \in A$.

Proof. The proof is similar to the proof of Proposition 4. □

Proposition 11. *Let* (U_1, U_2, π) *be the multigranulation generalized soft approximation space over two universes* U_1 *and* U_2 *and* $B, B_1, B_2 \in IF(U_2)$. *Then, the following properties for* $\underline{\Sigma_{i=1}^m \sigma_i}_o^B$, ${}^o\overline{\Sigma_{i=1}^m \sigma_i}^B$ *with respect the aftersets hold:*
(1) *If* $B_1 \subseteq B_2$ *then* $\underline{\Sigma_{i=1}^m \sigma_i}_o^{B_1} \subseteq \underline{\Sigma_{i=1}^m \sigma_i}_o^{B_2}$;
(2) *If* $B_1 \subseteq B_2$ *then* ${}^o\overline{\Sigma_{i=1}^m \sigma_i}^{B_1} \subseteq^o \overline{\Sigma_{i=1}^m \sigma_i}^{B_2}$;
(3) $\underline{\Sigma_{i=1}^m \sigma_i}_o^{B_1 \cap B_2} = \underline{\Sigma_{i=1}^m \sigma_i}_o^{B_1} \cap \underline{\Sigma_{i=1}^m \sigma_i}_o^{B_2}$;
(4) $\underline{\Sigma_{i=1}^m \sigma_i}_o^{B_1 \cup B_2} \supseteq \underline{\Sigma_{i=1}^m \sigma_i}_o^{B_1} \cup \underline{\Sigma_{i=1}^m \sigma_i}_o^{B_2}$;
(5) ${}^o\overline{\Sigma_{i=1}^m \sigma_i}^{B_1 \cup B_2} =^o \overline{\Sigma_{i=1}^m \sigma_i}^{B_1} \cup^o \overline{\Sigma_{i=1}^m \sigma_i}^{B_2}$;
(6) ${}^o\overline{\Sigma_{i=1}^m \sigma_i}^{B_1 \cap B_2} \subseteq^o \overline{\Sigma_{i=1}^m \sigma_i}^{B_1} \cap^o \overline{\Sigma_{i=1}^m \sigma_i}^{B_2}$.

Proof. The proof is similar to the proof of Proposition 5. □

Proposition 12. *Let* (U_1, U_2, π) *be the multigranulation generalized soft approximation space over two universes* U_1 *and* U_2 *and* $B, B_1, B_2 \in IF(U_1)$. *Then, the following properties for* ${}^B\underline{\Sigma_{i=1}^m \sigma_i}_o$, ${}^B\overline{\Sigma_{i=1}^m \sigma_i}^o$ *with respect the foresets hold:*
(1) *If* $B_1 \subseteq B_2$ *then* ${}^{B_1}\underline{\Sigma_{i=1}^m \sigma_i}_o \subseteq^{B_2} \underline{\Sigma_{i=1}^m \sigma_i}_o$;
(2) *If* $B_1 \subseteq B_2$ *then* ${}^{B_1}\overline{\Sigma_{i=1}^m \sigma_i}^o \subseteq^{B_2} \overline{\Sigma_{i=1}^m \sigma_i}^o$;
(3) ${}^{B_1 \cap B_2}\underline{\Sigma_{i=1}^m \sigma_i}_o =^{B_1} \underline{\Sigma_{i=1}^m \sigma_i}_o \cap^{B_2} \underline{\Sigma_{i=1}^m \sigma_i}_o$;
(4) ${}^{B_1 \cup B_2}\underline{\Sigma_{i=1}^m \sigma_i}_o \supseteq^{B_1} \underline{\Sigma_{i=1}^m \sigma_i}_o \cup^{B_2} \underline{\Sigma_{i=1}^m \sigma_i}_o$;
(5) ${}^{B_1 \cup B_2}\overline{\Sigma_{i=1}^m \sigma_i}^o =^{B_1} \overline{\Sigma_{i=1}^m \sigma_i}^o \cup^{B_2} \overline{\Sigma_{i=1}^m \sigma_i}^o$;
(6) ${}^{B_1 \cap B_2}\overline{\Sigma_{i=1}^m \sigma_i}^o \subseteq^{B_1} \overline{\Sigma_{i=1}^m \sigma_i}^o \cap^{B_2} \overline{\Sigma_{i=1}^m \sigma_i}^o$.

Proof. The proof is similar to the proof of Proposition 6. □

Proposition 13. *Let* (U_1, U_2, π) *be the multigranulation generalized soft approximation space over two universes* U_1 *and* U_2 *and* $B_1, B_2, B_3, \ldots B_n \in IF(U_2)$, *and* $B_1 \subseteq B_2 \subseteq B_3 \subseteq \ldots \subseteq B_n$. *Then, the following properties with respect the aftersets hold:*
(1) $\underline{\Sigma_{i=1}^m \sigma_i}_o^{B_1} \subseteq \underline{\Sigma_{i=1}^m \sigma_i}_o^{B_2} \subseteq \underline{\Sigma_{i=1}^m \sigma_i}_o^{B_3} \subseteq \ldots\ldots \subseteq \underline{\Sigma_{i=1}^m \sigma_i}_o^{B_n}$;
(2) ${}^o\overline{\Sigma_{i=1}^m \sigma_i}^{B_1} \subseteq^o \overline{\Sigma_{i=1}^m \sigma_i}^{B_2} \subseteq^o \overline{\Sigma_{i=1}^m \sigma_i}^{B_3} \subseteq \ldots\ldots \subseteq^o \overline{\Sigma_{i=1}^m \sigma_i}^{B_n}$.

Proof. The proof is similar to the proof of Proposition 5. □

Proposition 14. *Let* (U_1, U_2, π) *be the multigranulation generalized soft approximation space over two universes* U_1 *and* U_2 *and* $B_1, B_2, B_3, \ldots B_n \in IF(U_1)$, *and* $B_1 \subseteq B_2 \subseteq B_3 \subseteq \ldots \subseteq B_n$. *Then, the following properties with respect the foresets hold:*
(1) ${}^{B_1}\underline{\Sigma_{i=1}^m \sigma_i}_o \subseteq^{B_2} \underline{\Sigma_{i=1}^m \sigma_i}_o \subseteq^{B_3} \underline{\Sigma_{i=1}^m \sigma_i}_o \subseteq \ldots\ldots \subseteq^{B_n} \underline{\Sigma_{i=1}^m \sigma_i}_o$;
(2) ${}^{B_1}\overline{\Sigma_{i=1}^m \sigma_i}^o \subseteq^{B_2} \overline{\Sigma_{i=1}^m \sigma_i}^o \subseteq^{B_1} \overline{\Sigma_{i=1}^m \sigma_i}^o \subseteq \ldots\ldots \subseteq^{B_n} \overline{\Sigma_{i=1}^m \sigma_i}^o$.

Proof. The proof is similar to the proof of Proposition 6. □

5. Application in Decision-Making Problem

Decision making is a major area of study in almost all types of data analysis. To select effective alternatives from aspirants is the process of decision making. Since our environment is becoming changeable and complicated day by day and the decision-making process proposed by a single expert is no longer good, therefore, a decision-making algorithm based on consensus by using collective wisdom is a better approach. From imprecise multi-observer data, Maji et al. [71] proposed a useful technique of object recognition. Feng et al. [72] pointed out errors in Maji et al. [71] and rebuilt a framework correctly. Shabir et al. [63] presented MGRS model based on soft relations by using crisp sets and proposed a decision-making algorithm. Jamal and Shabir [64] presented a decision-making algorithm by using the OMGRS model in terms of FS based on soft relations. This paper extends Jamal's OMGFRS model and presents the decision-making method based on multi-soft relations by use of OMGIFRS.

The lower and upper approximations are the closest to approximated subsets of a universe. We obtain two corresponding values $\underline{\Sigma_{i=1}^{m}\sigma_i}_o^B(e_j)(x_k)$ and ${}^o\overline{\Sigma_{i=1}^{m}\sigma_i}^B(e_j)(x_k)$ with respect to the afterset to the decision alternative $x_k \in U_1$ by the IF soft lower and upper approximations of an IF $B \in IF(U_2)$.

We present Algorithms 1 and 2 for our proposed model here.

Algorithm 1: Aftersets for decision-making problem

(1) Compute the optimistic multigranulation lower IF soft set approximation $\underline{\Sigma_{i=1}^{m}\sigma_i}_o^B$ and optimistic multigranulation upper IF soft set approximation ${}^o\overline{\Sigma_{i=1}^{m}\sigma_i}^B$ of an IF set $B = \langle \mu_B, \gamma_B \rangle$ with respect to the aftersets;
(2) Compute the score values for each of the entries of the $\underline{\Sigma_{i=1}^{m}\sigma_i}_o^B$ and ${}^o\overline{\Sigma_{i=1}^{m}\sigma_i}^B$ and denote them by $\underline{S}_{ij}(x_i, e_j)$ and $\overline{S}_{ij}(x_i, e_j)$ for all i, j;
(3) Compute the aggregated score $\underline{S}(x_i) = \sum_{j=1}^{n} \underline{S}_{ij}(x_i, e_j)$ and $\overline{S}(x_i) = \sum_{j=1}^{n} \overline{S}_{ij}(x_i, e_j)$;
(4) Compute $S(x_i) = \underline{S}(x_i) + \overline{S}(x_i)$;
(5) The best decision is $x_k = \max_i S(x_i)$;
(6) If k has more than one value, say k_1, k_2, then we calculate the accuracy values $\underline{H}_{ij}(x_i, e_j)$ and $\overline{H}_{ij}(x_i, e_j)$ for only those x_k for which $S(x_k)$ are equal;
(7) Compute $H(x_k) = \sum_{j=1}^{n} \underline{H}_{kj}(x_k, e_j) + \sum_{j=1}^{n} \overline{H}_{kj}(x_k, e_j)$ for $k = k_1, k_2$;
(8) If $H(x_{k_1}) > H(x_{k_2})$, then we select x_{k_1};
(9) If $H(x_{k_1}) = H(x_{k_2})$, then select any one of x_{k_1} and x_{k_2}.

Algorithm 2: Foresets for decision-making problem

(1) Compute the optimistic multigranulation lower IF soft set approximation ${}^B\underline{\Sigma_{i=1}^{m}\sigma_i}_o^B$ and upper multigranulation IF soft set approximation ${}^B\overline{\Sigma_{i=1}^{m}\sigma_i}^o$ of an IF set $B = \langle \mu_B, \gamma_B \rangle$ with respect to the foresets;
(2) Compute the score values for each of the entries of the ${}^B\underline{\Sigma_{i=1}^{m}\sigma_i}_o^B$ and ${}^B\overline{\Sigma_{i=1}^{m}\sigma_i}^o$ and denote them by $\underline{S}_{ij}(x_i, e_j)$ and $\overline{S}_{ij}(x_i, e_j)$ for all i, j;
(3) Compute the aggregated score $\underline{S}(x_i) = \sum_{j=1}^{n} \underline{S}_{ij}(x_i, e_j)$ and $\overline{S}(x_i) = \sum_{j=1}^{n} \overline{S}_{ij}(x_i, e_j)$;
(4) Compute $S(x_i) = \underline{S}(x_i) + \overline{S}(x_i)$;
(5) The best decision is $x_k = \max_i S(x_i)$;
(6) If k has more than one value, say k_1, k_2, then we calculate the accuracy values $\underline{H}_{ij}(x_i, e_j)$ and $\overline{H}_{ij}(x_i, e_j)$ for only those x_k for which $S(x_k)$ are equal;
(7) Compute $H(x_k) = \sum_{j=1}^{n} \underline{H}_{kj}(x_k, e_j) + \sum_{j=1}^{n} \overline{H}_{kj}(x_k, e_j)$ for $k = k_1, k_2$;
(8) If $H(x_{k_1}) > H(x_{k_2})$ then we select x_{k_1};
(9) If $H(x_{k_1}) = H(x_{k_2})$ then select any one of x_{k_1} and x_{k_2}.

Now, we show the proposed approach of decision making step by step by using following example. The following example discusses an algorithm to make a wise decision for the selection of a car.

Example 3. *Suppose a multi-national company wants to select a best officer and there are 10 short-listed applicants which are categorized in two groups, platinum and diamond. The set* $U_1 = \{m_1, m_2, m_3, m_4, m_5, m_6\}$ *represents the applicants of platinum group and* $U_2 = \{c_1, c_2, c_3, c_4\}$ *represents the applicants of diamond group. Let* $A = \{e_1, e_2, e_3\}$ *={*e_1*=education,*e_2*=experience,* e_3*=computer knowledge} be the set of parameters. Let two different teams of interviewers analyze and compare the competencies of these applicants.*

We have $\sigma_1 : A \to P(U_1 \times U_2)$ represent the comparison of the first-interviewer team defined by

$$\begin{aligned}
\sigma_1(e_1) &= \{(m_1,c_1),(m_1,c_2),(m_2,c_2),(m_2,c_4),(m_4,c_2),(m_4,c_3),(m_5,c_3),(m_5,c_4),(m_6,c_1)\}, \\
\sigma_1(e_2) &= \{(m_1,c_1),(m_2,c_3),(m_4,c_1),(m_5,c_1),(m_6,c_2),(m_6,c_3)\}, \\
\text{and } \sigma_1(e_3) &= \{(m_1,c_1),(m_2,c_4),(m_3,c_1),(m_3,c_3),(m_4,c_1),(m_5,c_3),(m_5,c_4)\},
\end{aligned}$$

where $\sigma_1(e_1)$ compares the education of applicants, $\sigma_1(e_2)$ compares the experience of applicants, $\sigma_1(e_3)$ compares the computer knowledge of applicants.

Similarly, $\sigma_2 : A \to P(U_1 \times U_2)$ represent the comparison of the second-interviewer team defined by

$$\begin{aligned}
\sigma_2(e_1) &= \{(m_1,c_1),(m_1,c_2),(m_2,c_3),(m_3,c_4),(m_4,c_2),(m_5,c_2),(m_6,c_3)\}, \\
\sigma_2(e_2) &= \{(m_1,c_1),(m_3,c_2),(m_4,c_1),(m_6,c_4)\}, \\
\text{and } \sigma_2(e_3) &= \{(m_1,c_1),(m_1,c_3),(m_2,c_2),(m_2,c_3),(m_4,c_1),(m_5,c_4),(m_6,c_4)\},
\end{aligned}$$

where $\sigma_2(e_1)$ compares the education of applicants, $\sigma_2(e_2)$ compares the experience of applicants, $\sigma_2(e_3)$ compares the computer knowledge of applicants.

From these comparisons, we obtain two soft relations from U_1 to U_2. Now, the aftersets

$$\begin{aligned}
m_1\sigma_1(e_1) &= \{c_1,c_2\},\ m_2\sigma_1(e_1) = \{c_2,c_4\},\ m_3\sigma_1(e_1) = \emptyset, \\
m_4\sigma_1(e_1) &= \{c_2,c_3\},\ m_5\sigma_1(e_1) = \{c_3,c_4\},\ m_6\sigma_1(e_1) = \{c_1\} \text{ and} \\
m_1\sigma_1(e_2) &= \{c_1\},\quad m_2\sigma_1(e_2) = \{c_3\},\quad m_3\sigma_1(e_2) = \emptyset, \\
m_4\sigma_1(e_2) &= \{c_1\},\quad m_5\sigma_1(e_2) = \{c_1\},\quad m_6\sigma_1(e_2) = \{c_2,c_3\}, \text{ and} \\
m_1\sigma_1(e_3) &= \{c_1\},\quad m_2\sigma_1(e_3) = \{c_4\},\quad m_3\sigma_1(e_3) = \{c_1,c_3\}, \\
m_4\sigma_1(e_3) &= \{c_1\},\quad m_5\sigma_1(e_3) = \{c_3,c_4\},\ m_6\sigma_1(e_3) = \emptyset, \text{ and} \\
m_1\sigma_2(e_1) &= \{c_1,c_2\},\ m_2\sigma_2(e_1) = \{c_3\},\quad m_3\sigma_2(e_1) = \{c_4\}, \\
m_4\sigma_2(e_1) &= \{c_2\},\quad m_5\sigma_2(e_1) = \{c_2\},\quad m_6\sigma_2(e_1) = \{c_3\} \text{ and} \\
m_1\sigma_2(e_2) &= \{c_1\},\quad m_2\sigma_2(e_2) = \emptyset,\quad m_3\sigma_2(e_2) = \{c_2\}, \\
m_4\sigma_2(e_2) &= \{c_1\},\quad m_5\sigma_2(e_2) = \emptyset,\quad m_6\sigma_2(e_2) = \{c_4\}, \text{ and} \\
m_1\sigma_2(e_3) &= \{c_1,c_3\},\ m_2\sigma_2(e_3) = \{c_2,c_3\},\ m_3\sigma_2(e_3) = \emptyset, \\
m_4\sigma_2(e_3) &= \{c_1\},\quad m_5\sigma_2(e_3) = \{c_4\},\quad m_6\sigma_2(e_3) = \{c_4\},
\end{aligned}$$

where $m_i\sigma_j(e_1)$ represents all those applicants of the diamond group whose education is equal to m_i, $m_i\sigma_j(e_2)$ represents all those applicants of the diamond group whose experience is equal to m_i and $m_i\sigma_j(e_3)$ represents all those applicants of the diamond group whose computer knowledge is equal to m_i. In addition, foresets

$$
\begin{aligned}
\sigma_1(e_1)c_1 &= \{m_1, m_6\}, \quad \sigma_1(e_1)c_2 = \{m_1, m_2, m_4\}, \ \sigma_1(e_1)c_3 = \{m_4, m_5\}, \ \sigma_1(e_1)c_4 = \{m_2, m_5\}, \text{ and} \\
\sigma_1(e_2)c_1 &= \{m_1, m_4, m_5\}, \ \sigma_1(e_2)c_2 = \{m_6\}, \quad \sigma_1(e_2)c_3 = \{m_2, m_6\}, \ \sigma_1(e_2)c_4 = \emptyset, \text{ and} \\
\sigma_1(e_3)c_1 &= \{m_1, m_3, m_4\}, \ \sigma_1(e_3)c_2 = \emptyset, \quad \sigma_1(e_3)c_3 = \{m_3, m_5\}, \sigma_1(e_3)c_4 = \{m_2, m_5\}. \\
\sigma_2(e_1)c_1 &= \{m_1\}, \quad \sigma_2(e_1)c_2 = \{m_1, m_4, m_5\}, \ \sigma_2(e_1)c_3 = \{m_2, m_6\}, \ \sigma_2(e_1)c_4 = \{m_3\}, \text{ and} \\
\sigma_2(e_2)c_1 &= \{m_1, m_4\}, \quad \sigma_2(e_2)c_2 = \{m_3\}, \quad \sigma_2(e_2)c_3 = \emptyset, \quad \sigma_2(e_2)c_4 = \{m_6\}, \text{ and} \\
\sigma_2(e_3)c_1 &= \{m_1, m_4\}, \quad \sigma_2(e_3)c_2 = \{m_2\}, \quad \sigma_1(e_3)c_3 = \{m_1, m_2\}, \ \sigma_2(e_3)c_4 = \{m_5, m_6\},
\end{aligned}
$$

where $\sigma_j(e_1)c_i$ represents all those applicants of the platinum group whose education is equal to c_i, $\sigma_j(e_2)c_i$ represents all those applicants of the platinum group whose experience is equal to c_i and $\sigma_j(e_3)c_i$ represents all those applicants of the platinum group whose computer knowledge is equal to c_i.

$$
\begin{aligned}
\text{Define } B &= \langle \mu_B, \gamma_B \rangle : U_2 \to [0,1] \text{ which represents the preference of applicants given by a multi-national company such that} \\
\mu_B(c_1) &= 0.9, \ \mu_B(c_2) = 0.8, \ \mu_B(c_3) = 0.4, \ \mu_B(c_4) = 0 \text{ and} \\
\gamma_B(c_1) &= 0.0, \ \gamma_B(c_2) = 0.2, \ \gamma_B(c_3) = 0.5, \ \gamma_B(c_4) = 0.8. \\
\text{Define } B_1 &= \langle \mu_{B_1}, \gamma_{B_1} \rangle : U_1 \to [0,1] \text{ which represents the preference of applicants given by a multi-national company such that} \\
\mu_{B_1}(m_1) &= 1, \ \mu_{B_1}(m_2) = 0.7, \ \mu_{B_1}(m_3) = 0.5, \ \mu_{B_1}(m_4) = 0.1, \\
\mu_{B_1}(m_5) &= 0, \ \mu_{B_1}(m_6) = 0.4 \text{ and} \\
\gamma_{B_1}(m_1) &= 0, \ \gamma_{B_1}(m_2) = 0.2, \ \gamma_{B_1}(m_3) = 0.5, \ \gamma_{B_1}(m_4) = 0.7, \\
\gamma_{B_1}(m_5) &= 1, \ \gamma_{B_1}(m_6) = 0.5.
\end{aligned}
$$

Therefore, the optimistic multigranulation lower and upper approximations (with respect to the aftersets as well as with respect to the foresets) are given in Tables 8 and 9.

$$
\begin{aligned}
\underline{\sigma_1 + \sigma_2}_o^B &= (\underline{\sigma_1 + \sigma_2}_o^{\mu_B}, \underline{\sigma_1 + \sigma_2}_o^{\gamma_B}), \\
{}^o\overline{\sigma_1 + \sigma_2}^B &= ({}^o\overline{\sigma_1 + \sigma_2}^{\mu_B}, {}^o\overline{\sigma_1 + \sigma_2}^{\gamma_B}).
\end{aligned}
$$

Table 8. Optimistic multigranulation lower approximations of B.

$\underline{\sigma_1 + \sigma_2}_o^{\mu_B}, \underline{\sigma_1 + \sigma_2}_o^{\gamma_B}$	m_1	m_2	m_3	m_4	m_5	m_6
$\underline{\sigma_1 + \sigma_2}_o^{\mu_B}(e_1)$	0.8	0	0	0.4	0	0.4
$\underline{\sigma_1 + \sigma_2}_o^{\mu_B}(e_2)$	0.9	0.4	0.8	0.9	0.9	0
$\underline{\sigma_1 + \sigma_2}_o^{\mu_B}(e_3)$	0.4	0	0.4	1	0	0
$\underline{\sigma_1 + \sigma_2}_o^{\gamma_B}(e_1)$	0.2	0.8	0.8	0.5	0.8	0.5
$\underline{\sigma_1 + \sigma_2}_o^{\gamma_B}(e_2)$	0	0.5	0.2	0	0	0.8
$\underline{\sigma_1 + \sigma_2}_o^{\gamma_B}(e_3)$	0.5	0.8	0.5	0	0.8	0.8

Table 8 shows the exact degree of competency of applicant m_i to B in education, experience and computer knowledge.

Table 9 shows the possible degree of competency of applicant m_i to B in education, experience and computer knowledge.

In Table 10, $S(m_1) = S(m_4)$, so we calculate accuracy values for m_1 and m_4, as shown in Table 11.

Table 9. Optimistic multigranulation upper approximations of B.

${}^{o}\overline{\sigma_1+\sigma_2}^{\mu_B}, {}^{o}\overline{\sigma_1+\sigma_2}^{\gamma_B}$	m_1	m_2	m_3	m_4	m_5	m_6
${}^{o}\overline{\sigma_1+\sigma_2}^{\mu_B}(e_1)$	0.9	0	0	0.8	0	0
${}^{o}\overline{\sigma_1+\sigma_2}^{\mu_B}(e_2)$	0.9	0	0	0.9	0	0
${}^{o}\overline{\sigma_1+\sigma_2}^{\mu_B}(e_3)$	0.9	0	0	0	0	0
${}^{o}\overline{\sigma_1+\sigma_2}^{\gamma_B}(e_1)$	0	1	1	0.2	1	1
${}^{o}\overline{\sigma_1+\sigma_2}^{\gamma_B}(e_2)$	0	1	1	0	1	1
${}^{o}\overline{\sigma_1+\sigma_2}^{\gamma_B}(e_3)$	0	1	1	1	0.8	1

Table 10. Values of score function of applicants.

	$\underline{S}_{ij}(e_1)$	$\underline{S}_{ij}(e_2)$	$\underline{S}_{ij}(e_3)$	$\overline{S}_{ij}(e_1)$	$\overline{S}_{ij}(e_2)$	$\overline{S}_{ij}(e_3)$	$\underline{S}(x_i)$	$\overline{S}(x_i)$	$S(x_i)$
m_1	0.6	0.9	−0.1	0.9	0.9	0.9	1.4	2.7	4.1
m_2	−0.8	−0.1	−0.8	−1	−1	−1	−1.7	−3	−4.7
m_3	−0.8	0.6	−0.1	−1	−1	−1	−0.3	−3	−3.3
m_4	−0.1	0.9	0.9	0.6	0.9	0.9	1.7	2.4	4.1
m_5	−0.8	0.9	−0.8	−1	−1	−0.8	−0.7	−2.8	−3.5
m_6	−0.1	−0.8	−0.8	−1	−1	−1	1.7	−3	−4.7

Table 11. Values of accuracy function.

	$\underline{H}_{ij}(e_1)$	$\underline{H}_{ij}(e_2)$	$\underline{H}_{ij}(e_3)$	$\overline{H}_{ij}(e_1)$	$\overline{H}_{ij}(e_2)$	$\overline{H}_{ij}(e_3)$	H
m_1	1	0.9	0.9	0.9	0.9	0.9	5.5
m_6	0.9	0.9	1	1	0.9	1	5.7

It is shown in Table 11 that $H(m_6) = 5.7$ is maximum. Therefore, a multi-national company will select applicant m_6.

Therefore, the optimistic multigranulation lower and upper approximations (with respect to the foresets) are given in Tables 12 and 13.

$$
\begin{aligned}
{}^{B}\underline{\sigma_1+\sigma_2}_o &= ({}^{\mu_B}\underline{\sigma_1+\sigma_2}_o, {}^{\gamma_B}\underline{\sigma_1+\sigma_2}_o), \\
{}^{B}\overline{\sigma_1+\sigma_2}^o &= ({}^{\mu_B}\overline{\sigma_1+\sigma_2}^o, {}^{\gamma_B}\overline{\sigma_1+\sigma_2}^o).
\end{aligned}
$$

Table 12. Optimistic multigranulation lower approximations of B.

${}^{\mu_B}\underline{\sigma_1+\sigma_2}_o, {}^{\gamma_B}\underline{\sigma_1+\sigma_2}_o$	c_1	c_2	c_3	c_4
${}^{\mu_B}\underline{\sigma_1+\sigma_2}_o(e_1)$	0.4	0	0	0
${}^{\mu_B}\underline{\sigma_1+\sigma_2}_o(e_2)$	0	0.4	0.4	0.4
${}^{\mu_B}\underline{\sigma_1+\sigma_2}_o(e_3)$	0.1	0.7	0	0
${}^{\gamma_B}\underline{\sigma_1+\sigma_2}_o(e_1)$	0.5	1	1	1
${}^{\gamma_B}\underline{\sigma_1+\sigma_2}_o(e_2)$	1	0.5	0.5	0.5
${}^{\gamma_B}\underline{\sigma_1+\sigma_2}_o(e_3)$	0.7	0.2	1	1

Table 12 shows the exact degree of competency of applicant c_i to B in education, experience and computer knowledge.

Table 13 shows the possible degree of competency of applicant c_i to B in education, experience and computer knowledge.

Table 13. Optimistic multigranulation upper approximations of B.

$\mu_B\overline{\sigma_1+\sigma_2}^o, \gamma_B\overline{\sigma_1+\sigma_2}^o$	c_1	c_2	c_3	c_4
$\mu_B\overline{\sigma_1+\sigma_2}^o(e_1)$	1	0.1	0	0
$\mu_B\overline{\sigma_1+\sigma_2}^o(e_2)$	1	0	0	0
$\mu_B\overline{\sigma_1+\sigma_2}^o(e_3)$	1	0	0	0
$\gamma_B\overline{\sigma_1+\sigma_2}^o(e_1)$	0	0.7	1	1
$\gamma_B\overline{\sigma_1+\sigma_2}^o(e_2)$	0	1	1	1
$\gamma_B\overline{\sigma_1+\sigma_2}^o(e_3)$	0	1	1	1

It is shown in Table 14 that $S(c_1) = 1.3$ is maximum. Therefore, a multi-national company will select applicant c_1.

Table 14. Values of score function of colors of car.

	$\underline{S}_{ij}(e_1)$	$\underline{S}_{ij}(e_2)$	$\underline{S}_{ij}(e_3)$	$\overline{S}_{ij}(e_1)$	$\overline{S}_{ij}(e_2)$	$\overline{S}_{ij}(e_3)$	$\underline{S}(x_i)$	$\overline{S}(x_i)$	$S(x_i)$
c_1	−0.1	−1	−0.6	1	1	1	−1.7	3	1.3
c_2	−1	−0.1	0.5	−0.6	−1	−1	−0.6	−2.6	−3.2
c_3	−1	−0.1	−1	−1	−1	−1	−2.1	−3	−5.1
c_4	−1	−0.1	−1	−1	−1	−1	−2.1	−3	−5.1

6. Comparison

The RS describes a target set by a lower and upper approximation based on single granulation. However, the multiple granulation with approximations of a target set is needed in many real world problems as well. For example, Qian et al. [41,42] built a framework of OMGRS and PMGRS by getting inspiration of multi-source datasets and multiple granulation is needed by multi-scale data for set approximations [73]. Many things are different when comparing our work with existing theories. Mainly, we make a note on the differences of our work and existing ones, such as angle of thinking, MGRS environment and research objective. Our research with respect to the angle of thinking is different from other existing theories. For a comparative study, our proposed model transforms decision-making systems into a formal decision context. Our study is different from the existing ones in [41,63,74] in terms of MGRS because our work is about IFSs which are useful in dealing with uncertainty. In [63], Shabir et al. used crisp sets to present MGRS model based on soft relations. Later, they used a FS instead of a crisp set and presented OMGFRS [64]. We extended the OMGFRS model in terms of IFS and proposed OMGIFRS model based on soft binary relations to make better decision in decision making-problems. An IFS is better than a crisp set or a FS to discuss the uncertainty. In IFS, an element is described with membership degree as well as non-membership degree but in FS, an element is described with membership degree only. That is why our proposed model has more capability to reveal the uncertainty because of IFS. Furthermore, we have used soft relations which have many applications in dealing with uncertainty because of its parameterized collection of binary relations.

7. Conclusions

This paper proposes the MGRS model in terms of IFS based on soft binary relations over two universes. First of all, we defined granulation roughness based on two soft binary relations using IFSs with respect to the aftersets and foresets. In this way, we obtain two

IFSSs with respect to the aftersets and foresets. Some properties of OMGIFRS have been studied. Then, we generalized these concepts to granulation roughness of an IFS based on multi-soft relations and discussed their properties. We presented a decision-making algorithm regarding the aftersets and foresets with an example in practical decision-making problem. In IFS, the sum of membership degree, non-membership degree and hesitant degree of an element is less than or equal to 1. However, in some decision-making problems, the sum of membership degree, non-membership degree and hesitant degree of an element may be greater than 1. In this case, the Pythagorean fuzzy set which is an extension of IFS is the better set to deal with uncertainty. The Pythagorean fuzzy set extension makes better improvement in applicability and flexibility of IFS. Further work may be discussed about investigation of pessimistic MGRS of an IFS based on soft relations. Other OMGIFRS models with interval valued IFSs, uncertain linguistic FSs, basic uncertain information SSs, linguistic Z-number FSSs and Pythagorean FSs may be discussed in future.

Author Contributions: Conceptualization, S.B. and M.S.; methodology, M.Z.A.; software, S.B.; validation, M.G.A.; formal analysis, M.Z.A.; investigation, M.Z.A.; resources, M.G.A.; data curation, M.S.; writing—original draft preparation, M.Z.A.; writing—review and editing, S.B.; visualization, M.Z.A.; supervision, S.B.; project administration, M.Z.A.; funding acquisition, M.G.A. All authors have read and agreed to the published version of the manuscript.

Funding: This research received no external funding.

Institutional Review Board Statement: Not applicable.

Informed Consent Statement: Not applicable.

Data Availability Statement: We did not use any data for this research work.

Acknowledgments: The researchers would like to thank the Deanship of Scientific Research, Qassim University for funding the publication of this project.

Conflicts of Interest: The authors declare that they have no conflicts of interest.

References

1. Zadeh, L.A. Fuzzy sets. *Inform. Control* **1965**, *8*, 338–353. [CrossRef]
2. Atanassov, K.T. Intuitionistic fuzzy sets. *Fuzzy Sets Syst.* **1986**, *20*, 87–96. [CrossRef]
3. Atanassov, K.T. *Intuitionistic Fuzzy Sets: Theory and Application*; Springer: Berlin/Heidelberg, Germany, 1999.
4. Ejegwa, P.A.; Onoja, A.M.; Emmanuel, I.T. A note on some models of intuitionistic fuzzy sets in real life situations. *J. Glob. Res. Math. Arch.* **2014**, *2*, 42–50.
5. Molodtsov, D. Soft set theory—First results. *Comput. Math. Appl.* **1999**, *37*, 19–31. [CrossRef]
6. Pawlak, Z. Rough sets. *Int. J. Comput. Inform. Sci.* **1982**, *11*, 341–356. [CrossRef]
7. Khan, M.J.; Kumam, P.; Liu, P.; Ashraf, K. A novel approach to generalized intuitionistic fuzzy soft sets and its application in decision support system. *Mathematics* **2019**, *7*, 742. [CrossRef]
8. Shabir, M.; Ayub, S.; Bashir, S. Prime and Semiprime L-Fuzzy Soft Bi-Hyperideals. *J. Hyperstruct.* **2017**, *6*, 109–119.
9. Shabir, M.; Ayub, S.; Bashir, S. Applications of L-Fuzzy soft sets in semihypergroups. *J. Adv. Math. Stud.* **2017**, *10*, 367–385.
10. Maji, P.K.; Biswas, R.; Roy, A.R. Fuzzy soft sets. *J. Fuzzy Math.* **2001**, *9*, 589–602.
11. Maji, P.K.; Biswas, R.; Roy, A.R. Intuitionistic fuzzy soft sets. *J. Fuzzy Math.* **2001**, *9*, 677–692.
12. Xu, W.; Ma, J.; Wang, S.; Hao, G. Vague soft sets and their properties. *Comput. Math. Appl.* **2010**, *59*, 787–794. [CrossRef]
13. Ali, M.I. A note on soft sets, rough soft sets and fuzzy soft sets. *Appl. Soft Comput.* **2011**, *11*, 3329–3332.
14. Ali, M.I.; Davvaz, B.; Shabir, M. Some properties of generalized rough sets. *Inf. Sci.* **2013**, *224*, 170–179. [CrossRef]
15. Majumdar, P.; Samanta, S.K. Generalised fuzzy soft sets. *Comput. Math. Appl.* **2010**, *59*, 1425–1432. [CrossRef]
16. Xiao, Z.; Xia, S.; Gong, K.; Li, D. The trapezoidal fuzzy soft set and its application in MCDM. *Appl. Math. Model.* **2012**, *36*, 5844–5855. [CrossRef]
17. Yang, X.B.; Lin, T.Y.; Yang, J.Y.; Li, Y.; Yu, D.Y. Combination of interval-valued fuzzy set and soft set. *Comput. Math. Appl.* **2009**, *58*, 521–527. [CrossRef]
18. Agarwal, M.; Biswas, K.K.; Hanmandlu, M. Generalized intuitionistic fuzzy soft sets with applications in decision-making. *Appl. Soft Comput.* **2013**, *13*, 3552–3566. [CrossRef]
19. Feng, F.; Fujita, H.; Ali, M.I.; Yager, R.R. Another view on generalized intuitionistic fuzzy soft sets and related multi attribute decision making methods. *IEEE Trans. Fuzzy Syst.* **2018**, *27*, 474–488. [CrossRef]
20. Samanta, S.K.; Mondal, T.K. Intuitionistic fuzzy rough sets and rough intuitionistic fuzzy sets. *J. Fuzzy Math.* **2001**, *9*, 561–582.
21. Chakrabarty, K.; Biswas, R.; Nanda, S. Fuzziness in rough sets. *Fuzzy Sets Syst.* **2000**, *110*, 247–251. [CrossRef]

22. Sang, M.Y.; Seok, J.L. New approach to intuitionistic fuzzy rough sets. *Int. J. Fuzzy Log. Intell. Syst.* **2020**, *20*, 129–137.
23. Kumar, S.S.; Inbarani, H.H. Optimistic multi-granulation rough set based classification for medical diagnosis. *Procedia Comput. Sci.* **2015**, *47*, 374–382. [CrossRef]
24. Tsang, E.C.; Degang, C.; Yeung, D.S. Approximations and reducts with covering generalized rough sets. *Comput. Math. Appl.* **2008**, *56*, 279–289. [CrossRef]
25. Didier, D.; Prade, H. Rough Fuzzy sets and fuzzy rough sets. *Int. J. Gen. Syst.* **1990**, *17*, 191–209.
26. Cao, B.Y.; Yang, J.H. Advances in fuzzy geometric programming. In *Fuzzy Information and Engineering;* Advances in Soft Computing; Springer: Berlin/Heidelberg, Germany, 2007; Volume40, pp. 497–502.
27. Bashir, S.; Abbass, H.; Mazhar, R.; Shabir, M. Rough fuzzy ternary subsemigroups based on fuzzy ideals with 3-Dimensional congruence relation. *Comput. Appl. Math.* **2020,** *39*, 1–16. [CrossRef]
28. Sun, B.Z.; Ma, W.M.; Qian, Y.H. Multigranulation fuzzy rough set over two universes and its application to decision making. *Knowl. Based Syst.* **2017**, *123*, 61–74. [CrossRef]
29. Wong, S.K.; Wang, L.S.; Yao, Y.Y. Interval structure: A framework for representing uncertain information. In Proceeding of the UAI92, Stanford, CA, USA, 17–19 July 1992; pp. 336–343.
30. Wong, S.K.M.; Wang, L.S.; Yao, Y.Y. On modeling uncertainty with interval structures. *Comput. Intell.* **1995**, *11*, 406–426. [CrossRef]
31. Liu, G.L. Rough set theory based on two universal sets and its applications. *Knowl. Based Syst.* **2010**, *23*, 110–115. [CrossRef]
32. Pei, D.W.; Xu, Z.B. Rough set models on two universes. *Int. J. Gen. Syst.* **2004**, *33*, 569–581. [CrossRef]
33. Liu, C.H.; Miao, D.Q.; Zhang, N. Graded rough set model based on two universes and its properties. *Knowl. Based Syst.* **2012**, *33*, 65–72. [CrossRef]
34. Ma, W.M.; Sun, B.Z. Probabilistic rough set over two universes and rough entropy. *Int. J. Approx. Reason.* **2012**, *53*, 608–619. [CrossRef]
35. Tan, A.; Wu, W.Z.; Shi, S.; Zhao, S. Granulation selection and decision making with multigranulation rough set over two universes. *Int. J. Mach. Learn. Cybern.* **2018**, *10*, 2501–2513. [CrossRef]
36. Zadeh, L.A. Toward a theory of fuzzy information granulation and its centrality in human reasoning and fuzzy logic. *Fuzzy Sets Syst.* **1997**, *90*, 111–127. [CrossRef]
37. Rauszer, C. Rough logic for multi-agent systems. In *Logic at Work'92;* Masuch, M., Polos, L., Eds.; Springer: Berlin/Heidelberg, Germany, 1991; Volume 808, pp. 151–181.
38. Khan, M.A.; Banerjee, M.; Rieke, R. An update logic for information systems. *Int. J. Approx. Reason.* **2014**, *55*, 436–456. [CrossRef]
39. Khan, M.A. Formal reasoning in preference-based multiplesource rough set model. *Inf. Sci.* **2016**, *334–335*, 122–143. [CrossRef]
40. Qian, Y.H.; Liang, J.Y.; Yao, Y.Y.; Dang, C.Y. Incomplete mutigranulation rough set. *IEEE Trans. Syst. Man. Cybern. Part A* **2010**, *20*, 420–430. [CrossRef]
41. Qian, Y.H.; Liang, J.Y.; Yao, Y.Y.; Dang, C.Y. MGRS: A multigranulation rough set. *Inf. Sci.* **2010**, *180*, 949–970. [CrossRef]
42. Qian, Y.H.; Li, S.Y.; Liang, J.Y.; Shi, Z.Z.; Wang, F. Pessimistic rough set based decisions: A multigranulation fusion strategy. *Inf. Sci.* **2014**, *264*, 196–210. [CrossRef]
43. Lin, G.P.; Liang, J.Y.; Qian, Y.H. Multigranulation rough sets: From partition to covering. *Inf. Sci.* **2013**, *241*, 101–118. [CrossRef]
44. Kruse, R.; Gebhardt, J.; Klawonn, F. *Foundations of Fuzzy Systems;* Wiley: New York, NY, USA, 1994.
45. Lee, C.C. Fuzzy logic in control systems: Fuzzy logic controller, parts I and II. *IEEE Trans. Syst. Man Cybern.* **1990**, *20*, 404–418. [CrossRef]
46. Liu, C.H.; Miao, D.Q.; Qian, J. On multi-granulation covering rough sets. *Int. J. Approx. Reason.* **2014**, *55*, 1404–1418. [CrossRef]
47. Xu, W.; Sun, W.; Zhang, X.; Zhang, W. Multiple granulation rough set approach to ordered information systems. *Int. J. Gen. Syst.* **2012**, *41*, 475–501. [CrossRef]
48. Xu, W.H.; Wang, Q.R.; Zhang, X.T. Multi-granulation rough sets based on tolerance relations. *Soft Comput.* **2013**, *17*, 1241–1252. [CrossRef]
49. Xu, W.H.; Zhang, W.X. Measuring roughness of generalized rough sets induced by a covering. *Fuzzy Sets Syst.* **2007**, *158*, 2443–2455. [CrossRef]
50. Dou, H.L.; Yang, X.B.; Fan, J.Y.; Xu, S.P. The models of variable precision multigranulation rough sets. In Proceedings of the 7th International Conference on Rough Sets and Knowledge Technology, Chengdu, China, 17–20 August 2012; pp. 465–473.
51. Ju, H.R.; Yang, X.B.; Dou, H.L.; Song, J.J. Variable precision multigranulation rough set and attributes reduction. In *Transactions on Rough Sets XVIII;* Springer: Berlin/Heidelberg, Germany, 2014; pp. 52–68.
52. Feng, T.; Mi, J.S. Variable precision multigranulation decision-theoretic fuzzy rough sets. *Knowl.-Based Syst.* **2016**, *91*, 93–101. [CrossRef]
53. You, X.; Li, J.; Wang, H. Relative reduction of neighborhood-covering pessimistic multigranulation rough set based on evidence theory. *Information* **2019**, *10*, 334. [CrossRef]
54. Huang, B.; Wu, W.Z.; Yan, J.; Li, H.; Zhou, X. Inclusion measure-based multi-granulation decision- theoretic rough sets in multi-scale intuitionistic fuzzy information tables. *Inf. Sci.* **2020**, *507*, 421–448. [CrossRef]
55. Liang, M.; Mi, J.; Feng, T.; Xie, B. A dynamic approach for updating the lower approximation in adjustable multi-granulation rough sets. *Soft Comput.* **2020**, *24*, 15951–15966. [CrossRef]
56. Shao, Y.; Qi, X.; Gong, Z. A general framework for multi-granulation rough decision-making method under q-rung dual hesitant fuzzy environment. *Artifcial Intell. Rev.* **2020**, *53*, 4903–4933. [CrossRef]

57. Kong, Q.; Zhang, X.; Xu, W.; Xie, S. Attribute reducts of multi-granulation information system. *Artifcial Intell. Rev.* **2019**, *53*, 1353–1371. [CrossRef]
58. Sun, L.; Wang, L.; Ding, W.; Qian, Y.; Xu, J. Neighborhood multi-granulation rough sets-based attribute reduction using lebesgue and entropy measures in incomplete neighborhood decision systems. *Knowl-Based Syst.* **2020**, *192*, 105373. [CrossRef]
59. Huang, B.; Guo, C.; Zhuang, Y.L.; Li, H.; Zhou, X. Intuitionistic fuzzy multigranulation rough sets. *Inf. Sci.* **2014**, *277*, 299–320. [CrossRef]
60. Pang, J.; Guan, X.; Liang, J.; Wang, B.; Song, P. Multi-attribute group decision-making method based on multi-granulation weights and three-way decisions. *Int. J. Approx. Reason.* **2020**, *117*, 122–147. [CrossRef]
61. Sun, B.Z.; Ma, W.M. Multigranulation rough set theory over two universes. *J. Intell. Fuzzy Syst.* **2015**, *28*, 1251–1269. [CrossRef]
62. Xu, Z. Intuitionistic preference relations and their application in group decision making. *Inf. Sci.* **2007**, *177*, 2363–2379. [CrossRef]
63. Shabir, M.; Din, J.; Ganie, I.A. Multigranulation roughness based on soft relations. *J. Intell. Fuzzy Syst.* **2021**, *40*, 10893–10908. [CrossRef]
64. Jamal, D.; Shabir, M. Multigranulation roughness of a fuzzy set based on soft relations over dual universes and its application. 2021, submitted.
65. Chen, S.M.; Tan J.M. Handling multicriteria fuzzy decision-making problems based on vague set theory. *Fuzzy Sets Syst.* **1994**, *67*, 163–172. [CrossRef]
66. Hong, D.H.; Choi, C.H. Multi-criteria fuzzy decision-making problems based on vague set theory. *Fuzzy Sets Syst.* **2000**, *114*, 103–113. [CrossRef]
67. Xu, Y.; Sun, Y.; Li, D. Intuitionistic fuzzy soft set. In Proceedings of the 2nd International Workshop on Intelligent Systems and Applications IEEE, Wuhan, China, 22–23 May 2010; pp. 1–4.
68. Feng, F.; Ali, M.I.; Shabir, M. Soft relations applied to semigroups. *Filomat* **2013**, *27*, 1183–1196. [CrossRef]
69. Li, Z.; Xie, N.; Gao, N. Rough approximations based on soft binary relations and knowledge bases. *Soft Comput.* **2017**, *21*, 839–852. [CrossRef]
70. Anwar, M.Z.; Bashir, S.; Shabir, M. An efficient model for the approximation of intuitionistic fuzzy sets in terms of soft relations with applications in decision making. *Math. Probl. Eng.* **2021**, in press. [CrossRef]
71. Roy, A.R.; Maji, P.K. A fuzzy soft set theoretic approach to decision making problems. *J. Comput. Appl. Math.* **2007**, *203*, 412–418. [CrossRef]
72. Feng, F.; Jun, Y.B.; Liu, X.; Li, L. An adjustable approach to fuzzy soft set based decision making. *J. Comput. Appl. Math.* **2010**, *234*, 10–20. [CrossRef]
73. Wu, W.Z.; Leung, Y. Theory and applications of granular labelled partitions in multi-scale decision tables. *Inf. Sci.* **2011**, *181*, 3878–3897. [CrossRef]
74. Li, J.; Ren, Y.; Mei, C.; Qian, Y.; Yang, X. A comparative study of multigranulation rough sets and concept lattices via rule acquisition. *Knowl.-Based Syst.* **2016**, *91*, 152–164. [CrossRef]

Σ mathematics MDPI

Article

Applications of the Fractional Calculus in Fuzzy Differential Subordinations and Superordinations

Alina Alb Lupaş

Department of Mathematics and Computer Science, University of Oradea, 1 Universitatii Street, 410087 Oradea, Romania; dalb@uoradea.ro or alblupas@gmail.com

Abstract: The fractional integral of confluent hypergeometric function is used in this paper for obtaining new applications using concepts from the theory of fuzzy differential subordination and superordination. The aim of the paper is to present new fuzzy differential subordinations and superordinations for which the fuzzy best dominant and fuzzy best subordinant are given, respectively. The original theorems proved in the paper generate interesting corollaries for particular choices of functions acting as fuzzy best dominant and fuzzy best subordinant. Another contribution contained in this paper is the nice sandwich-type theorem combining the results given in two theorems proved here using the two theories of fuzzy differential subordination and fuzzy differential superordination.

Keywords: fuzzy differential subordination; fuzzy differential superordination; fuzzy best dominant; fuzzy best subordinant; fractional integral; confluent hypergeometric function

Citation: Alb Lupaş, A. Applications of the Fractional Calculus in Fuzzy Differential Subordinations and Superordinations. *Mathematics* **2021**, *92*, 2601. https://doi.org/10.3390/math9202601

Academic Editors: Sorin Nadaban and Ioan Dzitac

Received: 17 September 2021
Accepted: 14 October 2021
Published: 15 October 2021

1. Introduction

The concept of fuzzy differential subordination and its dual, the concept of fuzzy differential superordination were introduced in the last decade as a result of the trend for adapting the notion of fuzzy set to different topics of research. Even if the notion of fuzzy set didn't look promising when it was first introduced by Lotfi A. Zadeh in his paper published in 1965 [1], in the recent years it became part of many branches of science and scientific research. Mathematical sciences also aimed for introducing and using fuzzifications of the already established classical theories in different fields of research. The review paper published in 2017 [2] shows some parts of the history of the fuzzy set notion and how Zadeh's new concept has revolutionized Soft Computing and Artificial Intelligence as well as other fields of science and technique. Another review paper published as part of a special issue dedicated to celebrating Zadeh's birth centenary [3] shows other aspects from the development process of fuzzy logic based on the notion of fuzzy set.

Fuzzy sets theory was connected to geometric function theory in 2011 when the notion of fuzzy subordination was introduced [4] with as inspiration the theory of differential subordination initiated by Miller and Mocanu in 1978 [5] and 1981 [6]. The core of the theory of differential subordination was gradually adapted to fuzzy sets notions in the subsequent years [7–9] following the main lines of research as found in the monograph published in 2000 [10]. Obtaining fuzzy subordination and superordination results involving operators was a topic approached early in the study of fuzzy subordinations. Fuzzy differential subordinations were first obtained using a convolution product of Sălăgean operator and Ruscheweyh derivative in 2013 [11]. The study following this line of research was continued and, in 2015, fuzzy differential subordinations were investigated using Ruscheweyh operator [12], and then using a multiplier transformation in 2016 [13]. Fuzzy differential subordinations for prestar-like functions of complex order and some applications were published in 2017 [14] and in the same year, the dual notion of fuzzy differential superordination was introduced [15]. After this event, the two notions were used together in many investigations giving applications such as in [16] and adding operators to the research such as the Srivastava–Attitya operator in [17]. In recent years, theories of fuzzy differential subordination and fuzzy differential superordination developed nicely. New fuzzy differential subordinations associated with integral operators were investigated in [18,19] and a linear operator was considered for the study in [20]. New applications of Sălăgean and Ruscheweyh operators for obtaining fuzzy differential subordinations were investigated in [21] and fuzzy differential subordinations based upon the Mittag–Leffler-type Borel distribution also emerged [22].

The present paper considers for the investigation from a fuzzy point of view an operator obtained using fractional integrals of confluent hypergeometric function. The investigation of such an operator for obtaining new fuzzy differential subordinations and superordinations results is motivated by recent fuzzy related investigations that considered a fractional integral associated with generalized Mittag–Leffler function [23], a fuzzy Atangana–Baleanu fractional derivative operator in [24] and fuzzy differential subordinations obtained using a hypergeometric integral operator [25].

Fractional calculus has had tremendous development in recent years proving to have applications in many research domains such as physics, engineering, turbulence, electric networks, biological systems with memory, computer graphics, etc. As an example, the Korteweg–de Vries equation, developed to represent a broad spectrum of physics behaviors of the evolution and association of nonlinear waves, is studied using a new integral transform where the fractional derivative is proposed in the Caputo sense [26]. Related to biological systems, examples can be given as the new study on the mathematical modelling of the human liver with the Caputo–Fabrizio fractional derivative proposed in [27] and fractional calculus analysis of the transmission dynamics of the dengue infection seen in [28].

Numerous interesting approaches were taken into account regarding fractional integral calculus and new results have emerged in an impressive number of papers published lately. A study of fractional integral operators involving a certain generalized multi-index Mittag–Leffler function was conducted in [29]. Integral inequalities were investigated using integral operators defined with fractional integral of Gauss hypergeometric function [30] and using γ-convex functions and a generalized fractional integral operator based on Raina's function [31]. Subclasses of analytic functions defined using a fractional integral operator were introduced and studied in [32,33]. Applications of differential subordination theory to analytic and p-valent functions defined by a generalized fractional differintegral operator were presented in [34] and a new fractional integral operator is used in the study on the Mittag–Leffler-confluent hypergeometric function [35].

Following the research line in which fuzzy differential subordinations and superordinations are connected to operators, the interesting fractional integral of confluent hypergeometric function introduced and investigated in [36] using the classical theories of differential subordinations and superordinations is further considered from this new perspective involving fuzzy set theory notions. The purpose of the investigation is to present new fuzzy differential subordinations and superordinations which lead to interesting corollaries when using functions with remarkable geometric properties known from geometric function theory as fuzzy best dominant and fuzzy best subordinant, respectively.

2. Preliminaries

The study presented in this paper is done in the general context of geometric function theory.

The unit disc of the complex plane is denoted by $U = \{z \in \mathbb{C} : |z| < 1\}$ and the class of analytic functions in U by $\mathcal{H}(U)$. For n a positive integer and $a \in \mathbb{C}$, $\mathcal{H}[a, n]$ denotes the subclass of $\mathcal{H}(U)$ consisting of functions written in the form $f(z) = a + a_n z^n + a_{n+1} z^{n+1} + \ldots, z \in U$.

A function with beautiful applications in defining operators is the fractional integral of order λ given as:

Definition 1 ([37]). *The fractional integral of order λ ($\lambda > 0$) is defined for a function f by*

$$D_z^{-\lambda} f(z) = \frac{1}{\Gamma(\lambda)} \int_0^z \frac{f(t)}{(z-t)^{1-\lambda}} dt,$$

where f is an analytic function in a simply connected region of the z-plane containing the origin, and the multiplicity of $(z-t)^{\lambda-1}$ is removed by requiring $\log(z-t)$ to be real, when $(z-t) > 0$.

The definitions of the notions used in the present investigation are next recalled. Confluent (or Kummer) hypergeometric function is defined as:

Definition 2 ([10] p. 5). *Let a and c be complex numbers with $c \neq 0, -1, -2, \ldots$ and consider*

$$\phi(a,c;z) = {}_1F_1(a,c;z) = 1 + \frac{a}{c}\frac{z}{1!} + \frac{a(a+1)}{c(c+1)}\frac{z^2}{2!} + \ldots, z \in U. \tag{1}$$

This function is called confluent (Kummer) hypergeometric function, is analytic in $\mathbb{C}$ and satisfies Kummer's differential equation

$$zw''(z) + (c-z)w'(z) - aw(z) = 0.$$

The operator introduced in [36] using the fractional integral of confluent hypergeometric function is given in the following definition:

Definition 3 ([36]). *Let a and c be complex numbers with $c \neq 0, -1, -2, \ldots$ and $\lambda > 0$. We define the fractional integral of confluent hypergeometric function*

$$D_z^{-\lambda}\phi(a,c;z) = \frac{1}{\Gamma(\lambda)} \int_0^z \frac{\phi(a,c;t)}{(z-t)^{1-\lambda}} dt = \tag{2}$$

$$\frac{1}{\Gamma(\lambda)} \frac{\Gamma(c)}{\Gamma(a)} \sum_{k=0}^{\infty} \frac{\Gamma(a+k)}{\Gamma(c+k)\Gamma(k+1)} \int_0^z \frac{t^k}{(z-t)^{1-\lambda}} dt.$$

Remark 1 ([36]). *The fractional integral of confluent hypergeometric function can be written*

$$D_z^{-\lambda}\phi(a,c;z) = \frac{\Gamma(c)}{\Gamma(a)} \sum_{k=0}^{\infty} \frac{\Gamma(a+k)}{\Gamma(c+k)\Gamma(\lambda+k+1)} z^{k+\lambda}, \tag{3}$$

after a simple calculation. Evidently $D_z^{-\lambda}\phi(a,c;z) \in \mathcal{H}[0,\lambda]$.

For the concept of fuzzy differential subordination to be used, the following notions are necessary:

Definition 4 ([38]). *A pair (A, F_A), where $F_A : X \to [0,1]$ and $A = \{x \in X : 0 < F_A(x) \leq 1\}$ is called the fuzzy subset of X. The set A is called the support of the fuzzy set (A, F_A) and F_A is called the membership function of the fuzzy set (A, F_A). One can also denote $A = \text{supp}(A, F_A)$.*

Remark 2 ([38]). *If $A \subset X$, then $F_A(x) = \begin{cases} 1, \text{if } x \in A \\ 0, \text{if } x \notin A \end{cases}$.*

For a fuzzy subset, the real number 0 represents the smallest membership degree of a certain $x \in X$ to A and the real number 1 represents the biggest membership degree of a certain $x \in X$ to A.

The empty set $\emptyset \subset X$ is characterized by $F_\emptyset(x) = 0$, $x \in X$, and the total set X is characterized by $F_X(x) = 1$, $x \in X$.

Definition 5 ([4]). *Let $D \subset \mathbb{C}$, $z_0 \in D$ be a fixed point and let the functions $f, g \in \mathcal{H}(D)$. The function f is said to be fuzzy subordinate to g and write $f \prec_{\mathcal{F}} g$ or $f(z) \prec_{\mathcal{F}} g(z)$, if are satisfied the conditions:*

(1) $f(z_0) = g(z_0)$,

(2) $F_{f(D)}f(z) \leq F_{g(D)}g(z)$, $z \in D$.

Definition 6 ([8] Definition 2.2). *Let $\psi : \mathbb{C}^3 \times U \to \mathbb{C}$ and h univalent in U, with $\psi(a, 0; 0) = h(0) = a$. If p is analytic in U, with $p(0) = a$ and satisfies the (second-order) fuzzy differential subordination*

$$F_{\psi(\mathbb{C}^3 \times U)}\psi(p(z), zp'(z), z^2p''(z); z) \leq F_{h(U)}h(z), \quad z \in U, \tag{4}$$

then p is called a fuzzy solution of the fuzzy differential subordination. The univalent function q is called a fuzzy dominant of the fuzzy solutions of the fuzzy differential subordination, or more simply a fuzzy dominant, if $F_{p(U)}p(z) \leq F_{q(U)}q(z)$, $z \in U$, for all p satisfying (4). A fuzzy dominant $\widetilde{q}$ that satisfies $F_{\widetilde{q}(U)}\widetilde{q}(z) \leq F_{q(U)}q(z)$, $z \in U$, for all fuzzy dominants q of (4) is said to be the fuzzy best dominant of (4).

Definition 7 ([11]). *Let $\varphi : \mathbb{C}^3 \times U \to \mathbb{C}$ and let h be analytic in U. If p and $\varphi(p(z), zp'(z), z^2p''(z); z)$ are univalent in U and satisfy the (second-order) fuzzy differential superordination*

$$F_{h(U)}h(z) \leq F_{\varphi(\mathbb{C}^3 \times U)}\varphi(p(z), zp'(z), z^2p''(z); z), \quad z \in U, \tag{5}$$

i.e.,

$$h(z) \prec_{\mathcal{F}} \varphi(p(z), zp'(z), z^2p''(z); z), \quad z \in U,$$

then p is called a fuzzy solution of the fuzzy differential superordination. An analytic function q is called fuzzy subordinant of the fuzzy differential superordination, or more simply a fuzzy subordination if

$$F_{q(U)}q(z) \leq F_{p(U)}p(z), \ z \in U,$$

for all p satisfying (5). A univalent fuzzy subordination $\widetilde{q}$ that satisfies $F_{q(U)}q \leq F_{q(U)}\widetilde{q}$ for all fuzzy subordinate q of (5) is said to be the fuzzy best subordinate of (5). Please note that the fuzzy best subordinant is unique to a relation of U.

The purpose of this paper is to obtain several fuzzy differential subordination and superordination results, using the following known results.

Definition 8 ([8]). *Denote by Q the set of all functions f that are analytic and injective on $\overline{U} \backslash E(f)$, where $E(f) = \{\zeta \in \partial U : \lim\limits_{z \to \zeta} f(z) = \infty\}$, and are such that $f'(\zeta) \neq 0$ for $\zeta \in \partial U \backslash E(f)$.*

Lemma 1 ([8]). *Let the function q be univalent in the unit disc U and θ and ϕ be analytic in a domain D containing $q(U)$ with $\phi(w) \neq 0$ when $w \in q(U)$. Set $Q(z) = zq'(z)\phi(q(z))$ and $h(z) = \theta(q(z)) + Q(z)$. Suppose that*

1. Q is starlike univalent in U and

2. $Re\left(\frac{zh'(z)}{Q(z)}\right) > 0$ for $z \in U$.

If p is analytic with $p(0) = q(0)$, $p(U) \subseteq D$ and

$$F_{p(U)}\theta(p(z)) + zp'(z)\phi(p(z)) \leq F_{h(U)}\theta(q(z)) + zq'(z)\phi(q(z)),$$

then

$$F_{p(U)}p(z) \leq F_{q(U)}q(z)$$

and q is the fuzzy best dominant.

Lemma 2 ([11]). *Let the function q be convex univalent in the open unit disc U and ν and ϕ be analytic in a domain D containing $q(U)$. Suppose that*

1. $Re\left(\frac{\nu'(q(z))}{\phi(q(z))}\right) > 0$ for $z \in U$ and

2. $\psi(z) = zq'(z)\phi(q(z))$ is starlike univalent in U.

If $p(z) \in \mathcal{H}[q(0), 1] \cap Q$, with $p(U) \subseteq D$ and $\nu(p(z)) + zp'(z)\phi(p(z))$ is univalent in U and

$$F_{q(U)}\nu(q(z)) + zq'(z)\phi(q(z)) \leq F_{p(U)}\nu(p(z)) + zp'(z)\phi(p(z)),$$

then

$$F_{q(U)}q(z) \leq F_{p(U)}p(z)$$

and q is the fuzzy best subordinant.

3. Main Results

The first fuzzy subordination result obtained using the operator given by (2) is the following theorem:

Theorem 1. *Let the function q be analytic and univalent in U such that $q(z) \neq 0$ and $\left(\frac{D_z^{-\lambda}\phi(a,c;z)}{z}\right)^{\delta} \in \mathcal{H}(U)$, for all $z \in U$, where a, c be complex numbers with $c \neq 0, -1, -2, \ldots$ and $\lambda, \delta > 0$. Suppose that $\frac{zq'(z)}{q(z)}$ is starlike univalent in U. Let*

$$Re\left(\frac{zq''(z)}{q'(z)} - \frac{zq'(z)}{q(z)} + \frac{2\mu}{\beta}(q(z))^2 + \frac{\xi}{\beta}q(z) + 1\right) > 0, \tag{6}$$

for $\alpha, \xi, \mu, \beta \in \mathbb{C}$, $\beta \neq 0$, $z \in U$ and

$$\psi_{\lambda}^{a,c}(\delta, \alpha, \xi, \mu, \beta; z) := \alpha + \beta\delta\left[\frac{z(D_z^{-\lambda}\phi(a,c;z))'}{D_z^{-\lambda}\phi(a,c;z)} - 1\right] + \tag{7}$$

$$\mu\left[\frac{D_z^{-\lambda}\phi(a,c;z)}{z}\right]^{2\delta} + \xi\left[\frac{D_z^{-\lambda}\phi(a,c;z)}{z}\right]^{\delta}.$$

If q satisfies the following fuzzy subordination

$$F_{\psi_{\lambda}^{a,c}(U)}\psi_{\lambda}^{a,c}(\delta, \alpha, \xi, \mu, \beta; z) \leq F_{q(U)}\left(\alpha + \beta\frac{zq'(z)}{q(z)} + \mu(q(z))^2 + \xi q(z)\right), \tag{8}$$

for $\alpha, \xi, \mu, \beta \in \mathbb{C}$, $\beta \neq 0$, then

$$F_{D_z^{-\lambda}\phi(U)}\left(\frac{D_z^{-\lambda}\phi(a,c;z)}{z}\right)^{\delta} \leq F_{q(U)}q(z), \quad z \in U, \tag{9}$$

and q is the best dominant.

Proof. Define $p(z) := \left(\frac{D_z^{-\lambda}\phi(a,c;z)}{z}\right)^{\delta}$, $z \in U$, $z \neq 0$. Differentiating it we obtain $p'(z) = -\frac{\delta}{z}p(z) + \delta\left(\frac{D_z^{-\lambda}\phi(a,c;z)}{z}\right)^{\delta-1}\frac{(D_z^{-\lambda}\phi(a,c;z))'}{z}$. Then $\frac{zp'(z)}{p(z)} = \delta\left[\frac{z(D_z^{-\lambda}\phi(a,c;z))'}{D_z^{-\lambda}\phi(a,c;z)} - 1\right]$.

By setting $\theta(w) := \mu w^2 + \xi w + \alpha$ and $Q(w) := \frac{\beta}{w}$, it is evident that θ is analytic in $\mathbb{C}$, ϕ is analytic in $\mathbb{C}\backslash\{0\}$ and $\phi(w) \neq 0$, $w \in \mathbb{C}\backslash\{0\}$.

Considering $Q(z) = zq'(z)\phi(q(z)) = \beta\frac{zq'(z)}{q(z)}$ and $h(z) = Q(z) + \theta(q(z)) = \alpha + \beta\frac{zq'(z)}{q(z)} + \mu(q(z))^2 + \xi q(z)$, we deduce that Q is starlike univalent in U.

Differentiating we obtain $h'(z) = \xi + q'(z) + 2\mu q(z)q'(z) + \beta\frac{(q'(z)+zq''(z))q(z)-z(q'(z))^2}{(q(z))^2}$ and $\frac{zh'(z)}{Q(z)} = \frac{zh'(z)}{\beta\frac{zq'(z)}{q(z)}} = 1 + \frac{\xi}{\beta}q(z) + \frac{2\mu}{\beta}(q(z))^2 - \frac{zq'(z)}{q(z)} + \frac{zq''(z)}{q'(z)}$, which imply that $Re\left(\frac{zh'(z)}{Q(z)}\right) = Re\left(1 + \frac{\xi}{\beta}q(z) + \frac{2\mu}{\beta}(q(z))^2 - \frac{zq'(z)}{q(z)} + \frac{zq''(z)}{q'(z)}\right) > 0$.

We obtain $\alpha + \beta\frac{zp'(z)}{p(z)} + \mu(p(z))^2 + \xi p(z) = \alpha + \beta\delta\left[\frac{z(D_z^{-\lambda}\phi(a,c;z))'}{D_z^{-\lambda}\phi(a,c;z)} - 1\right] + \mu\left[\frac{D_z^{-\lambda}\phi(a,c;z)}{z}\right]^{2\delta} + \xi\left[\frac{D_z^{-\lambda}\phi(a,c;z)}{z}\right]^{\delta}$.

Using (8), we deduce $F_{p(U)}\left(\alpha + \beta\frac{zp'(z)}{p(z)} + \mu(p(z))^2 + \xi p(z)\right) \le F_{q(U)}\left(\alpha + \beta q(z) + \mu(q(z))^2 + \beta\frac{zq'(z)}{q(z)}\right)$.

By an application of Lemma 1 we obtain $F_{p(U)}p(z) \le F_{q(U)}q(z)$, $z \in U$, i.e., $F_{D_z^{-\lambda}\phi(U)}\left(\frac{D_z^{-\lambda}\phi(a,c;z)}{z}\right)^{\delta} \le F_{q(U)}q(z)$, $z \in U$ and q is the fuzzy best dominant. □

Corollary 1. *Let c, a be complex numbers with $c \neq 0, -1, -2, \ldots$ and $\delta, \lambda > 0$. Assume that (6) holds. If*

$$F_{\psi_\lambda^{a,c}(U)}\psi_\lambda^{a,c}(\delta,\alpha,\xi,\mu,\beta;z) \le F_{q(U)}\left(\alpha + \beta\frac{(A-B)z}{(1+Az)(1+Bz)} + \mu\left(\frac{1+Az}{1+Bz}\right)^2 + \xi\frac{1+Az}{1+Bz}\right),$$

for $\xi, \alpha, \beta, \mu \in \mathbb{C}$, $\beta \neq 0$, $-1 \le B < A \le 1$, and $\psi_\lambda^{a,c}$ is introduced in (7), then

$$F_{D_z^{-\lambda}\phi(U)}\left(\frac{D_z^{-\lambda}\phi(a,c;z)}{z}\right)^{\delta} \le F_{q(U)}\frac{1+Az}{1+Bz}, \quad z \in U,$$

and $\frac{1+Az}{1+Bz}$ is the fuzzy best dominant.

Proof. Consider in Theorem 1 $q(z) = \frac{1+Az}{1+Bz}$, $-1 \le B < A \le 1$. □

Corollary 2. *Let c, a be complex numbers with $c \neq 0, -1, -2, \ldots$ and $\delta, \lambda > 0$. Assume that (6) holds. If*

$$F_{\psi_\lambda^{a,c}(U)}\psi_\lambda^{a,c}(\delta,\alpha,\xi,\mu,\beta;z) \le F_{q(U)}\left(\alpha + \xi\left(\frac{1+z}{1-z}\right)^{\gamma} + \mu\left(\frac{1+z}{1-z}\right)^{2\gamma} + \beta\frac{2\gamma z}{1-z^2}\right),$$

for $\xi, \alpha, \beta, \mu \in \mathbb{C}$, $\beta \neq 0$, $0 < \gamma \le 1$, where $\psi_\lambda^{a,c}$ is introduced in (7), then

$$F_{D_z^{-\lambda}\phi(U)}\left(\frac{D_z^{-\lambda}\phi(a,c;z)}{z}\right)^{\delta} \le F_{q(U)}\left(\frac{1+z}{1-z}\right)^{\gamma}, \quad z \in U,$$

and $\left(\frac{1+z}{1-z}\right)^{\gamma}$ is the fuzzy best dominant.

Proof. Theorem 1 give Corollary for $q(z) = \left(\frac{1+z}{1-z}\right)^{\gamma}$, $0 < \gamma \le 1$. □

Theorem 2. *Let q be analytic and univalent in U such that $q(z) \neq 0$ and $\frac{zq'(z)}{q(z)}$ be starlike univalent in U. Assume that*

$$Re\left(\frac{\xi}{\beta}q(z) + \frac{2\mu}{\beta}(q(z))^2\right) > 0, \text{ for } \mu, \xi, \beta \in \mathbb{C},\ \beta \neq 0. \tag{10}$$

Let c, a be complex numbers with $c \neq 0, -1, -2, \ldots$ and $\delta, \lambda > 0$. If $\psi_\lambda^{a,c}(\delta, \alpha, \xi, \mu, \beta; z)$ is univalent in U and $\left(\frac{D_z^{-\lambda}\phi(a,c;z)}{z}\right)^\delta \in \mathcal{H}[0, (\lambda-1)\delta] \cap Q$, where $\psi_\lambda^{a,c}(\delta, \alpha, \xi, \mu, \beta; z)$ is introduced in (7), then

$$F_{q(U)}\left(\alpha + \beta\frac{zq'(z)}{q(z)} + \mu(q(z))^2 + \xi q(z)\right) \leq F_{\psi_\lambda^{a,c}(U)}\psi_\lambda^{a,c}(\delta, \alpha, \xi, \mu, \beta; z) \tag{11}$$

implies

$$F_{q(U)}q(z) \leq F_{D_z^{-\lambda}\phi(U)}\left(\frac{D_z^{-\lambda}\phi(a,c;z)}{z}\right)^\delta, \; z \in U, \tag{12}$$

and q is the fuzzy best subordinant.

Proof. Define $p(z) := \left(\frac{D_z^{-\lambda}\phi(a,c;z)}{z}\right)^\delta, z \in U, z \neq 0$.

Consider $\nu(w) := \mu w^2 + \xi w + \alpha$ and $\phi(w) := \frac{\beta}{w}$ it is evident that ν is analytic in $\mathbb{C}$, ϕ is analytic in $\mathbb{C}\backslash\{0\}$ and $\phi(w) \neq 0$, $w \in \mathbb{C}\backslash\{0\}$.

In this conditions $\frac{\nu'(q(z))}{\phi(q(z))} = \frac{q'(z)[\xi+2\mu q(z)]q(z)}{\beta}$, which imply $Re\left(\frac{\nu'(q(z))}{\phi(q(z))}\right) = Re\left(\frac{\xi}{\beta}q(z) + \frac{2\mu}{\beta}(q(z))^2\right) > 0$, for $\xi, \beta, \mu \in \mathbb{C}$, $\beta \neq 0$.

We obtain

$$F_{q(U)}\left(\alpha + \beta\frac{zq'(z)}{q(z)} + \mu(q(z))^2 + \xi q(z)\right) \leq F_{p(U)}\left(\alpha + \beta\frac{zp'(z)}{p(z)} + \mu(p(z))^2 + \xi p(z)\right).$$

Applying Lemma 2, we obtain

$$F_{q(U)}q(z) \leq F_{D_z^{-\lambda}\phi(U)}\left(\frac{D_z^{-\lambda}\phi(a,c;z)}{z}\right)^\delta, \quad z \in U,$$

and q is the fuzzy best subordinant. □

Corollary 3. *Let c, a be complex numbers with $c \neq 0, -1, -2, \ldots$ and $\delta, \lambda > 0$. Assume that (10) holds. If $\left(\frac{D_z^{-\lambda}\phi(a,c;z)}{z}\right)^\delta \in \mathcal{H}[0, (\lambda-1)\delta] \cap Q$ and*

$$F_{q(U)}\left(\alpha + \beta\frac{(A-B)z}{(1+Az)(1+Bz)} + \mu\left(\frac{1+Az}{1+Bz}\right)^2 + \xi\frac{1+Az}{1+Bz}\right) \leq F_{\psi_\lambda^{a,c}(U)}\psi_\lambda^{a,c}(\delta, \alpha, \xi, \mu, \beta; z),$$

for $\beta, \xi, \alpha, \mu \in \mathbb{C}$, $\beta \neq 0$, $-1 \leq B < A \leq 1$, where $\psi_\lambda^{a,c}$ is introduced in (7), then

$$F_{q(U)}\left(\frac{1+Az}{1+Bz}\right) \leq F_{D_z^{-\lambda}\phi(U)}\left(\frac{D_z^{-\lambda}\phi(a,c;z)}{z}\right)^\delta, \quad z \in U,$$

and $\frac{1+Az}{1+Bz}$ is the fuzzy best subordinant.

Proof. Theorem 2 for $q(z) = \frac{1+Az}{1+Bz}$, $-1 \leq B < A \leq 1$ give the corollary. □

Corollary 4. *Let c, a be complex numbers with $c \neq 0, -1, -2, \ldots$ and $\delta, \lambda > 0$. Assume that (10) holds. If $\left(\frac{D_z^{-\lambda}\phi(a,c;z)}{z}\right)^\delta \in \mathcal{H}[0, (\lambda-1)\delta] \cap Q$ and*

$$F_{q(U)}\left(\alpha + \beta\frac{2\gamma z}{1-z^2} + \mu\left(\frac{1+z}{1-z}\right)^{2\gamma} + \xi\left(\frac{1+z}{1-z}\right)^\gamma\right) \leq F_{\psi_\lambda^{a,c}(U)}\psi_\lambda^{a,c}(\delta, \alpha, \xi, \mu, \beta; z),$$

for $\beta, \xi, \alpha, \mu \in \mathbb{C}$, $0 < \gamma \leq 1$, $\beta \neq 0$, *where* $\psi_\lambda^{a,c}$ *is introduced in (7), then*

$$F_{q(U)}\left(\frac{1+z}{1-z}\right)^\gamma \leq F_{D_z^{-\lambda}\phi(U)}\left(\frac{D_z^{-\lambda}\phi(a,c;z)}{z}\right)^\delta, \; z \in U,$$

and $\left(\frac{1+z}{1-z}\right)^\gamma$ *is the fuzzy best subordinant.*

Proof. Theorem 2 for $q(z) = \left(\frac{1+z}{1-z}\right)^\gamma$, $0 < \gamma \leq 1$, give the corollary. □

Theorems 1 and 2 combined give the following sandwich theorem.

Theorem 3. *Let* q_1 *and* q_2 *be analytic and univalent in* U *such that* $q_1(z) \neq 0$ *and* $q_2(z) \neq 0$, *for all* $z \in U$, *with* $\frac{zq_1'(z)}{q_1(z)}$ *and* $\frac{zq_2'(z)}{q_2(z)}$ *being starlike univalent. Suppose that* q_1 *satisfies (6) and* q_2 *satisfies (10). Let* c, a *be complex numbers with* $c \neq 0, -1, -2, \ldots$ *and* $\delta, \lambda > 0$. *If* $\psi_\lambda^{a,c}(\delta, \alpha, \xi, \mu, \beta; z)$ *is as introduced in (7) univalent in* U *and* $\left(\frac{D_z^{-\lambda}\phi(a,c;z)}{z}\right)^\delta \in \mathcal{H}[0, (\lambda-1)\delta] \cap Q$, *then*

$$F_{q_1(U)}\left(\alpha + \beta\frac{zq_1'(z)}{q_1(z)} + \mu(q_1(z))^2 + \xi q_1(z)\right) \leq F_{\psi_\lambda^{a,c}(U)}\psi_\lambda^{a,c}(\delta, \alpha, \xi, \mu, \beta; z)$$

$$\leq F_{q_2(U)}\left(\alpha + \xi q_2(z) + \mu(q_2(z))^2 + \beta\frac{zq_2'(z)}{q_2(z)}\right),$$

for $\beta, \xi, \alpha, \mu \in \mathbb{C}$, $\beta \neq 0$, *implies*

$$F_{q_1(U)}q_1(z) \leq F_{D_z^{-\lambda}\phi(U)}\left(\frac{D_z^{-\lambda}\phi(a,c;z)}{z}\right)^\delta \leq F_{q_2(U)}q_2(z), \;\; z \in U,$$

and q_1 *and* q_2 *are respectively the fuzzy best subordinant and the fuzzy best dominant.*

For $q_1(z) = \frac{1+A_1z}{1+B_1z}$, $q_2(z) = \frac{1+A_2z}{1+B_2z}$, where $-1 \leq B_2 < B_1 < A_1 < A_2 \leq 1$, we obtain the following corollary.

Corollary 5. *Let* c, a *be complex numbers with* $c \neq 0, -1, -2, \ldots$ *and* $\delta, \lambda > 0$. *Assume that (6) and (10) hold. If* $\left(\frac{D_z^{-\lambda}\phi(a,c;z)}{z}\right)^\delta \in \mathcal{H}[0, (\lambda-1)\delta] \cap Q$ *and*

$$F_{q_1(U)}\left(\alpha + \beta\frac{(A_1-B_1)z}{(1+A_1z)(1+B_1z)} + \mu\left(\frac{1+A_1z}{1+B_1z}\right)^2 + \xi\frac{1+A_1z}{1+B_1z}\right) \leq F_{\psi_\lambda^{a,c}(U)}\psi_\lambda^{a,c}(\delta, \alpha, \xi, \mu, \beta; z)$$

$$\leq F_{q_2(U)}\left(\alpha + \beta\frac{(A_2-B_2)z}{(1+A_2z)(1+B_2z)} + \mu\left(\frac{1+A_2z}{1+B_2z}\right)^2 + \xi\frac{1+A_2z}{1+B_2z}\right),$$

for $\beta, \xi\alpha,, \mu, \in \mathbb{C}$, $\beta \neq 0$, $-1 \leq B_2 \leq B_1 < A_1 \leq A_2 \leq 1$, *where* $\psi_\lambda^{a,c}$ *is introduced in (7), then*

$$F_{q_1(U)}\left(\frac{1+A_1z}{1+B_1z}\right) \leq F_{D_z^{-\lambda}\phi}\left(\frac{D_z^{-\lambda}\phi(a,c;z)}{z}\right)^\delta \leq F_{q_2(U)}\frac{1+A_2z}{1+B_2z},$$

hence $\frac{1+A_1z}{1+B_1z}$ *and* $\frac{1+A_2z}{1+B_2z}$ *are the fuzzy best subordinant and the fuzzy best dominant, respectively.*

Corollary 6. *Let c, a be complex numbers with $c \neq 0, -1, -2, \ldots$ and $\delta, \lambda > 0$. Assume that (6) and (10) hold. If $\left(\frac{D_z^{-\lambda}\phi(a,c;z)}{z}\right)^{\delta} \in \mathcal{H}[0, (\lambda - 1)\delta] \cap Q$ and*

$$F_{q_1(U)}\left(\alpha + \beta\frac{2\gamma_1 z}{1-z^2} + \mu\left(\frac{1+z}{1-z}\right)^{2\gamma_1} + \xi\left(\frac{1+z}{1-z}\right)^{\gamma_1}\right) \leq F_{\psi_\lambda^{a,c}(U)}\psi_\lambda^{a,c}(\delta, \alpha, \xi, \mu, \beta; z)$$

$$\leq F_{q_2(U)}\left(\alpha + \xi\left(\frac{1+z}{1-z}\right)^{\gamma_2} + \mu\left(\frac{1+z}{1-z}\right)^{2\gamma_2} + \beta\frac{2\gamma_2 z}{1-z^2}\right),$$

for $\beta, \xi\alpha,, \mu, \in \mathbb{C}$, $\beta \neq 0$, $0 < \gamma_1, \gamma_2 \leq 1$, where $\psi_\lambda^{a,c}$ is introduced in (7), then

$$F_{q_1(U)}\left(\frac{1+z}{1-z}\right)^{\gamma_1} \leq F_{D_z^{-\lambda}\phi(U)}\left(\frac{D_z^{-\lambda}\phi(a,c;z)}{z}\right)^{\delta} \leq F_{q_2(U)}\left(\frac{1+z}{1-z}\right)^{\gamma_2},$$

hence $\left(\frac{1+z}{1-z}\right)^{\gamma_1}$ and $\left(\frac{1+z}{1-z}\right)^{\gamma_2}$ are the fuzzy best subordinant and the fuzzy best dominant, respectively.

4. Conclusions

The interesting operator presented in Definition 3 was previously defined and studied related to several aspects of differential subordination theory in [36] as a fractional integral of confluent hypergeometric function. In this paper, the study of the operator is continued using the recently introduced notions of fuzzy differential subordination and fuzzy differential superordination as a result of the preoccupation with adapting the classical notions of differential subordination and superordination to fuzzy sets theory. Fuzzy differential subordinations and fuzzy differential superordinations are presented in the original theorems giving their best fuzzy dominant and best fuzzy subordinant, respectively. Using particular functions, interesting corollaries are presented that could inspire future studies related to the univalence of the operator. A sandwich-type result is obtained in the last theorem combining the results proved using the two theories of fuzzy differential subordination and fuzzy differential superordination. Since the operator gives nice results in studies done with both theories, it could be used for introducing new fuzzy classes of analytic functions and performing studies on those classes using both theories.

Finding applications in other domains for the operator and for the results of the fuzzy investigation presented in this paper remains an open problem to which future interdisciplinary applications are desired.

Funding: This research received no external funding.

Institutional Review Board Statement: Not applicable.

Informed Consent Statement: Not applicable.

Data Availability Statement: Not applicable.

Conflicts of Interest: The author declares no conflict of interest.

References

1. Zadeh, L.A. Fuzzy Sets. *Inf. Control* **1965**, *8*, 338–353. [CrossRef]
2. Dzitac, I.; Filip, F.G.; Manolescu, M.J. Fuzzy Logic Is Not Fuzzy: World-renowned Computer Scientist Lotfi A. Zadeh. *Int. J. Comput. Commun. Control.* **2017**, *12*, 748–789. [CrossRef]
3. Dzitac, S.; Nădăban, S. Soft Computing for Decision-Making in Fuzzy Environments: A Tribute to Professor Ioan Dzitac. *Mathematics* **2021**, *9*, 1701. [CrossRef]
4. Oros, G.I.; Oros, G. The notion of subordination in fuzzy sets theory. *Gen. Math.* **2011**, *19*, 97–103.
5. Miller, S.S.; Mocanu, P.T. Second order-differential inequalities in the complex plane. *J. Math. Anal. Appl.* **1978**, *65*, 298–305. [CrossRef]
6. Miller, S.S.; Mocanu, P.T. Differential subordinations and univalent functions. *Michig. Math. J.* **1981**, *28*, 157–171. [CrossRef]

7. Oros, G.I.; Oros, G. Fuzzy differential subordination. *Acta Univ. Apulensis* **2012**, *3*, 55–64.
8. Oros, G.I.; Oros, G. Dominants and best dominants in fuzzy differential subordinations. *Stud. Univ. Babeş-Bolyai Math.* **2012**, *57*, 239–248.
9. Oros, G.I.; Oros, G. Briot-Bouquet fuzzy differential subordination. *An. Univ. Oradea Fasc. Mat.* **2012**, *19*, 83–87.
10. Miller, S.S.; Mocanu, P.T. *Differential Subordinations. Theory and Applications*; Marcel Dekker, Inc.: New York, NY, USA; Basel, Switzerland, 2000.
11. Alb Lupaş, A. On special fuzzy differential subordinations using convolution product of Sălăgean operator and Ruscheweyh derivative. *J. Comput. Anal. Appl.* **2013**, *15*, 1484–1489.
12. Venter, A.O. On special fuzzy differential subordination using Ruscheweyh operator. *An. Univ. Oradea Fasc. Mat.* **2015**, *XXII*, 167–176.
13. Alb Lupaş, A. A note on special fuzzy differential subordinations using multiplier transformation. *An. Univ. Oradea Fasc. Mat.* **2016**, *XXIII*, 183–191.
14. Wanas, A.K.; Majeed, A.H. Fuzzy differential subordinations for prestarlike functions of complex order and some applications. *Far East J. Math. Sci.* **2017**, *102*, 1777–1788. [CrossRef]
15. Atshan, W.G.; Hussain, K.O. Fuzzy Differential Superordination. *Theory Appl. Math. Comput. Sci.* **2017**, *7*, 27–38.
16. Ibrahim, R.W. On the subordination and superordination concepts with applications. *J. Comput. Theor. Nanosci.* **2017**, *14*, 2248–2254. [CrossRef]
17. Thilagavathi, K. Fuzzy subordination and superordination results for certain subclasses of analytic functions associated with Srivastava-Attitya operator. *Int. J. Pure Appl. Math.* **2018**, *118*, 921–929.
18. Oros, G.I. New fuzzy differential subordinations. *Commun. Fac. Sci. Univ. Ank. Ser. A1 Math. Stat.* **2021**, *70*, 229–240.
19. El-Deeb, S.M.; Alb Lupaş, A. Fuzzy differential subordinations associated with an integral operator. *An. Univ. Oradea Fasc. Mat.* **2020**, *XXVII*, 133–140.
20. El-Deeb, S.M.; Oros, G.I. Fuzzy differential subordinations connected with the linear operator. *Math. Bohem.* **2021**, 1–10.
21. Alb Lupaş, A.; Oros, G.I. New Applications of Sălăgean and Ruscheweyh Operators for Obtaining Fuzzy Differential Subordinations. *Mathematics* **2021**, *9*, 2000. [CrossRef]
22. Srivastava, H.M.; El-Deeb, S.M. Fuzzy Differential Subordinations Based upon the Mittag-Leffler Type Borel Distribution. *Symmetry* **2021**, *13*, 1023. [CrossRef]
23. Wanas, A.K.; Majeed, A.H. Fuzzy subordination results for fractional integral associated with generalized Mittag-Leffler function. *Eng. Math. Lett.* **2019**, *2019*, 10.
24. Rashid, S.; Ashraf, R.; Akdemir, A.O.; Alqudah, M.A.; Abdeljawad, T.; Mohamed, M.S. Analytic Fuzzy Formulation of a Time-Fractional Fornberg–Whitham Model with Power and Mittag–Leffler Kernels. *Fractal Fract.* **2021**, *5*, 113. [CrossRef]
25. Oros, G.I. Fuzzy Differential Subordinations Obtained Using a Hypergeometric Integral Operator. *Mathematics* **2021**, *9*, 2539. [CrossRef]
26. Rashid, S.; Khalid, A.; Sultana, S.; Hammouch, Z.; Shah, R.; Alsharif, A.M. A novel analytical view of time-fractional Korteweg-De Vries equations via a new integral transform. *Symmetry* **2021**, *13*, 1254. [CrossRef]
27. Baleanu, D.; Jajarmi, A.; Mohammadi, H.; Rezapour, S. A new study on the mathematical modelling of human liver with Caputo–Fabrizio fractional derivative. *Chaos Solitons Fract.* **2020**, *134*, 109705. [CrossRef]
28. Srivastava, H.M.; Jan, R.; Jan, A.; Deebai, W.; Shutaywi, M. Fractional-calculus analysis of the transmission dynamics of the dengue infection. *Chaos* **2021**, *31*, 53130. [CrossRef]
29. Srivastava, H.M.; Bansal, M.; Harjule, P. A study of fractional integral operators involving a certain generalized multi-index Mittag-Leffler function. *Math. Methods Appl. Sci.* **2018**, *41*, 6108–6121. [CrossRef]
30. Saxena, R.K.; Purohit, S.D.; Kumar D. Integral Inequalities Associated with Gauss Hypergeometric Function Fractional Integral Operators. *Proc. Natl. Acad. Sci. India Sect. A Phys. Sci.* **2018**, *88*, 27–31. [CrossRef]
31. Rashid, S.; Khalid, A.; Bazighifan, O.; Oros, G.I. New modifications of integral inequalities via γ-convexity pertaining to fractional calculus and their applications. *Mathematics* **2021**, *9*, 1753. [CrossRef]
32. Alb Lupaş, A. Properties on a subclass of analytic functions defined by a fractional integral operator. *J. Comput. Anal. Appl.* **2019**, *27*, 506–510.
33. Alb Lupaş, A. Inequalities for Analytic Functions Deffined by a Fractional Integral Operator. In *Frontiers in Functional Equations and Analytic Inequalities*; Anastassiou, G., Rassias, J., Eds.; Springer: Berlin, Germany, 2020; pp. 731–745.
34. Cho, N.E.; Aouf, M.K.; Srivastava, R. The principle of differential subordination and its application to analytic and p-valent functions defined by a generalized fractional differintegral operator. *Symmetry* **2019**, *11*, 1083. [CrossRef]
35. Ghanim, F.; Al-Janaby, H.F. An analytical study on Mittag-Lefler-confluent hypergeometric functions with fractional integral operator. *Math. Methods Appl. Sci.* **2020**, *44*, 3605-3614. [CrossRef]
36. Alb Lupaş, A.; Oros, G.I. Differential Subordination and Superordination Results Using Fractional Integral of Confluent Hypergeometric Function. *Symmetry* **2021**, *13*, 327. [CrossRef]
37. Srivastava, H.M.; Owa, S. An application of the fractional derivative. *Mud Jpn.* **1984**, *29*, 383–389.
38. Gal, S.G.; Ban, A.I. *Elemente de Matematică Fuzzy*; University of Oradea: Oradea, Romania, 1996.

MDPI
St. Alban-Anlage 66
4052 Basel
Switzerland
Tel. +41 61 683 77 34
Fax +41 61 302 89 18
www.mdpi.com

Mathematics Editorial Office
E-mail: mathematics@mdpi.com
www.mdpi.com/journal/mathematics

MDPI

www.ingramcontent.com/pod-product-compliance
Lightning Source LLC
LaVergne TN
LVHW070417100526
838202LV00014B/1479
* 9 7 8 3 0 3 6 5 5 5 8 7 4 *